Build To Order

Glenn Parry • Andrew Graves

Editors

Build To Order

The Road to the 5-Day Car

Springer

Glenn Parry, Senior Research Fellow
Andrew Graves, Chair for Technology Management

School of Management
University of Bath
Bath, BA2 7AY
UK

ISBN 978-1-84996-759-4   e-ISBN 978-1-84800-225-8

DOI 10.1007/978-1-84800-225-8

British Library Cataloguing in Publication Data
A catalogue record for this book is available from the British Library

Cover design: eStudio Calamar S.L., Girona, Spain

Printed on acid-free paper

9 8 7 6 5 4 3 2 1

springer.com

*It is amazing what you can accomplish
if you do not care who gets the credit.*

Harry S. Truman

Foreword

Over the past 100 years the European Automotive Industry has been repeatedly challenged by best practice. First by the United States, through the development of 'mass production' pioneered by Henry Ford and more recently by 'lean production techniques' as practised by the leading Japanese producers, particularly Toyota. It has consistently risen to these challenges and has shown it can compete and even outperform its competitors with world-class products. However, the European industry is now faced with growing competition and growth from new emerging low-cost countries and needs to re-define its competitive advantage to remain at the forefront of the sector. Automotive growth is driven by two factors, new markets and new technologies. Global competition is increasing, with technology and product differentiation becoming the most important sales factors, but with continued cost pressure. Within the market the winners will be more profitable and the losers will disappear.

The Automotive Industry makes a significant contribution to the socio-economic fabric of the European Union. Manufacturing output represents €700 billion and research and development spending €24 billion. European automotive suppliers number 5000 member companies and represent 5 million employees and generate €500 billion in revenues. These are significant figures that generate wealth and high value employment within the EU. European firms must consistently improve their competitive position to ensure that the industry does not migrate to growing new markets.

'Build To Order: The Road to the 5-Day Car' provides us with a vision for a sustainable future European Automotive Industry that is able to meet these challenges and win. Within this book Dr Parry and Professor Graves have been able to integrate the ideas and concepts from the key automotive centres of excellence across Europe. This provides both the technical depth and breadth required, giving a structured approach to Build to Order, from the view point of the market, product, supply chain and most important, delivering a vehicle that "the customer wants when they want it".

The challenge, set out in 'Build To Order: The Road to the 5-Day Car', is the transformation of the automotive industry in Europe from mass manufacturer to build to order. The opportunity is currently there, but needs to be acted upon quickly and before new entrants adopt these concepts. This book provides an essential guide to those in the automotive industry taking up the challenge.

Lars Holmqvist
CEO CLEPA
European Association of Automotive Suppliers

Acknowledgements

First and foremost we recognise the EU's ILIPT project, funded by both industry partners and the European Commission, who together have made this book possible. We would like to extend our thanks and appreciation to the many organisations and people who supported and contributed to the research both financially and with their time.

The editors would like to thank the authors for their hard work in producing the text. Additional thanks to Jens Roehrich, Gareth Stone, Janice Legge and Max Dickson for their contributions to editing the manuscripts. Thanks also to Dr Valerie Crute who has supported our work whilst we have been producing the book and to Dr Joe Miemczyk and Dr Mickey Howard for supporting the project within our centre.

In addition, we would like to thank all those colleagues from industry, government and academia who have helped formulate our thinking with regard to build-to-order. In particular, the foresight and encouragement offered by Daniel Jones, chairman of the Lean Enterprise Academy, and Malcolm Harbour, MEP, together with the enthusiasm and knowledge regarding the importance of manufacturing given by Sir John Harvey Jones.

Finally, we would like to thank our wives, Tsui Parry and Dr Claire Graves as we have found that organising and directing a significant part of the ILIPT programme as well as putting a book together is unavoidably intrusive on family life. Producing a book with so many authors has been a challenging experience, but has hopefully provided greater clarity with regard to the BTO vision. This would not have been possible without their ongoing support and occasional editing assistance.

Contents

Part III   Collaboration

Chapter 1
Introduction and Overview

Glenn Parry and Andrew Graves

School of Management, University of Bath, Bath, UK

Abstract. This book addresses the conceptual and practical aspects of the automotive industry's next goal – the delivery to the customer of a bespoke vehicle 5 days after placing the order.

We have brought together a selection of leading European automotive experts from across industry and academia to provide insights into the goals of producing cars to order. Whilst the "voices" are all different, the message is the same – build to order is the future of the European motor industry. A compelling need to change has been identified and viable methods to do so have been developed and are presented here. Whilst the research and findings are the result of many years' work in the automotive industry, most have application in many other sectors.

Many car companies are losing money. The mass-production business model of the automotive industry is flawed and perhaps becoming dysfunctional. The industry suffers from global overcapacity and rising stock levels and exhibits inherently low profitability. It has now been nearly two decades since "lean production" was documented and the auto industry in the West set out to employ Japanese best practice and close the productivity gap. While lean efforts have delivered improvements in manufacturing efficiency, they have been largely ineffective in increasing profitability, due to a myopic focus on factory processes. The automotive industry has optimised systems for mass production, but not tackled the problems of capacity and demand. We find ourselves in a position where, following leading practice, a car can be built from flat steel in a production facility within 11 h. A customer ordering a car in a dealership has to wait around 40 days to purchase their desired vehicle, or buy one from stock. How has this occurred?

1.1 Drivers and Followers

Toyota, the originator of lean, which was developed as the Toyota Production System (TPS), has moved ahead of its rivals and is now the global leader in manufacture. Toyota has been very open in letting other car manufacturers come and study its production techniques. In doing so it has ensured that many have spent their time seeking to catch up and following their lead instead of developing their own processes, something that may threaten Toyota's dominance. A spokesman for GM stated "Should the day come where GM is no longer the largest (globally) it will come out fighting the next day" (Kranz 2007). As any racer knows, it is very difficult to catch a moving vehicle from a standing start let alone get past them again, especially when they are world champion. Toyota has focussed on constantly improving its quality as it grows its market share. However, its approach is very different to that of GM, typified by their president, Katsuaki Watanabe "We want to be number one in terms of quality, but we don't pay much attention to catching up with General Motors". And whilst others have been catching up in terms of manufacturing efficiency, Toyota is rising to become the world's biggest car maker, overtaking the Ford Motor Company, Chrysler group and now challenging General Motors, who have held the position for 76 years. It is not so much that quality is lacking in the competition, but Toyota and the other Japanese manufacturers took the lead and hence the customer base. The car buying public is frequently brand-loyal, so a significant and valued differentiator will be required by any company trying to tempt new buyers into their dealerships. This is very difficult when those buyers' previous experience of the product was poor and their subsequent experience at a new dealer positive.

There is also a new and increasing pressure upon the incumbents from the rise of companies from newer automotives markets such as Korea, India and, most significantly. With a low-wage economy and a government thirst to create manufacturing sector work, the Chinese automotive industry is developing its automotive mass production capability. These companies are in their infancy and their products are currently behind the world's best. However, they are moving quickly to catch up. Illustrating the rate of development, the Nanjing Automotive Company (NAC) plant in Nanjing went from brown field to functioning factory in 2 years through purchase of the MG Rover assembly lines and models. They have begun with production of the "Rover 75", a vehicle first seen in Europe in 1998. Whilst this may be seen as old technology, their domestic market is very large and may well buy up all their annual 200,000 units' production. These sales will provide them with the revenue needed to develop their next generation of cars. NAC is not alone. Other Chinese manufacturers such as Chery and Shanghai (SAIC) have also been acquiring automotive know-how through partnerships with automotive companies around the world. Their domestic sales will provide them with the cash flows for future investment. The size of the Chinese home market, together with their speed of expansion and low cost labour market, could rapidly make them a creditable threat on the global stage.

The time is right for a change in the approach of the European Automotive industry. Build-to-order (BTO) systems appear to offer the solution. But what exactly do we mean by build-to-order?

1.2 Defining Build-to-Order

Build-to-order refers to a demand-driven production approach where a product is scheduled and built in response to a confirmed order received for it from a final customer. The final customer refers to a known individual owner and excludes all orders by the original equipment manufacturer (OEM), national sales companies (NSC), car dealers, fleet orders or other intermediaries in the supply chain. BTO excludes the order amendment function, whereby forecast orders in the pipeline are amended to customer requirements, as this is another level of sophistication for a build-to-stock (BTS) system. Build-to-stock is the dominant approach used today right across the automotive supply chain and refers to products that are built before a final purchaser has been identified, with production volume driven by historical demand information. This high stock level, endemic across the auto industry, allows some dealers to find an exact or very close match to the customer's desired vehicle within the dealer networks and supplier parks. The vehicle can then be delivered as soon as transport can be arranged. This has been used to justify stock levels. Whilst providing a rapid response to customer demand, the approach is expensive, mainly in terms of stock, but also transportation as finished goods are rarely where they are required.

A BTO system does not mean that all suppliers in the supplier chain should be producing only when a customer order has been confirmed. Clearly, it would not make economic sense for a manufacturer of windscreen wiper blades to employ BTO. These components should be built to a supplier order, effectively BTS. However, a large expensive item, such as an engine, could and possibly should be BTO. Part of the challenge in a BTO supplier network is in the identification of which suppliers should be BTO and which BTS. The point in the supply chain when this change occurs is called the "decoupling point". Currently, the majority of automotive supply chains lack a decoupling point and the dominant BTS approach has resulted in capital being tied up in stock in the supply chain and billions worth of finished automobiles.

1.3 Advantage of Build-to-Order

Significant advantages are to be gained by the automotive companies who would benefit from a considerable reduction in capital employed, reduced tooling, thus reduced fixed costs and a shift to variable costs. Consequently, the industry will have lower break-even volumes, allowing a greater responsiveness to changes in demand, fashion and regional preference. A leading manufacturer estimated its current inventory located at distribution parks in Europe to be worth €10 billion, reflecting the picture in the US. A press release in the US quantified the finished automobile stock held by the Detroit 3: GM retail inventory 94 days; Ford retail inventory 105 days; Chrysler retail inventory 126 days (Webster 2006). These figures suggest that stock levels are so high that, in theory at least, all production could halt completely for one-third of a year.

Building cars to customer order implies a considerable reduction in capital employed through the virtual elimination of stocks, as well as a reduction in the need to discount. Any company able to free this capital would improve their competitive position and gain access to capital for future product development. However, the implied ramifications for all key players in the system – component suppliers, logistics service providers, retailers and dealerships – have not yet been recognised by most companies; yet, the complexity of the automotive industry and the product itself make it imperative that not only the manufacturers change, but also that the entire supply network reaches the capability to support such drastic change.

The question for many automotive executives is not when, but how exactly will such a radically "different" business model operate?

1.4   The Road to Build-to-Order

It was clear by the late 1990s, that both US and European vehicle manufacturers(VMs) had significantly closed the performance gap on the leading Japanese producers through the adoption of a Lean Production philosophy, as outlined by MIT's International Motor Vehicle Programme (IMVP) in their enterprise benchmarking studies. There was a clear reduction in the variation of best practice performance across regions in final assembly, product development and supply chain delivery. However, it was also evident that most VMs had focussed upon optimising their assembly plant productivity and had developed "shop-floor myopia" at the expense of delivering cars that customers wished to purchase. Customers were increasingly being educated, through the internet and other sales channels, with regard to "instant gratification" or in other words, obtaining the products and services they required, without delay. Few automotive companies were profitable and customers were largely disaffected.

To explore these issues, a collaborative research programme was formed, funded by the UK Government (via the DTI and EPSRC) together with leading industrial partners, in order to develop an organisational and process framework within which a customers' need for a vehicle could be fulfilled in 3 days – from order placement, through manufacture and delivery (see Fig. 1.1). The team comprised leading academics and consultants led by: Professors Dan Jones and Peter Hines of the Lean Enterprise Research Centre at the University of Cardiff; Malcolm Harbour, MEP, of the International Car Distribution Programme; and Professor Andrew Graves, Co-Director of the IMVP, based at the University of Bath.

VMs:	Ford, Honda, Nissan, Peugeot, Vauxhall, Volkswagen
Suppliers:	GKN, Thyssen Krupp Automotive, TI Group (Bundy) Machine Tools Technology Association (MTTA)
Logistics / Distribution:	Axial, Institute of Logistics & Transport Wallenius Wilhelmsen
Retail / Dealers:	Inchcape, Lancaster, National Franchised Dealers Association (NFDA), Pendragon, Quicks
IT:	BEA Systems, Cap Gemini, Keane
Finance:	Goldman Sachs
Government:	DTI (UK), Engineering & Physical Sciences Research Council (EPSRC)

Fig. 1.1 3DayCar Partners (1999–2001)

Three days was set as a possible target for vehicle delivery as, at the time, it was unclear how long the process took. However, the programme's aim was in reality, an attempt to satisfy customer requirements by producing a car that the "customer wanted, when they wanted it".

Research at the core of the 3DayCar programme benchmarked the time taken for a customer's order to transit via the car dealership, sales channels, programme planning, operations and factory capability. From the factory, the finished vehicle was tracked via the distribution outlet and dealer, to the customer. For the first time, research utilising a value stream mapping methodology showed that the VMs took nearly 40 days on average to deliver a vehicle (Fig. 1.2).

• Order entry Dealer-Manufacturer	**3.8 days**
• Order bank	**9.8 days**
• Scheduled orders	**14.1 days**
• Sequenced orders held	**6.0 days**
• Physical production	**1.4 days**
• Loading at factory	**0.9 days**
• Distribution to dealer	**3.8 days**
' 39.8 Day Car ' Capability	

Fig. 1.2 Generic model for average delivery time for 3DayCar OEMs

It is clear from the data that physical production takes approximately only 1.4 days. Therefore, the increasing concentration of effort by VMs to improve factory operations, although important, was having little or no effect on the overall order to delivery time (OTD). Even when a theoretical "best practice" was investigated (i.e. best performance of each VM in each area), the OTD lead time was still 12 days – nowhere close to the 3DayCar target. In addition, it was also discovered that up to 50 days of stock was held in the overall system. A key number of problem areas were identified by the 3DayCar team, including:

- Vehicle complexity – the number and variety of models
- Capacity constraints – VMs vs suppliers
- Schedule/build unreliability
- Transparency – via IT legacy systems and lack of standards

The time from order entry to physical factory production was identified as taking a total of over 1 month! Possible solutions put forward by the 3DayCar team suggested a substantial re-engineering of the whole system with particular focus upon four key areas. These were:

- *Improved communications* – direct links between marketing, supply, logistics and product and process development
- *Alternative product strategies* – e.g. "plug & play" offering "Intel inside" solutions – i.e. "Siemens or Bosch inside"
- *New assembly strategies* – micro factories, changing the economy of scale for VMs.
- *Alternative body structures* – independent body panels (IBP), new materials

In conclusion, the 3DayCar programme found that the current "push" system of production was inherently inefficient and encouraged over-production and massive stock levels, while at the same time, failing to satisfy customers. The real focus of activity for the VMs, suppliers, logistic providers and dealers, for the 21st century was now to concentrate on real customer demand, rather than factory operations. Also, the need to "build to order" would become an essential tool in the return to profitability that would impact on every player from the customer to dealer, supplier and manufacturer.

1.5   The ILIPT Project

The findings of the 3DayCar programme pioneered in the UK created world-wide interest from both industry and governments with regard to developing the next competitive transformation for the automotive industry. In particular, the European Commission, through its "Intelligent Logistics for Innovative Product Technologies" (ILIPT) programme, proposed a pan-European research project to study the applicability of the 3DayCar findings across the European automotive sector. Of particular interest were the four problem areas identified by the 3DayCar team, as highlighted above, and the possible solutions in relation to European VMs, suppliers and logistics operators. The 3DayCar research results were therefore to be tested and analysed in the wider and more complex European arena. In addition, it was agreed to set a challenging target of 5 days from order to delivery for the European context. This would take into account the complexity of the European manufacturing and supply base and the research findings from the 3DayCar project, giving a theoretical "best practice" for BTO of a minimum of 12 days. To meet this new "5-Day Car" target the necessary improvement in productivity would require a radical restructuring across a broad spectrum of activities, as well as a possible revolutionary change with regard to its technological capacity.

This book therefore seeks to provide some of the answers to the "how to do BTO" question by drawing on expert knowledge generated by the BTO automotive expert participants who contribute to the ILIPT project. The project's aim is simple – to define, validate and operationalise processes, product structures and supply network structures that enable the purchase and on-time delivery of a new, customer-specified car, across Europe within a 5-day window. ILIPT is a 4-year joint European Commission and industry-funded project that will reach completion in the summer of 2008. This European Integrated Project is a consortium of 31 leading automotive-focussed partners with representatives from across Europe and throughout the supply chain. The auto-industry is a major contributor to prosperity throughout the EU and the ILIPT project has the potential to enhance European industry competitiveness, sustainability and will secure benefits for the next generation. The project extends the contribution of the International Motor Vehicle Program (IMVP), which produced the seminal Lean text *The Machine That Changed The World* and the 3DayCar initiative. ILIPT's re-invention of the industry required a global team including partners from eight EU Member States and participants from Russia, Switzerland and Brazil, comprising VMs, all supply tier levels, research establishments and academia.

The originating ILIPT project partners are:

- University of Bath
- BMW Motoren GmbH
- University of Cambridge
- Ceramicx Ireland Ltd
- CLEPA (European Association of Automotive Suppliers)
- DaimlerChrysler
- Dana Corporation
- De-Bonding Ltd
- E-Business und Prozessconsulting EBP GmbH
- EFTEC AG
- FEV Motorentechnik GmbH
- 4Flow AG

- Fraunhofer Institut für Materialfluss und Logistik (IML)
- Fraunhofer Institut für Produktionstechnik und Automatisierung (IPA)
- Fraunhofer-Institut für Arbeitswirtschaft und Organisation (IAO)
- Freeglass
- Hella Autotechnik s.r.o.
- Lear Automative EEDS, Spain S.L.
- University of Lueneburg
- MAN Ferrostaal
- University of Patras

- PLATOS
- Pontificia Universidade Catolica
  do Rio de Janeiro
- Saint-Gobain Sekurit
- Siemens AG
- Siemens VDO
- The St. Petersburg Institute for
  Informatics and Automation of

the Russian Academy of Science
(SPIIRAS)
- Technical University of Dresden
- ThyssenKrupp Automotive
- TRW Automotive GmbH
- VDI/VDE-IT

The industrial co-ordinator of the project is René Esser of ThyssenKrupp Automotive and it is administered by Helmut Kergel of VDI/VDE-IT. The ILIPT project is divided into three main work packages. The first package, ModCar, aims to develop a digital prototype of a 2015 passenger vehicle using an innovative system of flexible body frame panels and modules for rapid build to order. This prototype has been named ModCar. It is supported by the Technical Method Integrator, which examines novel IT tools, and methods needed to deliver "5-day car capable" design and production. Overall, the objective is to minimise complexity in design and production while optimising vehicle delivery and variety to the end-customer. The work package is led by Karl Josef Kerperin of Continental, previously Siemens VDO Automotive AG. The second work package, FlexNet, focuses on the concurrent development of new flexible processes, technologies and supply network organization structures. The work explores concepts of collaborative IT, supply chain planning and automated contract negotiation. It is led by Professor Bernd Hellingrath of the Fraunhofer Institute for Materials and Logistics. The objectives of the final theme, InterPro are to develop new network design tools and novel process simulation methods to test and validate concepts for implementation of a 5 day car process, as well as integrate key results and concepts from the other themes. In addition the work requires the development of an understanding of the transition path towards the future automotive BTO network, how to achieve a 5-day car between 2010 and 2015. This theme is led by Dr Glenn Parry of the University of Bath, School of Management.

The ILIPT project represents a vast European effort to develop new concepts. This book draws upon only a small selected subset of automotive experts from the ILIPT consortium to provide an overview of how the BTO vision may be achieved and provide Europe with a globally leading automotive industry.

1.6 Book Structure

This book's text is divided into five parts (Fig. 1.3).

In the first part we explore automotive industry dynamics. Dr Matthias Holweg from the University of Cambridge presents an overview of the evolution of competition within the automotive sector, and his colleague, Andreas Reichhart, then explores how BTO has evolved. Next, Alexandra Güttner and Thomas Sommer-

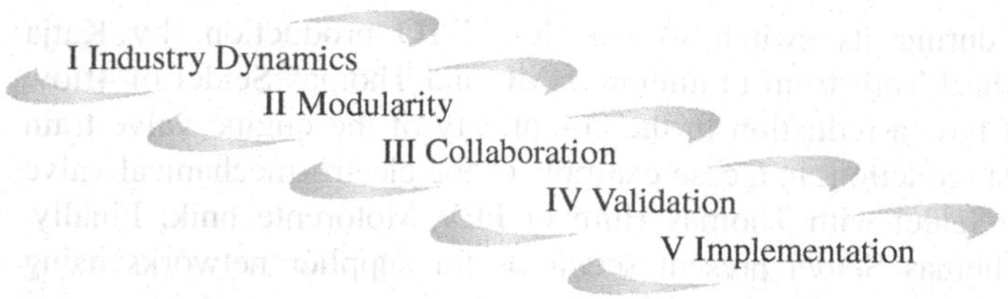

I Industry Dynamics
II Modularity
III Collaboration
IV Validation
V Implementation

Fig. 1.3 Book layout overview

Dittrich from Daimler AG discuss current issues at OEMs and suppliers, providing an insight from industry as to the practical realities of production and supply chain management. Building upon this, the management and practice of automotive outsourcing is then detailed by Jens Roehrich, University of Bath.

The second part describes the use of modularity. ILIPT's ModCar demonstrator prototype vehicle is used as the focus for exploring the practicalities and technical developments in car body modularity. The chapter begins with Philipp Gneiting and Thomas Sommer-Dittrich of Daimler AG providing the OEMs' perspective. Andreas Untiedt of ThyssenKrupp Automotive provides an insight into the development of a modular lightweight automotive body, which decouples the car structure from its surface, allowing short production times and reduced stock cost. This work is complemented by Dr Maik Gude and Professor Werner Hufenbach from TU Dresden, who explain the design of the car body shell that complements the modular car body architecture. Introducing modularity introduces complexity. The relationship between customer choices in car build combinations and the effects offering these has on production complexity and cost is presented by Jens Schaffer & Professor Heinrich Schleich, from the University of Lueneburg.

Professor Bernd Hellingrath begins the third part with an overview of the key principles for collaborative planning and execution of a network capable of delivering customised production. Siemens AG's Jan-Gregor Fischer and Philipp Gneiting from Daimler AG then describe flexible processes for inter-enterprise collaborative planning. Collaborative execution processes that reduce the time lots between customer order and vehicle assembly scheduling are presented by Jörg Mandel from Fraunhofer-IPA. The information communication technology challenges that need to be overcome to move towards the BTO concepts are discussed by Markus Witthaut and Michael Berger from Fraunhofer-IML with Jan-Gregor Fischer, Siemens AG. The work from this group is then illustrated by modelled scenario examples from Stefanie Ost and Joerg Mandel of Fraunhofer-IPA.

In Part IV we present the work done to validate the approach and tools developed to achieve the 5-day car. The work of EBP Consulting's Katja Klingebiel provides decision support for the transition to a more flexible and stable BTO strategy for all partners in an automotive supply chain, accomplished by the provision of a BTO reference model for high-level network and process design. Thomas Seidel of 4flow AG presents a method of integrating two supply chain modelling approaches into more rapidly produced feasible detailed supply chain scenarios. BMW Steyr's engine plant provides a case example of the implementation of the

ILIPT approaches during its switch to stockless BTO production, by Katja Klingebiel and Michael Toth from Fraunhofer-IML and Thomas Seidel of 4flow AG. An account of how a reduction in the complexity of the engine valve train leads directly to cost reduction, in a case example of the electro-mechanical valve train from Thomas Seidel with Thomas Huth of FEV Motorentechnik. Finally, Kati Brauer and Thomas Seidel present scenarios for supplier networks using a number of distribution concepts.

The final section looks at the broader picture. The motivations and barriers to forming electronic market places that will facilitate BTO systems are investigated by Dr Mickey Howard from the University of Bath. Dr Joe Miemczyk from Audencia Business School questions the received wisdom that that simply locating suppliers in close proximity to OEM assembly plants reduces delivery lead time and inventory. Finally, Gareth Stone and Dr Valerie Crute from the University of Bath present the challenges for transition.

Whilst the work cannot present a complete picture, hopefully it will provide both an overview and sufficient detail to engage the reader in the subject and make a persuasive case for BTO.

References

3DayCar: http://www.3daycar.com/

Kranz R (2007). Toyota: How high is up? *Automotive News*, January 8th

Webster SA (2006) Top dealer says there are more unsold cars than reported, Detroit Free Press (Online), 26th October

Part I
Industry Dynamics

Chapter 2
The Evolution of Competition in the Automotive Industry[1]

Matthias Holweg

Judge Business School, University of Cambridge

Abstract. At the dawn of the second automotive century it is apparent that the competitive realm of the automotive industry is shifting away from traditional classifications based on firms' production systems or geographical homes. Companies across the regional and volume spectrum have adopted a portfolio of manufacturing concepts derived from both mass and lean production paradigms, and the recent wave of consolidation means that regional comparisons can no longer be made without considering the complexities induced by the diverse ownership structure and plethora of international collaborations. In this chapter we review these dynamics and propose a double helix model illustrating how the basis of competition has shifted from cost-leadership during the heyday of Ford's original mass production, to variety and choice following Sloan's portfolio strategy, to diversification through leadership in design, technology or manufacturing excellence, as in the case of Toyota, and to mass customisation, which marks the current competitive frontier. We will explore how the production paradigms that have determined much of the competition in the first automotive century have evolved, what trends shape the industry today, and what it will take to succeed in the automotive industry of the future.

---

[1] This chapter provides a summary of research conducted as part of the ILIPT Integrated Project and the MIT International Motor Vehicle Program (IMVP), and expands on earlier works, including the book *The second century: reconnecting customer and value chain through build-to-order* (Holweg and Pil 2004) and the paper *Beyond mass and lean production: on the dynamics of competition in the automotive industry* (Économies et Sociétés: Série K: Économie de l'Enterprise, 2005, 15:245–270).

2.1 All Competitive Advantage is Temporary

The roots of today's motor industry can be traced back to Henry Ford, who, based on the inter-changeability of components and the use of the moving assembly line, laid the foundations for modern-day mass production techniques. Even the basic features of a car have not changed much since Ford's days: a car still has four wheels, is propelled by a gasoline engine and its body is still welded together from pressed metal parts. Despite the profound impact that Ford has had on the "industry of industries"[2], its competitive advantage was short-lived and Ford was soon overtaken by GM, which, based on the visions of Alfred P. Sloan, introduced a more decentralised organisational structure and offered customers the choice they wanted through a much broader product portfolio. While civilian production significantly shrunk during the years of the Second World War, the mass production of cars in the US leveraged the growth of the post-war period until the 1970s saw increasing competition from Japan, where companies like Toyota seemed to be able to offer better deals – in terms of quality and cost – to customers in the US and Europe.

The success story of lean production, leading to the difficult situation faced by the US and European manufacturers in the three decades since 1970, is well known and all major players in the industry have adopted the set of techniques that were first introduced at Toyota in Japan, the Toyota Production System (TPS), or "lean production" as it is more widely known. However, competitive forces are far from being static, and hence vehicle manufacturers can no longer rely on excellence in production only, especially since the performance gap between them has been closing (Holweg and Pil 2004). The automotive industry in the new millennium has seen the advent of three key challenges: regionalisation, saturation and fragmentation of markets, challenges that few manufacturers have addressed successfully to date. New capabilities are required to deal with this competitive situation and return to profitability. There is an increasing number of countries in the world today that have mastered the skills of producing cars with acceptable levels of quality, and often at a much lower cost compared with the US, Europe or Japan.

At the turn of the second automotive century the news from the automotive industry in the established regions is anything but encouraging: record losses are being reported in Detroit, and in Europe household names are, for the first time, being squeezed out of the market. Britain alone has seen the closure of five major car plants over as many years, and one might get the impression that for every factory that closes in the West, (at least) one is opening in Eastern Europe, India or China – suggesting that the days of the motor industry in the western world are numbered. In Japan, several corporate crises and even threats of bankruptcies have been averted, most prominently in the case of Nissan.

But painting a picture of gloom misses the point: the industry is mature, the barriers to entry are high and demand is growing – on average, global car production

---

[2] A term coined by Peter Drucker in 1946.

has increased by just below 2% annually since 1975, and major new markets in Asia and previously in Latin America have opened opportunities. The real conundrum is why successful strategies in this industry are so short-lived? Amidst a wealth of explanations pointing to legacy health care costs, to China's rise and to a perennial overcapacity, the real root cause is commonly overlooked: manufacturers have relied heavily on static business models, and have simply failed to adapt to a changing environment.

Revisiting the history of the industry soon shows to what extent fortunes have changed over the last century as companies failed to align their strategies to structural shifts in the marketplace. It was Henry Ford who built his empire based on his ability to mass produce vehicles at an unrivalled cost, albeit in "any colour as long as it is black". Ford's superiority was successfully challenged by Alfred P. Sloan at GM, who sensed the customers in maturing markets desired more variety than Ford was providing. Sloan offered "a car for every purse and purpose", and Ford soon lost its market leadership in 1927 – never to regain that position.

After the war all manufacturers soared on the seemingly insatiable demand that happily took every vehicle produced. Fortunes only changed when the oil crises increased demand for economical cars, which was met by increasing imports from Japan that threatened the heartland of the US and European manufacturers. Trade barriers were soon called for, but as Japanese transplant operations sprung up, this "invasion" could not be halted. This pattern of import competition entering the low segments of the market has replicated itself several times over since: in the 1970s Japanese imports threatened the US and European manufacturers, in the 1980s it was the growing South Korean motor industry that happily filled the space the Japanese vehicle manufacturers left as they moved upmarket, and there is little doubt that the Chinese manufacturers will lead the next wave of import competition by the end of the decade.

Initially, this success was achieved through leveraging their cost advantage, but today the Japanese and Koreans are competing on a level playing field – and thanks to superior manufacturing methods, have captured a 17% market share in Western Europe, and even 37% of the US car market. The real issue that drove this expansion was not labour cost, but the Western manufacturers' inability to adopt leaner manufacturing methods to meet the Eastern productivity and quality standards.

Instead, Western manufacturers sought salvation in size. The mantra of the 1990s was that an annual production of one million units and global market coverage ensured survival, and we are now seeing the fall-out from this single-minded pursuit of volume. Daimler-Benz was not the only one to get caught out: BMW equally failed in its venture with Rover, as did GM in its alliance with Fiat. The wider lesson here is that scale alone does not ensure survival. Those alliances, which do indeed provide economies of scale, crucially also feature a strong complementarity in terms of capabilities. Take Renault-Nissan for example: leveraging compatible product architectures, Nissan's manufacturing strongly complements Renault's design capabilities. This also applies to market coverage: Nissan is well represented in Asia and North America, where Renault has hardly any presence.

Renault is considerably stronger in Europe and South America, where Nissan plays a minor role. And for some, not allying with other firms makes perfect sense – of which Toyota and Honda are living proof – and even BMW does much better without a volume car division.

At present, all attention is on the growth in China and India, where combined vehicle production has grown to an equivalent of 44% of Western European output. Suppliers and manufacturers alike tremble at the thought of low-cost imports from this region, and the "China price" is an often-used menace in price negotiations. Frequently omitted though is that manufacturers and suppliers alike have benefited handsomely from the growth of the Chinese domestic market. Nonetheless, Western manufacturers have been responding with a steady migration into low-cost regions such as Eastern Europe – initially Poland, the Czech Republic and Hungary, and soon into Slovakia and Romania; yet, this strategy is short-lived at best: competing on cost alone is not only futile, it also misses the point.

Success in this mature industry neither has been, nor will it ever be decided on the basis of unit cost or scale alone. It is the ability of the manufacturer to sense trends in the market, and align its product range that determines success. And it is this stubborn refusal to accept these changes that poses the greatest threat to the Western motor industry: relying on high volumes of gas-guzzling SUVs in times of rising fuel prices and growing environmental concerns is as short-sighted as the European manufacturers' perennial love affair with luxury vehicles.

China and legacy costs are often portrayed as the main threats to the industry, but they are not the root cause of the woes we are feeling. In the long run, wages will rise even in China, as they have already done in Japan and Korea. And as one low-cost region develops, soon there will be another emerging. These mantras of scale and low unit cost might have worked in the past, but no longer suffice in today's dynamic world. Those who are able to adapt to shifts in market demand and to respond to customers' wishes will thrive, and rightfully so – if history teaches us one thing, it is that all competitive advantage is temporary.

In this chapter, we will explore in detail the past, present and future of competition in the automotive industry. How did production systems evolve that determined competition in the past, what are the present trends that shape this global industry, and what will it take to succeed in the future?

2.2   The Past: The Evolution of Production Systems

The motor industry has made a dramatic transition over the last century. From small workshops that had crafted customised vehicles for the affluent few, to Ford's mass-produced Model T, which made motoring available to the public at large, and to the Toyota Production System, which proved to the world that high productivity and high quality can be achieved at the same time. Many researchers have studied these drastic transitions in the motor industry, trying to understand how this drastic change could happen in such a short time. Historians such as

David Hounshell, Allan Nevins and Lawrence White, for example, debate the drivers and enablers of the change from the craft production of the late 19th century, which was prevalent at the time (Nevins 1954; Nevins and Hill 1957; White 1971; Hounshell 1984), and Womack et al. and Takahiro Fujimoto give a detailed account of the lean production paradigm as a contrast to the mass production approach (Womack et al. 1990; Fujimoto 1999).

At the start of the automotive industry were the craft producers of the likes of Panhard et Levassor, Duesenberg, and Hispano Suiza, which employed a skilled workforce to hand-craft single vehicles customised to the wishes of the few customers who could afford them. The core of the mass production logic, or the Fordist system, which was to turn the economies of the motor industry upside down from 1908 onwards, was not the moving assembly line, as many suspect, but in fact the inter-changeability of parts, and Ford's vision to maximise profit by maximising production and minimising cost. This notion was very different from the existing economies of the craft producers, where the cost of building one vehicle differed little if only a single car was made, or a thousand identical ones: since all parts were hand-made, and subsequently amended by the so-called "fitters", the amount of labour required per vehicle differed little, if at all. Furthermore, most vehicles at the time were customised to individual requirements, so standardising parts was not a priority.

It was this notion of the inter-changeability of parts that would become the critical enabler of Ford's mass production system, a concept that originally stems from the arms-making sector (Hounshell 1984). Initially proposed by Eli Whitney and later implemented by Samuel Colt, the ability to standardise parts meant that the assembly operation could be streamlined, and the entire job function of the "skilled fitter" was made redundant. The moving assembly line, however, implemented by Ford in his Highland Park factory in 1913 for the first time, is merely a logical evolution of the production concepts of flow production and standardisation of parts and job functions. As Robert Hall argues, "[...] there is strong historical evidence that any time humans have engaged in any type of mass production, concepts to improve the flow and improve the process occur naturally" (Hall 2004, personal communication). Historians to date disagree who actually invented the moving assembly line, whether it was within Ford or within the McCormick Harvesting Machine Company, and who within Ford made the critical changes (Nevins 1954; Hounshell 1984). In my view this debate is hardly relevant: it was Ford's vision to produce the most vehicles at the lowest possible cost that became the imprint of mass production, and the foundation stone of motor industry economics of the 20th century (White 1971; Rhys 1972). In the same way, it has been argued that Ford was influenced by the Taylorist approach of Scientific Management, which was proposed at a time when Ford's mass production model was still being crafted (Taylor 1911). However, there is no evidence that this influence actually happened, and indeed Ford never referred to Taylor as such in any official documentation (Hounshell 1984). Instead, the concomitant standardisation of work practices and the product itself, the inter-changeability of components, flow production, and the moving assembly line should be seen as tools that allowed Henry Ford to turn his

Model T Production, Sales Volume and Retail Price

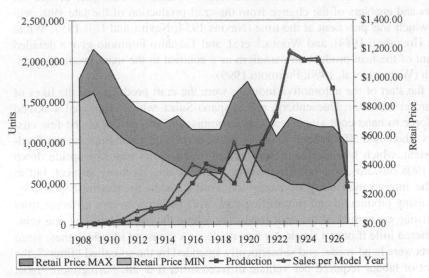

Fig. 2.1 Model T production, sales and retail price band 1908–1927. Source of data: Benson Ford Research Center. Nominal monetary values by year

vision into reality, rather than as the essence of mass production. As Peter Drucker puts it: "[...] The essence of the mass production process is the reversal of conditions from which the theory of monopoly was deduced. The new assumptions constitute a veritable economic revolution" (Drucker 1946).

Henry Ford had the vision that literally changed the face of the planet – to produce large volumes of cars in order to reduce the cost per unit, and make the cars available to the masses. And his new "mass production", first called as such in an article in the 1925 edition of the Encyclopaedia Britannica, worked well for almost two decades. Ford was able to reduce the labour hours for assembly of the vehicle from 750 h in 1913 to 93 h in 1914, and the entry-level sales price for a Model T could be reduced from $1,200 in 1909 to $690 in 1914 (see Fig. 2.1). With the introduction of the moving assembly line came labour challenges. The new type of work was not well received by the work force, and staff turnover soared to unsustainable levels (Hounshell, 1984). And although sometimes misinterpreted as a philanthropic move by Henry Ford, the famous "five-dollar-day" was primarily geared at making the workplace attractive for workers to stay, and as a secondary effect also meant that his own workers soon became able to buy these cars, so demand was stimulated.

The demise of the pure mass production logic came suddenly, and as a surprise to Henry Ford: when for the first time in 1927 more customers bought their second cars than bought their first, it soon became clear that the outdated Model T (which from 1914 to 1926 was indeed only available in one colour, black) could not offer the level of specifications expected by the customers. It was at this time that Alfred P. Sloan at General Motors could finally compete against Ford. By offering "a car for every purse and purpose", GM was able to offer customers the choice they desired, and the possibility to move up from the mass brands such as Chevrolet, to prestige brands, such as Cadillac – all within the realms of the GM brand portfolio. Ford's market share dwindled from 55% in 1921 to 30% in 1926, and it took Ford a long time to develop the replacement model for the Model T, the Model A, in order to be able to compete against GM. In the view of many historians, Sloan complemented Ford's mass production model by marrying the mass production logic with the need to offer choice and a brand portfolio to the customer. As a key element of constant innovation, or the "search for novelty", Sloan also introduced the "model year" in the 1930s, which involved cosmetic updates to each vehicle each year – a practice that persists today.

Hounshell (1984) refers to this stage as "flexible mass production", although one should be clear that the increasing levels of product variety led to just the opposite – factories found it difficult to cope with the product and part variety, so components and vehicles were made in large batches to make the economies of scale so critical to mass production. Consequently, lead times and inventory levels soon rose in those factories that Womack et al. describe as typical mass producers in their seminal work *The machine that changed the world*, which marked the second major turning point for the auto industry of the 20th century (Womack et al. 1990; for a comprehensive review see also Holweg 2007).

Womack et al. described the Toyota Production System (TPS), which had been developed at Toyota in Japan as an alternative way of manufacturing cars. Taiichi Ohno and Saiichi Toyoda, the intellectual fathers of the approach, had borrowed many ideas from Ford's original flow production system at Highland Park: tightly synchronised processes, short changeovers that allowed for small-batch production, machines that stopped in the event of a defect, and a social system designed around workforce empowerment and continuous improvement (Pil and MacDuffie 1996). For a detailed discussion of the evolution of the Toyota Production System see Cusumano 1985, and Fujimoto 1999. Further inspired by quality gurus such as Deming, this lean production system, a term coined by MIT researcher John Krafcik (Krafcik 1988), soon proved to the world that the notion of trading quality against productivity was invalid. Prior to this, the assumption was that high quality levels could be achieved only if more labour was used to correct the quality problems, and vice versa, so that higher productivity would invariably compromise product quality.

This "Japanese manufacturing model" had been known as "just-in-time" in the Western world since the early 1980s, but surprisingly little notice was taken (Schonberger 1982; Hall 1983; Monden 1983; Ohno 1988). It was only in the late 1980s, when Japanese imports captured an increasing portion of the US auto market, that the Western auto industry became concerned. Henry Ford II even called the Japanese imports an "economic Pearl Harbour". Initially, attempts were made to restrict imports through voluntary trade agreements (Altshuler et al., 1984), but it soon transpired that the Japanese possessed a unique ad-vantage. And it was not until researchers of the MIT International Motor Vehicle Program showed that – taking the differences in vehicle size into account – the best Japanese were almost twice as productive as their American counterparts. The Japanese took an average 16.8 h to build a car, the US makers 24.9, and the European 35.5 h (Womack et al. 1990). At the same time, Japanese vehicles showed much higher levels of product quality, and thus could disprove the common belief of a general trade-off between productivity and quality in manufacturing. Although known for almost a decade in the West, the *Machine* book brought the lean production paradigm into the Western world by showing its superiority in the global comparison – and all at a time, when the Japanese exports posed the greatest threat to their Western counterparts.

Since then, most manufacturers have adopted lean manufacturing techniques in their operations. Although for political reasons often not called "just-in-time" or "lean", initiatives like the "Ford Production System" and its counterparts at the other Western manufacturers are clear evidence that key features of the lean production paradigm have been implemented (to a varying extent) by most manufacturers in the US and Europe. Also, starting with the opening of Honda's factory in Marysville, Ohio in 1984, the Japanese carmakers established a strong local manufacturing presence in the US, Europe, and emerging markets through their transplant operations, further aiding the diffusion of lean manufacturing techniques into the component manufacturing bases in the Western world. These operations, in particular in the US, were established to circumvent import tariffs, but played a key role in disseminating the knowledge of lean production (Krafcik 1986; MacDuffie and Pil 1994; Pil and MacDuffie 1999).

2.3 The Present: Shifts in the Competitive Landscape

At the start of its second century, the automotive industry is undergoing a period of drastic change: we have seen both record profits and bankruptcy of global suppliers and manufacturers, some of the largest industry mergers and de-mergers, and – largely thanks to emerging new markets – an ever increasing global demand for automobiles. If one looked at the present news coverage of the automotive industry across the globe, the obvious conclusion would be that this is an industry in deep trouble. In its last year of being the largest vehicle manufacturer on the planet, GM posted a loss of $8.6bn dollars, and the combined job cuts announced in early 2006 by GM and Ford totalled 60,000, with no less than 26 plants to be closed in North America by 2008. Unsurprisingly, US employment in automotive manufacturing has steadily fallen from 1.3 million in 2000, to 1.1 million in 2005.

In Europe, the situation is hardly more comforting. In January 2006, Volkswagen tuned in with a further 20,000 job cuts, and Mercedes and parent company Daimler-Chrysler announced a combined 14,500 – in addition to the 40,000 Chrysler jobs lost after the 1988 merger. And while Fiat posted profits in 2005, this came after 17 consecutive quarters of operational losses and after employing almost as many CEOs over that period. And, last but not least, in April of 2005, MG Rover ceased operation, ending a century of the British volume car industry. To put this into perspective, British Leyland (the former name of Rover) was nothing less than the

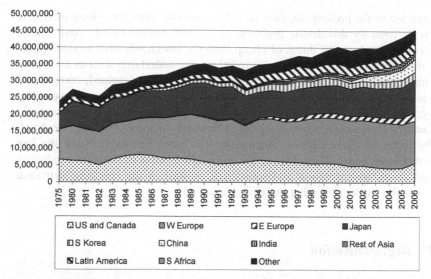

Fig. 2.2 World passenger car production by region, 1975–2006. Source: Ward's Yearbooks

fourth-largest vehicle manufacturer in the world in 1970, with a production volume of close to 1 million units per annum.

In Asia, where the perception is generally that the management techniques of manufacturing companies are superior, a similar picture emerges. With the possible exception of Toyota, we have seen the near bankruptcy and foreign takeover of Nissan, and major crises at Mitsubishi, Daewoo and Proton.

Overall, the automotive industry is not a happy place at the start of its second century. However, there is a paradox to this malaise: despite the depressing news, we are building more motor vehicles than ever. In 2004, global production of passenger cars totalled 42.5 million units, to be complemented by 21.2 million commercial vehicles, which added to the global total of 837 million vehicles in operation that need to be maintained and serviced. On average, the production of automobiles has been growing by 2.2% (1.8% for passenger cars) every year since 1975 (see Fig. 2.2). So why is it that this mature industry, with its high barriers of entry, that clearly finds customers for its products, finds it so hard to create a profitable and sustainable business proposition?

The answer is not as straightforward as some of the simplistic answers that have been suggested: legacy healthcare costs, overcapacity, and of course, the cheap imports from China. There is an element of truth in each, but none can explain what is fundamentally going on in the industry. Indeed, the legacy costs for some manufacturers like GM are calculated at $1,525 for each vehicle sold, but if a UAW worker earns $60,000 plus benefits, this cannot come as a surprise. This adjustment in labour cost should have come much earlier on, as GM was essentially still living in the good times of the past. Overcapacity, in 2004 estimated at approximately 20 million annual units globally, is a similar issue: the developments that have led to the present situation have been on the cards for a long time, and there is no need for this drastic problem, as we will discuss below. And finally, China constitutes both the threat of cheap imports, but largely also a huge opportunity due to the domestic demand. Nonetheless, the underlying shift in the manufacturing footprint, together with the persisting overcapacity created, competition in the automotive industry is fierce. Plagued by legacy costs and increasing product variety, vehicle manufacturers are constantly seeking ways to compete in a world that features increasingly demanding and impatient customers on the one hand, and the threat of cheap Chinese imports on the other. Let us examine the key trends that have, and still are shaping the competitive arena of the motor industry: regionalisation, fragmentation and saturation, as well as the resulting structural changes in the supply chain that these have invoked.

2.3.1 Regionalisation

Over the past few decades, we have observed several distinct shifts in the manufacturing footprint that has shaped the industry's structure as it is today. As demand in the established regions has been stagnating, we have seen several major waves of investment in emerging markets. In 1970, the vehicle production of the US, Western Europe and Japan combined accounted for 91% of the world's 22.5 million car production. Back then, the US and Western Europe in particular were large net exporters, while Japan was still on a steep curve of increasing both production and export volumes. By 2004, the picture had changed considerably. Of the 42.8 million units that were built, only 70% came from the three established regions, USA, Europe and Japan. The number of assembly plants had grown from 197 to 460, of which only 44% were located in North America, Western Europe and Japan. What had happened was that the industry had distributed its manufacturing base: whereas previously largely knock-down operations (CKD or SKD) were used in emerging markets, the growth of their respective domestic demand now justified full-scale assembly plants. The increase in demand in Latin America in the 1990s, for example, sparked a wave of investment in the motor industry in those countries. From 1980 to 2000, the combined vehicle production in Argentina, Brazil and Mexico nearly doubled to just under 4 million units. Yet, the experience obtained in Latin America also serves as a warning signal, as the demand in Brazil and Argentina collapsed sharply after currency devaluation. Exchange rate uncertainty remains an issue, today more than ever, with respect to the most recent wave of expansion in China, and the artificially pegged Yuan.

The opening of the Chinese domestic market, in conjunction with a strict growth policy, has seen the dramatic rise of the Chinese automotive industry. With virtually no passenger car production before 1980, China produced 2.32 million cars (total vehicles: 5.1 million) in 2004. Of these, 90% were made by the joint venture companies of the large foreign manufacturers, and virtually all have been (so far) sold domestically. Even by the later parts of this decade more than 90% of China's

Table 2.1 Share of world car production by region, 1971–2003. Source: Centre for Competitiveness and Innovation, University of Cambridge

		1971	1980	1990	1995	1997	2000	2001	2002	2003
World car production	(in million units)	26.45	28.61	36.27	36.07	38.45	41.23	39.97	41.22	41.78
Industrialised countries	Percentage of world car production	90.85	89.90	87.84	81.98	73.44	74.85	75.27	72.26	70.12
	Percentage of growth (based on previous period)	–	7.03%	23.87%	–7.19%	–4.5%	9.28%	–4.28%	0.83%	–1.62%
Newly industrialised countries	Percentage of world car production	5.14%	7.65%	8.66%	15.05%	17.31%	17.22%	18.00%	21.36%	23.45%
	Percentage of growth (based on previous period)	–	61.03%	43.38%	72.93%	22.6%	6.19%	–0.08%	24.61%	11.3%

production is used to meet growing domestic demand, and thus does not yet pose an import threat of the kind that Japan and South Korea did, and maybe still do.

What one can observe here is not what is commonly referred to as *globalisation*, but what is much better described as *regionalisation* of the industry. The net export balance that fostered the growth of the automotive industry in the industrialised world over much of the last century is gradually being replaced with an infrastructure that builds vehicles locally, close to the customer. The immediate result for the established regions has been a necessary yet painful capacity adjustment, and the closure of plants like Luton, Dagenham and Longbridge in the UK are likely to be followed by others in Western Europe. In the USA, the overcapacity situation is even more pronounced, and further Big Three plant closures in addition to those already announced are expected.

Lower labour costs are generally stated as the main reason for the increase in decentralising global production into countries with low labour costs, and comparing the nominal hourly remunerations, there are indeed stark differences (see Table 2.2). But how significant are labour costs? First of all, in the overall cost structure, the approximate production cost of a vehicle from the customer's point of view breaks down as follows: 31% of the list price is accounted for by distribution and marketing costs, as well as dealer and manufacturer margins; the 69% ex-factory costs split into 48% for procured parts and materials, 9% overheads, and only 13% is related to the vehicle production operation. Here, labour represents the largest component, alongside capital investment depreciation of the production assets. When one compares the above to the hourly rates a worker earns then it is obvious that labour cost is indeed a significant competitive factor in the lower segments of the market; yet, it does play a decreasing role in the higher market segment, where firms do not compete on cost alone, but on technological innovation, design and brand image.

Table 2.2 Average hourly remuneration for production workers in manufacturing. Data for 2003, Source: Bureau of Labour Statistics 2004 & Economist 2005

Germany	$29.91	South Korea	$10.28
USA	$21.97	Czech Republic	$4.71
UK	$20.37	Brazil	$2.67
Japan	$20.09	Mexico	$2.48
Spain	$14.96	PR China	$1.30

2.3.2 Fragmentation of Markets

The second key trend is one that is relatively easy to observe; namely, the implosion of traditional vehicle segments, in favour of cross-over and niche vehicles. The traditional segments of small cars (B-segment, e.g. Polo or Fiesta), compact cars (C-segment, e.g. Golf and Focus), family cars (D-segment, e.g. Passat and Mondeo), and executive class (E-segment, such as E-class and 5-series) have been joined by SUVs, MPVs, UAVs, and the like. In quantitative terms, this trend can be easily seen: across Europe, in 1990 a total of 187 models were offered, which increased to a total of 315 models in 2003. This increase is not only due to the new segments, such as MPVs and SUVs, but also to model line expansions in existing segments. The B-segment of the Corsa and Fiesta, for example, saw an increase from 16 to 31 models over that time period.

The increase in model range is accompanied by a general shortening of product life cycles. While the average time a product stayed in the market was around 7 years in 1970, this average has been reduced to 5 years – a trend consistent across the US and Western Europe. In Japan, life cycles have traditionally been much shorter, and some companies like Toyota have coped by building two generations on one platform, before changing both design and platform with the third generation.

Together, the increase in model range and the reduction in life cycles have a drastic impact on the economies of scale that can be achieved. The volume sold per model has been significantly reduced over time, which gives the manufacturers less and less opportunity to recover their considerable development cost. As a reaction, manufacturers are trying to increase the component sharing and platform usage across as many models as possible. Table 2.3 illustrates the overall shifts in volume per model, and the use of platforms in Europe.

This development is, and will continue to be, a major challenge for vehicle manufacturers. While the large players are currently working on leveraging their resources across their brands, for smaller companies this is not so easy. One reason why MG Rover failed was the need to cover the growing new market segments, while volumes were shrinking in the traditional segments in which it was offering products. Ultimately, its volumes were too small to finance the required product development programmes, and with an ageing line-up in limited segments, sales continued to fall.

Table 2.3 Platform usage in the European automotive industry. Source: Pil and Holweg (2004)

	1990	1995	1996	1997	1998	1999	2000	2001	2002
Number of platforms in use (Europe)	60	60	57	56	53	51	45	45	48
Number of body types offered (Europe)	88	137	139	148	157	162	170	178	182
Average number of body types per platform	1.5	2.3	2.4	2.6	3.0	3.2	3.8	4.0	3.8
Average production volume by platform (in 1,000s)	190	171	185	194	199	215	249	272	258
Average production volume by body type (in 1,000s)	129	75	76	73	67	68	66	69	68

2.3.3 Saturation and Overcapacity

The third key trend is a malaise that is entirely self-inflicted: as a result of the failure to adjust capacity to demand, the auto industry suffers from a global overcapacity that at this point is estimated at 20 million units – equivalent to the combined installed capacity in Western Europe! The basic reason for the overcapacity is an asymmetry: it is much easier to add capacity than it is to reduce it. With an average level of employment of 5,000 workers per assembly plant and an additional job multiplier of up to four jobs in the supply chain, governments encourage, and most often also subsidise, the building of new vehicle assembly plants. For the same reason, closing a plant when demand drops is difficult and quickly becomes a political issue.

The main consequence of the overcapacity is that manufacturers – in their quest to keep capacity utilisation high – produce into the growing inventories of unsold cars (around 1.5–2 months in most markets), and then employ sales incentives, such as discounts, high trade-in prices, free upgrades, and the like, to maintain their market share. Initially, the problem was confined to the North American market, which after the recession of 2001 has seen an increasing "war of attrition" between the manufacturers. Average incentives then and to-day range between $2,000 and $6,000 per vehicle. That way, the Big Three have indeed managed to maintain their market share, yet their position is not sustainable, as the respective 2005 losses of Ford's and GM's automotive businesses graphically illustrate.

The root cause here is a chronic inability to adjust output to demand and link the production schedule to actual customer orders. While Henry Ford founded the industry on the premise of making vehicles as efficiently and inexpensively as possible, this mass production "volume-push" approach is no longer viable in current settings of saturated markets, where one has to deal with increasingly demanding customers. At times when Dell illustrates that one can order a customised product that is built to order within only a few days, the established automotive business model seems obsolete. Several manufacturers have understood the need to link production to customer demand and have successfully initiated "build-to-order" (BTO) programmes, such as Renault, Nissan, BMW and Volvo. Their success has illustrated that one can indeed build a car to customer order within 3 weeks or less, and operate without the costly finished vehicle inventories and the incentives needed to clear the overproduced cars from dealer stock. Most other manufacturers recognise the need to get closer to their customers, but implementation often lags behind what the press releases state. One could argue that while there is widespread intellectual acceptance, there is an equally widespread institutional apathy.

2.3.4 Structural Changes in the Supply Chain

The pressures outlined above faced by the manufacturers have led to internal changes (such as increased platform usage across models), but the wakes are equally felt in the supply chain – most prominently at the interface with the first-tier suppliers. The main changes here are a general reduction of supplier numbers per vehicle assembly plant, the re-distribution (i.e. outsourcing) of value-added activities, and the increase in globally sourced components and materials.

The reduction in supplier numbers, shown in Table 2.4, is driven by two strategies: first, in order to develop longer term, collaborative (Japanese-style) relationships, vehicle manufacturers focus on a few key partners, rather than change suppliers opportunistically based on unit price only. Second, the increasing product variety means that vehicle manufacturers have to rely more and more on their suppliers to provide the design and assembly of key vehicle systems and modules. This drive towards outsourcing required a re-tiering of the supply chain, whereby several previous first-tier suppliers became "0.5-tier" module or systems suppliers, now sourcing components from their previous first-tier peers. For the vehicle manufacturer, outsourcing was also a means of harnessing the lower labour costs at suppliers ($17 versus $23/h in the USA) and to reduce transaction costs by dealing with fewer suppliers at the same time.

Unlike the components we see in the computer sector that feature standardised interfaces, a motor vehicle features a largely integral product architecture that

Table 2.4 Number of suppliers across Japan, Europe, the USA and new entrant countries. Source: Holweg and Pil (2004)

Region	1990	1994	2000
Japanese OEMs in Japan	170	173	206
European OEMs in Europe	494	357	341
US OEMs in North America	534	457	376
New entrant countries	409	615	201

renders the outsourcing of modules difficult. The drive towards modularity also called for a geographical change in the supply chain. With the need to provide sequenced parts deliveries at short notice, "supplier parks" were created in the 1990s that housed primarily module and systems suppliers in the immediate vicinity of the car assembly plant. And within these parks, logistics companies often took on tasks such as component sequencing and minor assembly tasks. In a general sense, considerable value added by the manufacturer was outsourced to component suppliers, and to a lesser extent, logistics service providers.

Interestingly, the structural changes in the auto supply chain do show a stark dichotomy. On the one hand, increasing outsourcing requires physical proximity to enable a fast response time to manufacturers' call-off signals. On the other hand, manufacturers are increasingly sourcing components from distant regions with low labour costs, such as Eastern Europe, Mexico or China, which induces long logistics lead times. Despite the hype, China was still a net importer of components in 2004, but it is widely estimated that this balance will shift towards increased component export in 2008. Sourcing from China, however, creates operational tensions, in particular where customised or configured components are sourced from abroad. For example, the wiring harness is generally specific to a particular vehicle, yet very labour intensive, which poses a constant temptation to source it from low-cost regions. With a logistics lead time of as many as 6 weeks, this means that the build schedule has to be set for these 6 weeks in advance, which severely limits manufacturing flexibility and makes a rapid response to an impatient customer almost impossible.

The auto industry is undergoing considerable change, and it is in particular the structure of the supply chain that is changing. Caught between a rock and a hard place, manufacturers are trying to become more responsive to customer needs and avoid the costly inventories and sales incentives that cut into their profitability at present. At the same time, they are trying to reduce cost by outsourcing tasks, and by sourcing components from low-cost regions, and in some cases, even relocating their vehicle assembly operations to these regions. While the current competitive pressures in the motor industry are not likely to subside for the time being, logistics companies on both the inbound and the outbound side can harness these for their growth. Bridging the gap between distant component suppliers, and coordinating a supply chain that increasingly is not only measured on cost alone, but also on how fast it can deliver the product to the customer, is a task that neither vehicle manufacturers nor suppliers are particularly well set up to do. Here, logistics companies have the unique ability to integrate their core transportation business with additional value-added services that can include anything from component sequencing and the management of supplier parks, to the late configuration of entire vehicles (Reichhart and Holweg 2008). In an industry that features intense competition and a global stage at the same time, logistics companies are the connecting element in the system, and now have the chance to advance as an enabler of a supply chain that is both cost-efficient and responsive to customer needs.

28 M. Holweg

2.4 The Future: Competing in the Second Automotive Century

At the turn into the second automotive century, the automotive industry finds itself in a complex competitive situation, and one that is hard to explain with the current notions of "craft, mass and lean producers" The reason is that the competitive landscape is much less clearly divided than it had been for most of the first automotive century. Boyer et al. (1998) illustrate this fact well by showing that – instead of a universal best practice – auto companies have developed individual forms of work organisations and production systems that are shaped by their respective national environments and business histories.

First, the persistent overcapacity in the industry has resulted in an unprecedented wave of mergers and acquisitions in the industry. Coupled with the financial crises in Asia and considerable mismanagement in many Japanese industrial conglomerates, the keiretsus, this has led to the situation that – apart from Toyota and Honda – all Japanese carmakers were at least partially owned by a Western vehicle manufacturer at the end of the 20th century. Also, most Western manufacturers have joined forces with others in order to achieve higher economies of scale in purchasing and product development, to develop a global brand portfolio, and to gain access to emerging markets. Many of these mergers have a rather troubled history, such as DaimlerChrysler and Mitsubishi, are far from delivering the financial returns that were hoped for, and have not led to the reduction in global overcapacity that had been hoped for (Holweg and Pil 2004).

Second, since almost all vehicle manufacturers across global regions have adopted lean manufacturing techniques, the competitive advantage of the Japanese has been considerably reduced. The results from the global assembly plant survey of the MIT International Motor Vehicle Program show that the gap between the US and Japan has been reduced to duration of build. As shown in Fig. 2.3, the average vehicle build takes 16.6 h in the US, compared with 12.3 h in Japan and 21.3 h in Europe (Holweg and Pil 2004). Equally, product quality has improved considerably since 1990. In fact, the quality has improved so much that JD Power

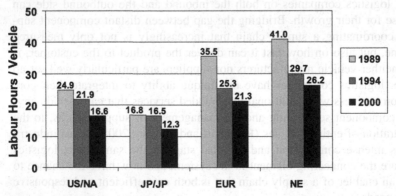

Fig. 2.3 Labour productivity across US, Europe, Japan and new entrant countries. Source: Holweg and Pil (2004)

(the institution that collects the customer quality data) had to tighten up their measurements in 1997, as most vehicles simply scored "zero defects". Overall, our current benchmarking studies found strong evidence that the "message of lean" had indeed been heard in the Western world, and although Japan is still in the lead, the competitive situation is far less drastic than it was in 1990.

Third, the globalisation and wave of mergers in the 1990s also meant that a global identity is far less obvious to establish. The same applies to the brand image. Is Volvo still Swedish, or is Saab now American? Not only has the ownership of many "national" producers changed, some of their vehicles may also not even be produced in their "home countries" any longer in the future. This raises further questions as to whether any regional comparisons (the "Japanese" model against the "Western" model, for example), still make any sense. This is furthermore problematic as a strong local manufacturing presence dilutes the incentives for policy-makers. In fact, the Big Three have continuously been losing their market share in the US, and in 2002 even lost their majority in the US passenger car market, down from a market share of more than 90% in the 1950s. Accordingly, the assembly capacity that is being added to the US market is almost exclusively thanks to new Japanese, Korean and European transplants, whereas the Big Three show a consistent net loss of capacity and employment. Thus, the transplants make an attractive proposition to policy makers, and are generally being subsidised by the respective local governments. Building automobiles remains the world's largest manufacturing activity, and the industry directly or indirectly employs one in every seven people (Sako 2002).

While the fortunes in the industry have changed drastically over the last century, the way we sell and distribute cars has not. In fact, Henry Ford's legacy equally lies in the way we run factories, and sell the vehicles that have been made by our mass production factories. Craft producers used to build all vehicles to customer order in the 1900s. Henry Ford made his Model T entirely to forecast and sold the cars from dealer stock, which allowed him to run the factories as efficiently as possible. His reasoning was that running higher volumes at the factory would reduce unit cost, and thus the sales price. Lower sales prices in turn would increase demand, and therefore sales. This logic was fine when demand exceeded supply, but in today's market, where increasingly demanding customers require customised vehicles at short lead times, this forecast-driven model is flawed (Holweg and Pil 2004). Yet, to date, most manufacturers drive their production by long-term sales forecasts, and then hope to sell their vehicles from dealer stock thereafter. As can be seen in Table 2.5, the majority of vehicles are still built to forecast across regions. The basic underlying problem of increasing the content of vehicles built to order (thus avoiding the costly inventory and sales incentives) are the long lead time it takes to build and deliver a vehicle to order. In Europe, the average order-to-delivery (OTD) lead time is 41 days, yet customers are generally only willing to wait 2–3 weeks (with the exception of few very patient customers, and the German market, where build-to-order has a long tradition). Thus, in order not to lose any sales to competitors with better availability, manufacturers produce vehicles against a sales forecast, and sell vehicles from stock, where they are

Table 2.5 Sales sourcing in major volume markets. Source: Shioji (2000), Williams (2000)

Sales source	Europe	United Kingdom	Germany	United States	Japan (Toyota)
Cars built to customer order (BTO) (%)	48	32	62	6	60
Sales from central stock (distribution centres) or transfer between dealers (BTF) (%)	14	51	8	5	6
Sales from dealer stock (BTF) (%)	38	17	30	89	34

instantly available to the customers. Supply is driven by the production forecast, and demand is adjusted by using sales incentives.

In a world of global overcapacity and fashion-conscious consumers, the results of this mass production logic are disastrous: vehicle manufacturers use increasing amounts of sales incentives to sell off their overproduction, and thereby not only erode their brand image, but also put serious strain on the residual values of their brands and models. This in turn hurts the (currently still) very profitable leasing operations of the vehicle manufacturers (Holweg and Pil 2001). In fact, manufacturers such as GM or Ford currently derive considerably more profits from their leasing and finance arms than from manufacturing cars in the first place. If the current make-to-forecast practice and the current levels of incentives persist, that situation may well change in the future. Since the start of the new millennium, the Big Three in particular have been fighting a war of attrition on the levels of incentives, and by 2004 levels of $3,000 per vehicle were consistently observed as average across the US market, and exceeded $5,000 for individual models. More recently, these incentives have also affected markets such as Europe, and surprisingly, the new entrant market, China, where the developing overcapacity is taking its toll.

2.5 What Next?

The question arises: what is to come next? What new concept might follow the implementation of lean production, increasing scale through platform-sharing, global mergers and collaborations, and build-to-order strategies? Where is the competitive realm going to shift after mass customising products? As could be observed in other sectors, the offer of services around the product could provide further differentiation. For example, one could think of providing a complete "mobility service" to the customer, rather than simply selling a vehicle. Yet, even if such advanced service offerings were to become mainstream in the near future, in terms of manufacturing strategy, however, such a shift would have little impact. Manufacturers would still build vehicles to customer specification, even if the customer does not own the vehicle any more, but simply remunerates a service subsidiary of the manufacturer for using the vehicle. Others argue that the internet will drastically

alter the way we market and sell cars, in the same way as telematics offers radically new ways of redefining vehicles as communication platforms (Sako 2002;
Fine 2003). The largely unfulfilled promises of the e-commerce and internet applications, as well as the slow establishment of telematics applications in vehicles,
however, cast serious doubts over their potential to radically alter competition in
the automotive industry.

In my view, the next major change in the competitive realm is going to be triggered by a major shift in technology, i.e. the advent of a "disruptive technology"
(Christensen 1997). Such radically new technology would then reset the competitive dynamics back to the days of Henry Ford – completely new technology will
require considerable changes to current practices and change existing economies,
as did mass production to the automotive industry at the time. Initially, manufacturers will seek to boost production volumes to achieve better economies of scale.
The speed of adoption is critical, as the "chicken-and-egg" dilemma (high product
price due to low production volume on the one hand, and low sales due to the high
price of the product on the other) needs to be overcome quickly in order to reach
market acceptance. Thus, as in the case of Henry Ford in 1908, the focus will be
on minimising production costs and increasing the market share in order to establish new technology. Only once the market matures will the competition shift
away from mere cost-driven strategies, towards variety, diversification, and customisation. The double helix dynamic that establishes itself, as shown in Fig. 2.4,
is one that mirrors the developments in many other markets and industries, and
one that has been used to describe the evolution of product architecture and other
management processes (Fine and Whitney 1996; Fine 1998). Although we have
seen these dynamics in many sectors, such as electronics and communication,
many times over, the striking fact is that technology in the motor industry has not
yet changed radically, and that we are on the verge of seeing the double helix
complete with in the next few decades. And this change might, for example, be
catalysed by alternative propulsion technologies entering the mass market.

It is not within the remit of this chapter to speculate about the adoption of advanced powertrains in the automotive industry. What is clear, though, is that environmental needs and the price of fossil fuels will require changes to the current

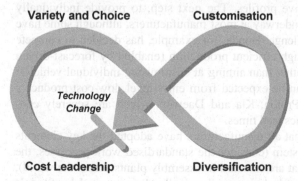

Fig. 2.4 Helix dynamics of competition in the automotive industry

powertrain technology. None of the options at hand has established itself as the dominant design or technology as yet – once this has happened, however, the dynamics of competition would run through the second cycle, with an initial focus on scale and cost leadership, moving towards greater variety and choice, and on to diversification, and ultimately, product customisation.

2.6 Conclusion

The competitive realm of the auto industry is dynamic, and has been throughout the past century. However, contrary to the past, the strategies adopted by firms are far less distinctly defined than they used to be. Over the last century we have witnessed the evolution from craft production to mass production under Henry Ford, to Sloan's policy of brand and product variety, to lean production, and more recently, to build-to-order initiatives at both volume and luxury vehicle manufacturers. Along the way, most manufacturers have adopted a wide range of mass and lean production tools and techniques, as well as Sloan's concept of a brand portfolio. Thus, today we see elements of all these approaches across manufacturers: the moving assembly line, the product and brand portfolio, model years, and lean production techniques are common at most manufacturers, even at those luxury makers that traditionally were seen to be "craft producers". In the process, the competitive realm has shifted considerably, and the main basis on which companies are competing has changed.

In this chapter, the dynamics of the competitive realm in the motor industry have been laid out over time, and four generic phases could be identified: cost leadership, variety and choice, diversification, and customisation. At present, most companies are at the diversification and customisation stages of this model, although it could be argued that Ford and GM in North America have remained at the "variety and choice" stage, competing on both cost and model variety, whereas others, such as BMW, Volkswagen, Toyota, Audi, and Renault, have found their diversifying feature: brand image, innovative design, leading product technology or manufacturing excellence provide the basis on which these companies have established individual competitive profiles. The next step, to provide individually customised vehicles, is well underway at most manufacturers, although some have chosen to opt out of this challenge. Honda, for example, has decided to compete on the basis of low cost through efficient production (enabled by forecast-driven strategies and low variety), rather than aiming at customising individual vehicles. Similar low-cost strategies can be expected from entry-level, low-cost producers such as Hyundai, Daihatsu, Proton, Kia and Daewoo, which are severely constrained by their import logistics lead times.

What is clear, though, is that all manufacturers have adopted the key elements of Ford's mass production system (consider the standardised work processes, the moving assembly lines etc. that are standard in assembly plants across the world), the need to provide variety and choice so drastically demonstrated by Sloan's success at GM, and the lean production paradigm that laid the foundation for Toyota's persisting success. Thanks to the implementation of lean production techniques, the way we manufacture vehicles has changed considerably – the way we sell vehicles, however, has changed little since the days of Henry Ford. Large vehicle stocks and sales incentives are the inevitable by-products of the forecast-driven production and sales strategies still pursued by most manufacturers. Few companies have realised that the new competitive battle, in a setting of global overcapacity, increasing dynamic variety and customers demanding customised products, is how to overcome this second legacy of the mass production system: forecast-driven production planning and vehicle supply. Early adopters of BTO strategies such as Volvo (Hertz et al. 2001) and Renault ("Project Nouvelle Distribution") have the objective of linking the mass production facility to customer demand. Early adopters will undoubtedly face challenges; yet, most will likely also benefit the most from adopting BTO, whereas the remaining companies are likely to be forced to follow suit, or to continue on their mass production path and become the providers of low-cost, entry-level cars in a segment that will continually be challenged by low-cost import competition. Truly sustainable competitiveness in tomorrow's automotive industry can only be found in developing customer-responsive supply systems that respond to both demanding customer needs, as well an increasing product and model variety that has invoked considerable changes in the economic foundations of the global automotive industry.

References

Altshuler A, Anderson M, Jones DT, Roos D, Womack J (1984) The future of the automobile. MIT Press, Cambridge

Boyer R, Charron E, Jürgens U, Tolliday S (eds) (1998) Between imitation and innovation: the transfer and hybridization of productive models in the international automobile industry. Oxford University Press, Oxford

Christensen C (1997) The innovator's dilemma. Harvard Business School Press, Boston

Cusumano MA (1985) The Japanese automobile industry: technology and management at Nissan and Toyota (Harvard East Asian Monographs, No 122). Harvard University Press, Boston

Drucker P (1946) The concept of the corporation. Day, New York

Fine CH (1998) Clockspeed – winning industry control in the age of temporary advantage. Perseus, Reading

Fine CH (2003) Opportunities when value networks collide: telematics at the intersection of automotive and telecommunications. RIETI Policy Symposium: The 2003 RIETI-HOSEI-MIT IMVP Meeting. Hosei University, Tokyo

Fine CH, Whitney D (1996) Is the make-or-buy process a core competency? Working paper, Center for Technology, Policy, and Industrial Development, Massachusetts Institute of Technology, Cambridge

Fujimoto T (1999) The evolution of a manufacturing system at Toyota. Oxford University Press, Oxford

Hall RW (1983) Zero inventories. McGraw Hill, New York

Hertz S, Johannsson JK, de Jager F (2001) Customer-oriented cost cutting: process management at Volvo. Supply Chain Manag 6(3):128–141

Holweg M (2007) The genealogy of lean production. J Oper Manag 25(2):420–437

Holweg M, Pil FK (2001) Successful build-to-order strategies start with the customer. Sloan Manag Rev (Fall):74–83

Holweg M, Pil FK (2004) The second century: reconnecting customer and value chain through build-to-order. MIT Press, Cambridge

Hounshell DA (1984). From the American system to mass production 1800–1932: the development of manufacturing technology in the United States. John Hopkins University Press, Baltimore

Krafcik J (1986) Learning from NUMMI. IMVP working paper. Massachusetts Institute of Technology, Cambridge

Krafcik JF (1988) The triumph of the lean production system. Sloan Manag Rev (Fall):41–52

MacDuffie JP, Pil FK (1994) Transferring Japanese human resource practices: Japanese auto plants in Japan and the US. Paper presented at the IMVP Research Briefing Meeting, June. Massachusetts Institute of Technology, Cambridge

Monden Y (1983) The Toyota production system. Productivity, Portland

Nevins A (1954) Ford: the times, the man, the company. Scribner, New York

Nevins A, Hill FE (1957) Ford: expansion and challenge 1915–1933. Scribner, New York

Ohno T (1988) The Toyota production system: beyond large-scale production. Productivity, Portland

Pil FK, Holweg M (2004) Linking product variety to order fulfilment strategies. Interfaces 34(5):394–403

Pil FK, MacDuffie JP (1996) The adoption of high-involvement work practices. Ind Relat 35:423–455

Pil FK, MacDuffie JP (1999) What makes transplants thrive? J World Bus 34(4):372–391

Reichhart A, Holweg M (2008) Co-located supplier clusters: forms, functions and theoretical perspectives. Int J Oper Prod Manag 28(1):53–78

Rhys DG (1972) The motor industry: an economic survey. Butterworth, London

Sako M (2002) The automobile industry. Warner M (ed) The international encyclopedia of business and management. Thomson, London

Schonberger RJ (1982) Japanese manufacturing techniques. Free Press, New York

Shioji H (2000) The order entry system in Japan. International Symposium on Logistics, Morioka, Japan

Taylor FW (1911) Scientific management. In: Pugh DS (ed) (1997) Organization theory – selected readings. Penguin, London

White LJ (1971) The automobile industry since 1971. Harvard University Press, Cambridge

Williams G (2000) Progress towards customer pull distribution. Research Paper 4/2000, The International Car Distribution Programme, Solihull

Womack JP, Jones DT, Roos D (1990) The machine that changed the world. Rawson Associates, New York

Chapter 3
Build-to-Order: Impacts, Trends and Open Issues

Andreas Reichhart and Matthias Holweg

Judge Business School, University of Cambridge

Abstract. The promise of BTO has been widely discussed: lower finished vehicle stocks and higher margins on the one hand have to be balanced against the higher cost of providing flexibility in manufacturing on the other. In this chapter we discuss the impacts of BTO on the OEM as well as the component supply and distribution chain, and outline the key challenges that OEMs face on their way to implementing BTO, including approaches for mitigating these.

3.1 Introduction

The automotive industry is a difficult market in which to compete. The competitive pressure and dynamics that firms face are both intense and complex. Competition takes place in multiple dimensions: firms need to develop appealing car models, build strong brands, and excel in operations simultaneously (Fujimoto 2006). For the past 15 years, the last dimension, "operational excellence", has been a synonym for lean production; yet, more recently, a number of OEMs have aimed to increase their responsiveness to end-customers. Such responsiveness is commonly associated with reductions in order-to-delivery (OTD) lead times and an increase in the percentage of cars that are built to customer order.

While some firms have achieved higher levels of flexibility, others have failed, or have not yet tried. The obstacles, or disadvantages, encountered by OEMs on their way to achieving such flexibility are inter alia increasing costs in vehicle assembly and component supply, as vehicle assembly plants are less well shielded from demand variability in the market. It has been argued that these cost increases are likely to be outweighed by cost reductions and revenue increases in the distribution system (Holweg and Pil 2004), and in a later section of this chapter,

Sect. 3.4, we will show that a higher build-to-order (BTO) content can indeed reduce finished vehicle stocks.

Despite the strategic importance of BTO in the automotive industry, a holistic assessment of the key obstacles along the automotive supply chain that prevent firms from achieving high BTO rates and short lead times is still missing. In order to provide further insight into this complex, yet important topic, we will focus on four areas. In the first section, we will discuss the penetration of BTO in the automotive industry. Subsequently, we will explain why the adoption of BTO strategies has been slow, highlighting the obstacles that firms face when implementing a BTO system. This section should also be of interest to executives who believe that their firms have already achieved a high level of customer responsiveness. As our research has shown most executives have an overly optimistic view of their firm's performance, mainly due to ill-defined key performance indicators, such as the firm's BTO content. In this sense, we also attempt to provide a foundation for critical reflection. The third section will highlight and quantify some of the benefits of BTO, thereby providing initial incentives for firms to overcome the discussed obstacles. Finally, we will recommend some changes to the current operating system in the automotive industry that can help mitigate the adverse cost impacts of increasing the flexibility of the supply chain. The insights provided are based on interviews with more 80 executives from ten different vehicle manufacturers, their suppliers and new vehicle dealers. In addition, we have visited more than 20 vehicle assembly plants and numerous supplier facilities. For Sect. 3.4, we have further analysed a large quantitative dataset provided by one of the OEMs in order to quantify and highlight some of the measurable impacts of BTO.

3.2 BTO: Where Is the Industry?

Most major OEMs have trialled BTO initiatives with the aim of increasing the share of cars that are built to customer specifications, while at the same time reducing their order-to-delivery (OTD) lead times. Table 3.1 provides an overview over a number of selected BTO programmes. As one can see, the majority of these programmes aimed at an OTD target of around 2–3 weeks, yet most OEMs were far from achieving this target. In the first comprehensive benchmarking of the automotive industry's BTO capabilities, Holweg and Pil (2001) found that the industry – despite good intentions – was far from achieving such short delivery lead times. The key reason identified for this performance gap was a convoluted order booking and scheduling process.

With their existing processes in place, even the best performing OEM could not achieve an OTD time of less than 21 days, because of all the steps required from receiving a customer order at the dealership to assigning a production slot to the

Table 3.1 Build-to-order programmes in the automotive industry

OEM	Programme name	OTD target
BMW	KOVP (customer-oriented sales and production process)	10 days
DaimlerChrysler	FastCar/global ordering	15 days
Ford	Order-to-delivery	15 days
General Motors	Order-to-delivery	20 days
Renault	Project Nouvelle Distribution (PND)	21 days (initially 14 days)
Nissan	SCOPE (Europe), ANSWER I+II (Japan), ICON (USA)	14 days
Volkswagen	Kunde-Kunde ("Customer-to-Customer")	14 days
Volvo	Distribution 90, COP (Customer Ordered Production)	14 days (initially 28 days)

car, and assembling and delivering that car to the customer. Some of the worst performing OEMs even had minimum OTD lead times (the so-called system capability) of close to 100 days. In order to include the recent advances in the industry, we have updated Holweg and Pil's results by studying the processes at four additional vehicle manufacturers. The revised figures indicated that now – at least in theory – there are OEMs that have the capability to deliver a BTO car in 10 days (see Fig. 3.1), while the demonstrated best practice (DBP) was reduced from 10 days to only 4 days[1].

While these findings appeared to support the progress in the industry towards the adoption of BTO, the interviews conducted – unfortunately – told a different story: although OEMs have made significant progress towards the reduction of their system capability, few manufacturers have achieved these in practice, often building far less than 100% of their cars to customer order. As a matter of fact, few OEMs in our sample achieved a BTO content of more than 50%, even in European markets. In addition, a number of OEMs had shifted their focus away from further reducing their OTD times towards increasing the reliability of their delivery dates. A number of high-ranked interviewees stated that an OTD capability of around 4 weeks was perceived as enough as long as this time window was met by the majority of orders, and the delivery date given to the end-customer upon placing the order was met without exception. Unfortunately, a more detailed comparison of the vehicle manufacturers' actual performance remained difficult for at least two key reasons. First, the measures used (e.g. actual OTD times and BTO content) differed across all vehicle manufacturers and a BTO content of 50% according to manufacturer A may have only been equivalent to 30% using manufacturer B's definition (see also Sect. 3.3.1.3). Second, surprisingly many OEMs still lacked the capability to measure these key performance indicators across all

---

[1] DBP is a theoretical measure achieved by combining the best parts of the order processing systems across all vehicle manufacturers.

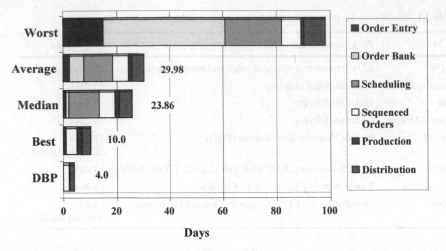

Fig. 3.1 Order-to-delivery system capability ($n = 10$)

of their markets. With this in mind, the following section will focus on explaining the key difficulties faced by the vehicle manufacturers studied on their way to becoming more customer-orientated.

3.3 The Obstacles on the Way to Build-to-Order

The slow adoption of BTO in the automotive industry was difficult to explain, as it was caused by a multitude of inter-related factors. Based on our research we have come to believe that these could be grouped into three main categories. The first set of barriers that firms faced were mostly internally related. Here, firms frequently failed to build a "business case" for BTO and convince management across all functions that BTO could indeed improve their profitability. A further obstacle was posed by market-related factors, or consumer preferences. At this point even the more advanced firms still failed to fully understand their end customers' preferences and the extent to which certain customer groups could be educated to appreciate the advantages of BTO. Finally, there remained some physical constraints in the automotive supply chain (i.e. supply-related obstacles) that inhibited the customer-responsive production of cars or individual components, often associated with the high cost of flexibility in production across the various tiers. It was these supply-related obstacles that were commonly used as arguments against BTO strategies. Yet, there were many firms that failed at the organisational level, or that built up the wrong types or levels of flexibility, and did not meet market requirements. Each of the three categories will be discussed in turn.

3.3.1 Organisational Obstacles

Organisational obstacles were the key group of factors that generally prevented a vehicle manufacturer from implementing the changes required to achieve a high rate of BTO production. As with any organisational change with severe implications for multiple departments, top-level management support was required for its implementation; yet, even top-level support did not guarantee a smooth and effective transition. In particular, there were three key problems inhibiting BTO at the organisational level:

1. The complexity of the business case,
2. Adverse impacts on various functions combined with misaligned performance measures, and
3. Incorrect BTO definitions.

3.3.1.1 The Business Case

Few, if any, large organisations will approve a project without a solid "business case", a projected income and cost statement to show what the project's impact on the firm's bottom line will be. The same applied to BTO programmes in most of the firms. However, presenting a solid (i.e. unchallengeable) business case for the implementation of BTO was difficult, if not impossible. First, there were numerous primary and secondary factors that had an impact on the associated increases in cost and revenues, and second, most of these factors could be estimated easily. Primary factors had a direct impact on cost or revenues and could be linked directly to changes in the order-to-delivery strategy (e.g. production costs, sales discounts), while secondary factors tended to show an effect with substantial time delay and may only have had an indirect impact on cost or revenues (e.g. customer satisfaction, brand value). Table 3.2 lists the factors identified during the research. A preliminary assessment will be presented later in the Sect. 3.4.

None of the vehicle manufacturers studied had successfully conducted a holistic assessment of the net impact of BTO, not even those that have implemented BTO strategies and could – theoretically – have had access to most of the required data. However, some of these factors could simply not be quantified precisely. For example, the impact that BTO had on brand value could only be roughly estimated. Also, the assessment of potential reductions in administrative overheads was difficult, especially as the managers involved were unlikely to agree that they were essentially performing tasks that would not be required in a BTO environment, suggesting that they were not adding value. On the other hand, the impacts that increasing flexibility requirements had on production operations in terms of cost increases inter alia due to spare capacity and smaller batch sizes could be estimated more closely by the manufacturing arm of the vehicle manufacturer, leading to a stark imbalance regarding the reliability of information used to assess BTO. Thus, BTO opponents within the organisation could present more convincing (i.e. quantifiable) evidence

Table 3.2 The complex business case for build-to-order

Order	Factor	Explanation	Proposed impact on profits
Primary	Decrease finished goods inventory	In a BTO system, no stock (apart from demonstration models and in-transit vehicles) will be required in the distribution system, significantly reducing current inventory levels of around 80 days of demand.	+
	Decrease sales discounts	When moving towards BTO, dealers will no longer need to grant sales discounts to customers (e.g. by reducing the sales price, through cash-back schemes or option upgrades) to take a vehicle from stock.	+
	Shorten model changes (i.e. product phase-outs)	Currently, it takes more than 80 weeks to deplete the stock of old models (at one OEM studied), once the production of the new model started, causing significant costs in terms of additional sales discounts and inventory holding.	+
	Increase option content	When ordering stock cars, dealers and NCS tend to be more cautious and do not order vehicles with expensive options (also due to the exponential impact on product variety), although an end-customer may have paid for this option.	+
	Reduce administrative overhead	Many MTF companies spend significant managerial effort planning production, managing finished vehicle stocks and negotiating sales targets with dealers. Successful BTO adopters have significantly reduced this overhead.	+
	Increase production cost	The volume and mix flexibility required has adverse impacts on production, which will need to cope with demand variability and loosen line balancing constraints.	-
	Increase component cost	The flexibility requirements will also be felt in the component supply chain, causing price increases for purchased components.	-
	Lost sales (due to price)	Some customers may choose to buy a vehicle from a different manufacturer, because their priority is to obtain an inexpensive car instead of one that meets their preferences exactly, expecting sales discounts.	-
	Lost sales (due to vehicle availability)	This factor may turn either way: Customers who want instant gratification may be lost when eliminating finished vehicle stocks (pro-MTF). On the other hand, it has been shown (see also Sect. 3.4) that an MTF system increases lead times for customer orders, thus driving away customers who want a BTO car with a short lead time (pro-BTO).	?
Secondary	Improve understanding of customer needs	Eliminates the disconnection between vehicle manufacturer and customer. Thus, vehicle manufacturers will be more sensitive to changes in customer demand.	+
	Increase customer satisfaction	Getting the customer exactly what they want with short lead times is likely to increase customer satisfaction.	+
	Increase brand value	Eliminating sales discounts is likely to increase brand value.	+
	Increase residual values	Eliminating sales discounts has been shown to increase the residual value of the respective car model, increasing the profitability of the manufacturer's leasing arm.	+

against the proposed changes. To quote a senior executive from one of the OEMs that has successfully introduced a BTO programme: "If we have had to quantify the net benefits from lean distribution [i.e. BTO] before going down that path, we would not have done it, yet we are glad we did." However, even vehicle manufacturers with high BTO rates faced challenges in their daily operations, especially when demand was low. They often reverted to pushing stock cars onto the market, because again they failed to comprehend the impact of such behaviour on their bottom line. To put it in the words of another senior executive from a different manufacturer: "Theoretically, we need to maximise profits. But there is almost no way to really understand how profitability is driven on a day-by-day basis. So often, we simply try to keep variability away from our [vehicle] assembly plants."

3.3.1.2 Adverse Impacts and Misaligned Performance Measures

The introduction of BTO required change and these changes created additional costs in functions like manufacturing and materials control. As cars were no longer built to forecast, the manufacturing function faced significant variability, further requiring it to relax the various assembly line and component supply constraints. While these requirements already created resistance against implementing BTO, simply because change was seldom appreciated at the outset, they were amplified by misaligned performance measures. As a matter of fact, misaligned performance measures combined with functional "silos" were the key reason behind the slow adoption of BTO across the vehicle manufacturers studied. For example, who can blame a manufacturing manager for opposing variability in production, if the manager is measured based on the number of vehicles produced per day or the unit cost in production? Even if the executives could be convinced of the overall benefits of BTO, they would not support changes with direct adverse impacts on their own performance (and often remuneration). However, performance measurement systems across most OEMs were still based on the principles of traditional mass production, with production volume, unit cost and market share being by far the most significant measures.

A similar problem existed with regard to the sales and distribution function. The strict use of volume (and often wholesale instead of retail) targets encouraged the distribution network to push vehicles onto the market when demand was low. Instead of assessing the overall profitability of a sale, dealers and distribution managers reverted to placing stock orders (when there were wholesale targets) and pushing stock cars at customers using sales discounts. As one of the dealers interviewed reported: "Some years ago, when [the OEM] introduced BTO, it was great. We could sell more profitable cars without carrying vehicle stocks. However, then [the OEM] started to introduce very high [wholesale] targets and we had to order stock cars again. It became so bad that we started pushing cars onto customers again just to clear our stocks, because new cars keep on arriving every month. All we would need is a couple of months without [wholesale] targets to clear our stocks and we could start selling BTO cars again."

3.3.1.3 Incorrect Build-to-Order Definitions

Comparing the BTO content across multiple vehicle manufacturers was far from a like-for-like comparison with very few OEMs applying strict measures. For example, one of the OEMs proclaimed that they only produced cars to order and hence had a BTO content of 100%, with most managers within the organisation convinced that they had reached their goal of ultimate responsiveness towards the end-customer. Instead, the OEM had simply defined that every car that was produced based on an order from one of its national sales companies (NSCs) was a BTO car. As the manufacturing arm would not produce cars without such orders, by definition, the OEM produced only BTO cars; yet, far from 100% of these orders had end-customers waiting for the cars at the time the order was placed or even at the time the car was produced. Similarly, most of the other vehicle manufacturers included certain orders in their BTO content, even though no end customer was associated with the order. Explanations for using such flawed definitions ranged from "[…] but we know that we will find a customer for this type of car soon" to "[…] if our dealers place these orders, they must know that their customers want these specifications."

Using such lenient definitions for BTO brought two major problems that could seriously damage an organisation's goal to become customer-driven. First, it prevented the vehicle manufacturer from accurately assessing its BTO capabilities, and second, such measures created a wrong (i.e. overly optimistic) perception of these capabilities. Both substantially impeded the organisation's ability to detect problems and improve its operations. If the majority of executives in an organisation believed that they produced a significant percentage of cars (if not all) to customer order, they were unlikely to engage in an open dialogue with the few colleagues who were sufficiently close to the market to know that the vehicle manufacturer needed to increase its flexibility.

3.3.2 Market-Related Obstacles

Despite the proposed positive impact that BTO can have on profitability, one key question has remained mostly unaddressed, and substantial discussion remains on the actual level and type of flexibility desired by end customers. For example, the interviewees consistently emphasised that the US market has traditionally had a very low percentage of BTO cars with vehicles being sold almost exclusively from stock, because US customers desired instant gratification and were not willing to wait even a few weeks for their car to be delivered. Despite the advances in reducing order-to-delivery lead times for BTO cars across the industry, even the most advanced vehicle manufacturers did not achieve average lead times of less than 2–3 weeks, a figure that did not provide instant gratification. An additional factor that needed to be considered was the sales discounts that were used to sell cars. There were no reliable figures on the percentage of customers who preferred receiving a "good deal" on a car as opposed to paying the full price for a car that meets their exact requirements. Rather, one could observe that few vehicle manufacturers were willing to eliminate special promotional campaigns or specific discounting techniques, although they would be capable of producing more cars to order than they did.

As a result there are two open issues that need to be answered before BTO can be fully implemented:

1. How to deal with customers that insist on instant gratification
2. What percentage of customers prefer shopping for a "good deal" (i.e. look for the largest sales discount)?

Addressing these questions is complex, because it is difficult to assess to what extent customer behaviour is a result, rather than the cause, of the current business model. Hence, it is hard to predict whether market education can change customer behaviour. There is no definitive answer to these questions so far, but those OEMs that tried to educate their dealers and change the way in which they sold cars to consumers reported good results. On the other hand, those that tried to push BTO, but neglected dealer training had a less successful story to tell. As one of the Vice Presidents interviewed reported: "We have spent far too little effort on educating our dealers. When we started to implement BTO, we expected our dealers to embrace it and start to use the new system. Now [7 years later], we know that this was a mistake."

Even for those customers who want a car built to their specifications, there is considerable ambiguity regarding the expected lead times. Apparently, sooner is not always better. A number of OEMs reported that long lead times were still regarded as a sign of exclusivity, and some customers even expected to receive a discount on a BTO car that was delivered in 1 week or less, assuming that it must have come from stock. Thus, the target order-to-deliver lead time for a number of OEMs was around 3–4 weeks; yet, for the majority of cars, actual lead times were much longer. The main reason was a lack of medium-term volume flexibility across different vehicle models combined with substantial differences in customer demand over the models' life cycle. Once a new model was introduced, demand generally exceeded supply, leading to long order queues. Towards the end of the life cycle, supply generally exceeded demand: while customers could have a BTO car without having to wait for weeks, OEMs produced a large percentage of cars to stock to keep their production operations at full capacity. Thus, there was generally only a short period in the middle of the model's life cycle, where a large percentage of cars were built to order within a 3- to 4-week window. To overcome these life-cycle-related problems, OEMs need to focus on increasing their medium-term flexibility, as outlined in the next section.

3.3.3 Supply-Related Obstacles

Key obstacles to the implementation of BTO are the physical constraints related to the production and logistics flows of components and vehicles along the supply chain. One of the key principles of lean production, as it has been adopted by virtually all major vehicle manufactures over the past few decades, is called "heijunka" or level-scheduling. To fully utilise the available capacity and enable kanban links between the individual work stations and supply chain partners, variability between subsequent production periods (e.g. days or weeks) needs to be minimised. In a make-to-forecast environment, production smoothing is achieved through finished goods inventory buffers that protect the vehicle assembly plant and its upstream supply chain against variability in end-customer demand. Thus, vehicles are produced to forecast and sold from stock. Forecast-based production further eliminates the uncertainty inherent to customer demand, allowing the vehicle assembly plant and its suppliers to know the exact production plans a couple of weeks (or even months) in advance. When moving towards BTO, both variability and uncertainty emerge as a problem for vehicle and component production, increasing production costs.

One possibility to cope with the inherent variability and uncertainty is to manage end-customer demand pro-actively, as is done in many service industries where stock production is not an option. Here, for example, the airline and travel industry use time-based pricing as a strategy to level demand. Although full time-based pricing may not be the best solution for the auto industry, OEMs could use promotional campaigns more pro-actively. For example, instead of using price discounts to clear vehicle stocks, they could use promotions to increase demand for BTO cars, levelling production plans while at the same time preventing the build-up of a finished goods inventory.

Despite these options for the pro-active management of demand variability, end-customer demand will never be fully level, and at the same time, demand uncertainty will remain. Thus, a key question is how the required flexibility can be obtained in the light of the various (often cost-related) constraints, in particular in times when the physical distances between supply chain partners and associated logistics lead times are increasing. Partly due to the organisational importance of the manufacturing function at vehicle manufacturers, these concerns have prevented (or at least postponed) the implementation of BTO initiatives at most of the firms studied, as discussed in the previous section; yet, more recently OEMs have found ways to mitigate these impacts and increase the flexibility in production and component supply. Of course, flexible production still increased production costs, but these increases did not necessarily have as big an impact as had been foreseen (see also Sect. 3.5).

3.4   The Impact of Build-to-Order on the Bottom Line

Build-to-order requires drastic change in the factory, the supply chain and the overall operating system (e.g. performance measurement), while leading to cost increases caused by increased flexibility requirements. Thus, it is important to assess the potential of BTO to reduce costs elsewhere and/or increase revenues (see also Sect. 3.3.1.1). During our research we found that surprisingly few OEMs had ever attempted to quantify the cost reductions achieved, and a holistic assessment was still lacking. There were two key reasons for the lack of empirical evidence within organisations: first, the required data were often not available, or distributed across a number of information systems, which prevented a meaningful analysis, and second, conducting the analyses was time consuming and required an in-depth understanding of both the dynamics of the business and the evaluation methods needed. For these reasons, such analyses had to be supported by senior management.

Together with one vehicle manufacturer, who was amongst the first OEMs to adopt a BTO strategy, we attempted to conduct a preliminary analysis of the impact of BTO on the firm's bottom line. While the impacts observed made a strong case for BTO, only a few of the factors shown in Table 3.2 could be assessed[2], mainly due to problems with data availability. The three areas in which sufficient data were available were:

1. Stock levels
2. Product phase-outs
3. Order-to-delivery lead times

These will be discussed in turn.

3.4.1   The Impact of Build-to-Order on Inventory Levels

When the OEM introduced its new BTO system in 1993, it proved to have a direct impact on its stock levels of finished cars (see Fig. 3.2). As one executive (who was part of the initial project) commented, the initial change had "[...] brought down inventory levels across Europe almost overnight". The first key process change that led to this development was the introduction of a direct order booking process, i.e. dealers could henceforth send orders directly to the factory and were asked to do so only once they had an end-customer order for the car. The second process change meant that the OEM abandoned the use of so-called "dummy orders" and market allocations, i.e. their production planning system would henceforth only contain real orders, while every market was allowed to place as many orders as they wished on a "first come, first served" basis.[3]

To quantify the exact relationship between the increase in BTO content and the decrease in finished vehicle stocks, the OEM supplied a detailed dataset containing market- and model-specific production and inventory data spanning the period from January 2003 to December 2005. Comparing the BTO content for each model and market combination and the respective inventory levels through a multiple

---

[2] Due to the confidentiality of this information, the data presented are disguised using indices. For a more comprehensive discussion of the research, please refer to Reichhart and Holweg (2007).
[3] For a comprehensive discussion of the order scheduling process in the automotive industry, please refer to Holweg et al. (2005) and Meyr (2004).

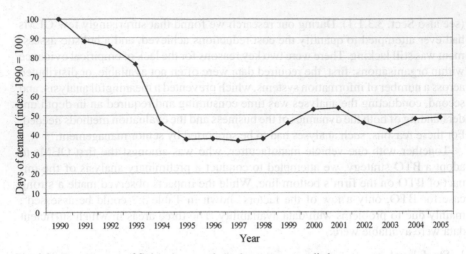

Fig. 3.2 Development of finished car stocks at the company studied

regression analysis revealed two interesting findings. First, the BTO content was in fact the strongest single factor to impact on inventory levels, and second, its impact could be quantified accurately by comparing the results from three differ- ent regression models (see Table 3.3). It could be shown that an increase in the BTO content of 1% reduced the vehicle stock in the markets by around 0.67 days of demand. This means that encouraging an increase in the BTO content by only 2% has the potential to eliminate inventory that is at least equal in value to all of the vehicle manufacturer's component stock inside the vehicle assembly plants.

Table 3.3 Impact of build-to-order content on inventory levels

Independent variables/model parameters	Model 1	Model 2	Model 3
BTO content	−0.69	−0.69	−0.67
	(11.26)**	(12.22)**	(12.04)**
1/(average sales per week)	N/A	222.29	182.19
		(5.06)**	(3.53)**
Average number of sales outlets per 1,000 sales in market	N/A	N/A	0.88
			(1.45)
Constant	91.96	84.16	81.79
	(22.80)**	(21.59)**	(19.44)**
Observations	94	94	94
R-squared	0.58	0.67	0.68

Absolute value of t statistics in parentheses: * = significant at 5%, ** = significant at 1%

A number of additional factors were tested for their impact on the finished car inventory, yet the BTO content remained by far the most important single factor: it alone could explain 58% of the variation in inventory levels across the 94 market–model combinations that were included in the analysis (i.e. all mainstream models sold in European markets). However, further analysis also revealed that simply pushing dealers into encouraging end-customers to have a car built to their specifications was not the solution. The underlying dynamics of the supply chain actually led to an initial increase in inventory when the BTO content was increased, because the outflow of cars out of dealer stock declined, while the inflow into dealer stock from the factory remained unchanged. In order to reduce inventory levels dealers must be allowed to temporarily reduce their wholesale order volume to clear stocks, before the new BTO strategy can have the desired impact. As a matter of fact, wholesale targets should be replaced by retail targets altogether for a BTO strategy to work successfully (see also Sect. 3.3.1.2).

3.4.2 Product Phase-outs

One aspect of a high finished car inventory that is often neglected in analyses are the discounts required to clear stocks when new vehicle models are introduced and natural demand for the old model decreases sharply. While the data did not allow for quantifying the cost associated with this aspect of make-to-forecast (MTF) production, it revealed very interesting results with regard to the time it took to clear the remaining stock of the old model. The dataset included information on the complete phase-out (i.e. last week of production until the sale of the last car) of three car models. Despite the three models covering different market segments, the phase-out times and patterns were remarkably similar. For all three models, it took more than 18 months to clear the vehicle stock in the market once production had stopped. Of course, the majority of cars were sold within the first couple of weeks after production stopped due to large discounting schemes like special promotional packages; yet, a significant percentage of cars remained in stock for longer than half a year (see Fig. 3.3).

Unfortunately, it was not possible to quantify the discounts given to customers for buying a car of the previous model, because the IT systems only captured historic data on list prices as opposed to the actual selling prices. Despite this lack of quantitative evidence, the interviews conducted confirmed that the "old" cars were unlikely to make a significant contribution to the vehicle manufacturer's profits. Further considering the capital and inventory costs of storing these cars for such a long period of time, it was more likely that the vehicle manufacturer was selling these cars at a loss.

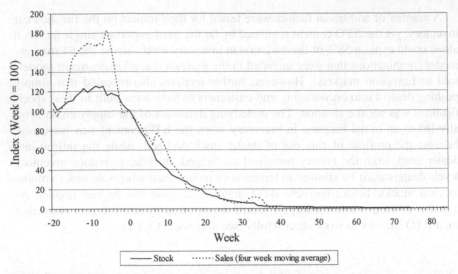

Fig. 3.3 Product phase-out for one car model (week 0: first week without production)

3.4.3 The Impact of Build-to-Order on Order Lead Times

The interviews conducted suggested that the majority of customers wanted their cars to be delivered in around 3–4 weeks; yet, as outlined in Sect 3.2, few vehicle manufacturers achieve such short delivery times. Although the case OEM had a theoretical system capability of less than 10 days for customer orders, very few models achieved such short delivery lead times in practice. Long order queues, i.e. other orders that had to be produced and delivered first, existed for most models, preventing short delivery lead times. In addition, the OEM's management had observed that markets with a higher BTO content and a lower finished car inventory offered shorter lead times to their customers for cars that were not sold from stock. Unfortunately, the history of customer lead times was not captured by the OEM's IT systems, but the regular snapshots of current order-to-production times taken from the system confirmed this observation. For example, Fig. 3.4 shows the relationship between the BTO content and the order-to-production times for new orders (for confidentiality reasons, actual figures were disguised by using indices). The inverse relationship between the two measures in this snapshot was – according to the interviews – representative of the underlying pattern.

The finding that increased BTO content reduced order-to-delivery lead times for BTO cars was to some extent counter-intuitive. It could only be explained by considering the dynamics of the order booking process and the complex supply and manufacturing constraints required for production scheduling. Dealers in markets with a low BTO content submitted stock orders to the OEM, which were slotted into the production schedule. Once an end customer ordered a car, these dealers tried to change a previously submitted stock order to suit the customer's specifications.

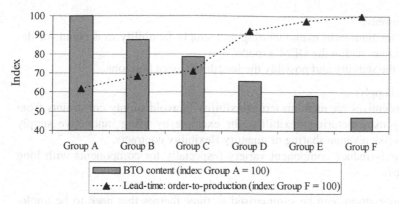

Fig. 3.4 Impact of build-to-order content on order-to-production times by market groups

However, the updated order was likely to violate some of the system constraints and was generally moved to a later production period behind stock orders for the same dealer. Therefore, the OEM produced cars that ended up in dealer stock, even though an end-customer from the same dealership was waiting for a car. This created a vicious circle because dealers who had to wait longer for a BTO car relied more on sales from stock in order not to lose sales to customers who were not prepared to wait. This in turn increased their lead times even further.

3.5 Coping Mechanisms

In the previous sections we highlighted the positive and negative impacts of BTO, as well as the challenges of implementing a BTO system. To support companies on their journey to achieving higher levels of customer responsiveness, we will now outline several mechanisms that can help mitigate the downside of BTO. Based on our research, we developed a series of recommendations for each of the three tiers. Surprisingly, we found that – apart from the at times still convoluted order processing system – the main improvement potential was outside of the vehicle assembly pant, both in component supply and vehicle distribution:

Vehicle distribution:
- Reduce the length of sales periods (e.g. use weekly instead of monthly sales targets).
- Use retail instead of wholesale targets.
- Use order targets (for customer orders) instead of delivery targets.
- Introduce continuous ordering instead of periodic batch processes.
- Establish central holding compounds based on a Pareto analysis of end-customer demand (i.e. high-runners vs low-runners) when stock production is required to keep manufacturing stable.

Vehicle assembly:
- Increase medium-term manufacturing and supply flexibility to account for demand changes within the OEM's model mix.
- Increase the stability and possibly the length of frozen horizons.

Component supply:
- Create incentives for medium-term flexibility to avoid supply constraints from hindering manufacturing flexibility, for example by using innovative supply contracts like revenue sharing or quantity flexibility contracts.
- Significantly reduce component variety (especially for components with long lead times).

The recommendations can be summarised as three themes that need to be implemented consistently across the three tiers:

1. Promote BTO throughout the supply chain by aligning all performance targets.
2. Reduce self-inflicted uncertainty and variability to allow all supply chain members to focus on the flexibility that is actually required by the end customer.
3. Create this flexibility across all tiers.

For example, dealers who were measured based on wholesale targets inevitably reverted to pushing cars onto the market. Manufacturing managers who were measured against unit costs in vehicle assembly tried to keep variability away from their operations at all cost, while purchasing executives with the aim of reducing purchasing costs had no incentive to reward supplier flexibility.

The second theme was equally important; for example, Fig. 3.5 shows an example of the so-called hockey stick effect, i.e. demand variability created by artificial reporting periods, in this case monthly and quarterly sales targets. The large spikes in demand every 4–5 weeks created a high level of variability for the vehicle assembly plant that was mainly caused by dealer behaviour. As a result – during the last week of every month when their month-end targets were in sight – dealers

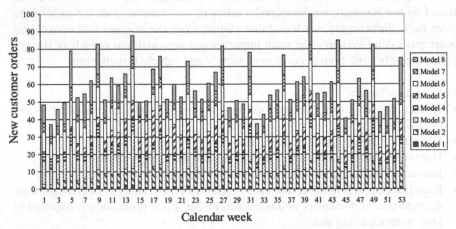

Fig. 3.5 New customer orders received at dealerships at one OEM (index: week 40 = 100)

Table 3.4 Coping mechanisms in vehicle assembly

Volume variability and uncertainty	Mix variability and uncertainty
Order buffers with fleet sales, employee sales and advanced stock orders Labour flexibility (hour banks; add or remove entire shifts) Ability to adjust "takt time" Produce at least two models at different stages in their life cycles in the same plant	Assemble multiple models per production line Flexible models that are produced in more than one assembly plant Flexible line balancing and supply constraints

incentivised their sales, while during the first week of every month, "[...] they simply relax and try to recover from the stressful week before", as one of the distribution managers pointed out.

Similarly, first-tier suppliers often faced a high level of planning uncertainty caused by uncertainties in the vehicle assembly plant's paint shop or other sources internal to the OEM's operations. Addressing these uncertainties caused significant costs in the supplier operations, often requiring some form of co-location[4]; yet, the required flexibility did not add any value to the end-customer, but rather inhibited the supply chain from focussing on the types of flexibility that were demanded by the market.

The final theme that needs to be considered is the provision of flexibility. Here, we found that a number of mitigation strategies – in particular in vehicle assembly – could be used to reduce the impact of flexibility on operational costs. While the mechanisms provided in Table 3.4 to counter the demands placed upon vehicle assembly by increased volume and mix flexibility requirements are not exhaustive, they provide an overview of the key best practice techniques found across the leading BTO manufacturers.

However, flexibility in vehicle assembly was often constrained by flexibility in component supply unless this issue was addressed simultaneously. Therefore, a number of vehicle manufacturers started to work at the supply chain interfaces to align the flexibility across the partners involved. For example, some OEMs have started to accept and plan for medium- to long-term uncertainty in customer demand by using innovative supply contracts, such as quantity flexibility contracts or revenue-sharing contracts. The traditional fixed-quantity contracts used across the industry yielded unsatisfactory results under high uncertainty, because they did not provide any guidelines and incentives for creating flexibility in supplier operations. Thus, by acknowledging the inherent uncertainty in the industry and by building flexibility incentives into supply arrangements, OEMs could play a proactive role in managing flexibility. For example, a number of OEMs suffered from supply constraints affecting their engine mix. Because the suppliers of diesel engine blocks could keep up with the recent increase in demand for diesel engines, the effective flexibility of the OEMs was limited, and they could not provide exactly

---

[4] For a more comprehensive discussion of co-location, please refer to Reichhart and Holweg (2008)

what their customers wanted. Therefore, many OEMs pushed petrol cars onto the market using sales incentives, a step that could have been avoided by building flexibility into the component supply chain.

3.6 The Way Forward

Despite the recent advances towards the adoption of BTO, the automotive industry has a long journey ahead to achieve the optimal balance between the responsive supply of new cars and the associated cost increases in manufacturing and component supply. The two unknown factors in this equation are the level of flexibility that is actually demanded by consumers and the extent to which the cost increases following BTO can be mitigated. We have presented a number of challenges that firms have been facing with regard to the adoption of BTO: while organisational obstacles prevent many firms from embarking on the journey in the first place, the difficulties of determining actual customer requirements further complicate reaching agreement on the right level of BTO within the organisation. Despite the remaining uncertainty about the actual level of flexibility required by consumers, we have shown that increases in flexibility lead to quantifiable improvements in the supply chain, such as a reduction in finished vehicle inventory, and we have discussed mitigating mechanisms to address the adverse impacts of flexibility on operating costs.

Because every vehicle manufacturer starts the journey towards BTO with a different organisational set-up and varying operational capabilities, it is difficult to prescribe a standard way to determine the appropriate level of customer responsiveness. While we have discussed a number of key characteristics of the required operating system, the two key factors that determine whether any effort has a chance to succeed are *transparency* and *alignment/purpose*. The key problem found was the prevalence of conflicting perceptions about the firm's performance with regard to BTO-related key performance indicators, such as BTO content and delivery reliability. This was followed by performance measures that were misaligned with the overall goal of increasing the BTO content and reducing lead times. A combination of these related issues undermined any serious efforts to improve, while preventing the vehicle manufacturer from understanding actual customer requirements. Therefore, an OEM needs to realistically assess its current performance and align its performance measures to support its order-to-delivery strategy, before it can start building the required operating system to support BTO.

The best performing manufacturers were those that could accurately answer questions like: "What is the BTO content (as defined by the percentage of orders placed with a factory that have an end-customer waiting for the car) for a given market?" or "What is the current stock level (including dealer stock) of model X in market Y?" Only those OEMs that could answer these questions were in the position to monitor their progress; yet, even they were often not able to leverage the advantages of BTO production. Therefore, a transparent and aligned performance measurement system must be accompanied by open dialogue across the involved functions in order to determine the right level of flexibility for the supply chain, an effort that must consider both the component supply system, because it often constituted a flexibility bottleneck for the supply chain, and the distribution system, where dealer behaviour frequently undermined the OEM's intentions to promote BTO.

Implementing BTO requires complex changes along the supply chain. While most commonly the perception is that these need to occur inside the factory, our research shows that the far more important changes need to occur at the organisational level in terms of the performance measurement system, and the way the OEM interacts with its supply and distribution system. Implementing BTO is neither fast nor easy, and certainly not a journey the OEM can embark upon without its partners along the supply chain; yet, a range of benefits – from a reduced finished vehicle inventory to higher margins per vehicle sold – make it an attractive option for remaining profitable in such a competitive industry.

From the current state of the industry, it remains difficult to predict how its OTD capabilities will develop: the extent to which pure BTO production will remain an unreachable goal depends on the advances in product and process technologies to decrease the cost of flexibility. Within the current set-up, some share of MTF production appears to be required to supply cars at reasonable cost. However – as we have shown – even with the existing technologies OEMs can choose from a wide range of hybrid strategies, once they overcome the initial organisational obstacles. For the time being, the most advanced OEMs aim to achieve OTD times of around 3–4 weeks with very high delivery reliability, based on a balanced mix of MTF and BTO production; yet, so far no manufacturer has accomplished such OTD times consistently across its entire model range.

References

Fujimoto T (2006) Toyota and its evolution since "The Machine". IMVP and WZB Sponsor Industry Workshop, Berlin

Holweg M, Pil FK (2001) Successful build-to-order strategies start with the customer. MIT Sloan Manag Rev 43(1):74–83

Holweg M, Pil FK (2004) The second century: reconnecting customer and value chain through build-to-order. MIT Press, Cambridge

Holweg M, Disney S, Hines P, Naim M (2005) Towards responsive vehicle supply: a simulation-based investigation into automotive scheduling systems. J Oper Manag 23(5):507–530

Meyr H (2004) Supply chain planning in the German automotive industry. OR Spectr 26(4):447–470

Reichhart A, Holweg M (2007) Lean distribution: concepts, contributions, conflicts. Int J Prod Res 45(16):3699–3722

Reichhart A, Holweg M (2008) Co-located supplier clusters: forms, functions and theoretical perspectives. Int J Oper Prod Manag 28(1):53–78

Chapter 4
Current Issues at OEMs and Suppliers

Alexandra Güttner and Thomas Sommer-Dittrich

Daimler AG; Group Research & Advanced Engineering;
Materials, Manufacturing and Concepts

Abstract. The European automotive industry has been seen to be confronted by the consequences of advanced market saturation in the core markets of Western Europe, North America and Japan, and increasing competitive pressure from the Far East. Stagnating or partly declining sales figures, growing overcapacity and costs due to increasing model variety and individualisation of the products, as well as the development of new markets in the boom regions of the world characterise the challenges of the automotive industry at the beginning of the 21st century. Rising regulatory requirements in the areas of safety and environmental protection further increase pressure on automotive manufacturers and suppliers. Modular product design is intended to make the advancing variety of options controllable for companies and affordable for the customers. New collaborative planning methods are being developed to deal with the complexity of the multistage supply chains of the industry and to maintain its capacity to act. The approaches proposed by the work presented in this book to achieve build-to-order may enable companies to meet these challenges.

4.1 New Challenges in Product Policy

The saturation of the automotive market requires OEMs to differentiate and individualise their product. The automotive manufacturers have, in recent years, reacted with a massive expansion of their model range and equipment options. For example, the optional equipment in vehicle manufacturing in the last 20 years has posted an increase of more than 200%, while product variety in the past decade has actually increased by more than 400% (Becker 2007, p. 110; Gehr and Hellingrath 2007, p. VII, 9; Gromer 2006, p. 80).

Since the 1980s, market power has been on the side of the customer, whose demands have increased significantly and this has been felt most in the premium segment of the automotive market. The rapid market introduction of customer-orientated innovations, which in the luxury class segment is easier to implement due to the higher cost tolerance of the buyers, plays an increasingly important role. When product innovations prove successful and where learning curves and economies of scale allow them to be produced more cost-efficiently after a model cycle, they are also offered in the medium-class and later, if required, in the compact and subcompact segments (Becker 2007, p. 33, 108 et seqq.; Gehr and Hellingrath 2007, p. 8; Diez 2005, p.134; Gromer 2006, p. 79).

Premium brands traditionally play an important role in the introduction of product innovations since they define themselves in terms of their constant innovative power, their outstanding product quality and, accordingly, a corporate image based on traditional values. A premium brands characteristic is an ability to demand margins above those of products with comparable functions and thus similar technical uses. These volume brands are not held in as high esteem with regard to image as innovation leaders, their quality of workmanship or the psychological effect of their products. Among the most well-known traditional premium European brands are Mercedes-Benz, BMW, Audi and Volvo. In comparison to the premium segment, luxury brands base their status far more on purely emotional elements such as their reputation, extravagance, luxurious details in their interior equipment (as, for example, a Champagne bar in the luggage compartment), but provide no additional use compared with the base product. The key characteristic in the luxury segment is the price, which far exceeds the rational product value. Names such as Lamborghini, Maserati, Maybach, or Rolls-Royce are among the European luxury brands. Classic volume brands such as VW, Renault, Peugeot, Opel or Ford instead confine themselves to fulfilling the real needs of the customers in their lifestyle environment with products of good quality at affordable prices (Diez 2005, p.125 et seqq.; Rosengarten and Stürmer 2005, p. 26 et seqq., 36 et seq., 177; Sanz et al. 2007, p. 256).

In the last few years, safety-relevant technical innovations such as airbags, ABS and ESP have increasingly become part of the standard equipment of new vehicles and have also gained ground in lower market segments. The majority of car buyers even respond to functions that are for convenience only. Navigation systems, for example, are already an integral part of the interior equipment in nearly one fifth of all new vehicles, with this trend rising significantly. Development departments are increasingly focussing on improving ergonomics and the air conditioning of the passenger compartment. After seat heating and air conditioning, already long established equipment options, attention is now turning to seat cooling or massage functions as well as the ergonomic form of seats and the feel of the operating elements (Becker 2007, p. 112 et seqq.; Gromer 2006, p. 81 et seq.).

Regarding vehicle electronics, the focus today is on developing entertainment and telecommunication devices, as well as safety components. Their ongoing popularity is not least due to high traffic density in cities and the risk of traffic jams on highways. Options such as internet access, television, and DVD players are available to passengers, while drivers must restrict themselves to hands-free car kits or Bluetooth technology, as well as navigation systems and car radios (Becker 2007, p. 113 et seq.).

In future, the driver will be supported by comprehensive driver assistance systems that are not only able to perceive their surroundings, but also evaluate and react to them. Those systems working today with radar sensors only will be replaced by communicative systems consisting of cameras, ultrasonic and radar sensors. In addition, data from navigation systems and satellites will be integrated into the assistance system so as to have precise information on the vehicle position available at any time. Warnings to be issued in future include deviation from a lane, information about road obstacles and, in addition, even the driver's reactions will be automated (Schlott 2007, p. 86; Wolters et al. 1999, p. 9 et seqq.).

These functions are developed further by vehicle-to-vehicle communication, where information about the driver's performance as well as traffic impediments or obstacles on the route are transmitted to the systems of other vehicles. These can then generate the corresponding mechanical reactions or issue warnings to their occupants. Other functions being developed today are the automatic identification of road signs, which can, where necessary, issue warnings to the driver, or a control unit for intelligent lighting systems, adapting the brightness of the headlights to road conditions, opposing traffic and vehicles in front. These functions will enhance the adaptive lighting systems already positioned in the market today (Schlott 2007, p. 88).

Technical development, forced by regulatory specifications, also focuses on preserving natural resources and improving the protection of the environment. Whereas in the past attempts were made to reduce the emission of pollutants with downstream units like catalytic converters or filters, today the entire powertrain technology is put to the test. Innovative vehicle concepts using hybrid-electric, purely electric or natural-gas engines and alternative fuels such as biodiesel or bioethanol are already on the market today. Great expectations for the future rest on the breakthrough of fuel-cell technology and the development of a commercially attractive and non-explosive hydrogen engine (Becker 2007, p. 114; Wolters et al. 1999, p. 218 et seqq.).

Regulatory guidelines not only demand innovations from the automotive industry, they also boost their market penetration and thus contribute to safeguarding the future for innovative companies. As, for example, making safety concepts initiated by the high price segment, like the safety belt, a standard accessory required for all vehicles. Some strongly benefit-orientated must-have technologies, supported by regulations or driven by strong customer needs, were accepted in the market within only one product generation, while upscale, nice-to-have technologies, such as convenience functions like air conditioning, auxiliary heating or central locking, require approximately two product life-cycles to penetrate middle and lower market segments. The third category of innovative technologies, so-called niche technologies such as memory functions for electrically adjustable seats or automatic opening of the luggage compartment only attain a small prevalence rate even decades after their introduction (Becker 2007, p. 80 et seq.; Gromer 2006, p. 81 et seq.).

Due to demanding market conditions, the rapid roll-out of innovations, be it as pioneer or as early successor, are of particular importance. Success depends crucially on developing the right strategy. It consists of a mixture of market push (aligning development activities to the market and consumer needs) and technology pull (discovering completely new and previously unknown product features). With technology-based inventions especially, it is imperative to keep an eye on the demand potential of the invention before deciding whether to also introduce the invention as an innovation onto the market (Becker 2007, p. 112, 180 et seqq.; Sanz et al. 2007, p. 396; Wolters et al. 1999, p. 50 et seqq.).

Customers purchasing a new vehicle today desire shorter delivery times, guaranteed delivery dates and the option to be able to demand changes at short notice and have these implemented before delivery. This is on top of the required variety of options and the possibility to assemble their vehicle individually via the internet or at the customer service centre. Customers in the volume segment above all, are not prepared to pay for the additional costs caused by increasing flexibility with regard to their requirements. Experts estimate that it will only be possible to achieve an increase in price of six percent, while actual manufacturing costs will increase by an average of 17% (Becker 2007, p. 33; Gehr and Hellingrath 2007, p. 8 et seqq.; Scholz 2007, p. 55).

Individualised finance offers will be increasingly important for both manufacturers and customers. Well-packaged finance offers enhance the attraction of offers for new cars, such as maintenance and service offers or inexpensive vehicle insurance rates (Knauer 2007, p. 12).

Producing a new vehicle contributes 39% to the business volume, but only eight percent to the profit of the automotive manufacturer. On the other hand, financing and insurance contributes 46% of the total profit and 30% of the business volume. Besides the car-fleet corporate customers, a major part of these earnings is contributed by recent strong growth in the leasing sector, with an increasingly large percentage of private customers. In future, individualised leasing offers could further enhance the importance of the leasing sector. The customers will receive the appropriate vehicle depending on the time of the year or their transport requirements: be it an off-road vehicle for a mountain tour in the winter, the coupé convertible for the summer weekend trip for two, or the station wagon or minivan for daily trips with the whole family (Becker 2007, p. 88; Gromer 2006, p. 164 et seq.; Wolters et al. 1999, p. 10).

4.2 Increased Pressure from Competitors

In future, significant quantitative growth of the automotive market is only expected in the boom regions of Southeast Asia, other emerging markets and Eastern

Europe. As a result of advanced market saturation, the established markets of Western Europe, North America and Japan have been stagnating or even shrinking for a number of years. Within this highly competitive triad, which will remain the most important market for automobiles for a long time to come, vendors can only grow either at the expense of competitors or by offering better quality. This means that not more, but qualitatively better and more expensive vehicles with high-quality equipment options will be sold (Becker 2007, p. 2, 12 et seq., 89 et seqq.; Gromer 2006, p. 79 et seqq.; Rosengarten and Stürmer 2005, p. 29; Sanz et al. 2007, p. 251; Wolters et al. 1999, p. 14 et seq.).

Saturation trends in the automobile market thus offer new prospects for manufacturers of premium vehicles in particular, as long as they employ their innovative power consistently and efficiently. Once product innovations no longer satisfy customer requirements and rising higher production costs incurred by them no longer contribute to market requirements, product characteristics such as quality and reliability become more important. Volume vendors offer that at significantly lower costs (Becker 2007, p. 81, 184).

Inexpensive vehicles, so-called low-cost cars, are a new phenomenon on the automobile market. They are driven by people with low income levels in Southeast Asia aspiring to be mobile. The new local automotive manufacturers like Brilliance of China or Tata of India, strengthened by cooperating with established manufacturers, have shown that fully-fledged cars can be built even for customers who are not financially strong.

Since prosperity in Europe is not developing uniformly, a market for a low-cost vehicle with significantly reduced comfort features, but sufficient safety features is also opening up here. Analogous to the distribution of income, the market structure had exhibited the same classic pyramid form for many decades: few buyers in the upper segment, an increased demand for middle class vehicles and the highest sales figures in the lower market segment. For some time now, the middle class has been eroded so that the shape of the market structure now resembles an hourglass. This means growing sales volumes in the premium range and the upper middle class, a larger market slump in the middle volume segment and a further increase in already high sales figures for the lower segments (Becker 2007, p. 30 et seq.; Diez 2005, p. 131 et seq.).

An example of the penetration of a European volume manufacturer into the low-cost segment is the Dacia Logan model from Renault. But such low-cost vehicles will not be able to contribute much to consolidated profits since the gross profit margin is obviously set very low. If manufacturers as well as suppliers master the challenge of streamlining their processes so that they can handle the enormous cost pressure in the low-cost segment, they can transfer their technical expertise and optimisation approaches to other vehicle segments in their product portfolio and thus create competitive advantages (Becker 2007, p. 33; Scholz 2007, p. 55).

Low-cost companies do not restrict themselves to the classic low-cost compact cars only, but now also offer inexpensive sedans, so they are competing for a market share in other segments. This is because fierce competition permits growth only at the expense of other market participants and possibly other market segments. Moreover, these low-priced vehicles will no longer come from Eastern European factories belonging to European vendors but – to an increasing degree – will be imported from China, and later from India. Brilliance is already risking entry into the European market in 2007, aiming to import up to 20,000 vehicles, depending on the demand of the market (Gromer 2006, p. 149 et seq.; Da Ke 2007, p. 42).

After the poor crash test marks of a Chinese vehicle, a Brilliance BS6, in Germany, it seemed that the European automotive industry, and the German industry in particular, could breathe a sigh of relief. However, viewed more closely, the test result deserved admiration rather than sneering. The Chinese automotive industry has only been built up over the last 20 years, and only during the last 9 years have they intensified development of their own automobiles. This was a step forward from the Landwind model from a Chinese competitor, Jiangling, tested 2 years previously, as Brilliance captured one of five possible stars in the test. What would have been a scandal for European vehicles is seen positively in China and speeds up further technical development of the products (Da Ke 2007, p. 42).

In the long run, Chinese competition for European vehicle manufacturers could prove to be as threatening as Japanese and later Korean manufacturers have become during recent years, since Chinese automakers can operate from a strong economic position. The automobile boom in China is earning substantial profits for local companies and joint ventures with Western manufacturers provide the necessary technical background. However, there is hope that the Chinese market share in Europe will grow at the expense of other foreign brands (Becker 2007, p. 120 et seq.).

Similar to the Chinese today, the Japanese started their entry into the European market at the beginning of the 1970s with an aggressive low price strategy. After growing continuously but moderately during the initial years, Japanese automakers caused quite a stir when they were able to almost triple their market share in Germany from 1978 to 1980. Subsequently, Japanese percentages in sales of new cars declined, rose and declined again during the following years. At the beginning of the new millennium they were again able to rise strongly, achieving a market share of 13.5%, growing at a rate of up to 10%. The success of Japanese brands, to a large extent achieved by Toyota, is all the more remarkable, since the European market during the last few years has shrunk at times, and some European mass producers have had to accept significant declines in sales (Becker 2007, p. 16 et seqq.; Gromer 2006, p. 118 et seqq.).

Only the Korean car manufacturers were able to achieve higher growth rates of, on average more than 20% in the years 2002 to 2004, gaining a market share of more than 4%. This was after their entry into the European market at the beginning of the 1990s, with previous growth of about 3.5% in 2000 that was followed by a major drop in their sales figures (Becker 2007, p. 17 et seq., 105).

The latest successes of Japanese manufacturers are largely the result of their realisation that building up a significant position in a market requires local capital expenditure beyond just setting up their own sales offices. They had initially built their own manufacturing plants in the USA, mainly to bypass import restrictions, but this enabled them to expand their position in the American market. They applied this concept to the European market and founded their so-called transplants there. The share of world-wide production by Japanese companies sold in Europe grew from about 3% in 1994 to 8% in 2004. On the one hand, the Japanese increased their acceptance by the European population by creating jobs and, on the other hand, were able to align their product policy closer to local tastes by establishing design and development departments. Recently, the Korean automakers followed a similar re-localisation strategy in order to be able to adapt their product designs to suit European tastes (Becker 2007, p. 69 et seqq., 115 et seqq.; Gromer 2006, p. 194; Wolters et al. 1999, p. 16).

For some time now Japanese and Koreans have been able to achieve considerable success by combining high quality and reliability with a good price–performance ratio, employing only medium levels of innovation. They have thus laid the foundation for a promising attempt to enter the premium segment. In this area, the Japanese automotive company Toyota has taken the lead with the Lexus brand. Since it was founded in the mid 1980s, it has been able to pick up a considerable share of the American market from the local automotive companies in the premium sector, whereas to date it has not been able to show any noteworthy sales figures in Europe. Introducing new product strategies aimed at market demands has made an impact on the Europeans in recent years, increasing sales of the Lexus models by more than 10%, not least due to forced implementation of innovations such as the hybrid drive. Should Toyota be able to gain a respectable market share with Lexus in Europe in the near future, it would then only be a matter of time until other Japanese or Korean manufacturers follow Toyota to the premium segment and also compete outside the European volume market segment (Becker 2007, p. 11, p. 119 et seq. p. 233; Sanz et al. 2007, p. 251).

4.3 Reactions from the European Automotive Industry

4.3.1 Product Strategy

The changing power positions in the automobile market are characterised by the car buyers' decreasing brand loyalty. The convergence of manufacturers' levels of quality and equipment on competing models means a customer's decision is based on almost identical equipment choices. Decisions are made on additional benefits such as a preference for equipment variants, for status symbols or follow a cost–use benefit analysis. A car manufacturer will only be able to attain a satisfactory level of customer loyalty if they offer a sufficient range of model variants in the broader market segments. If a broad selection of niche models is available to the consumer existing customers are not tempted to look to the competition and simultaneously opportunities are developed to win the competition's customers (Becker 2007, p. 28 et seq., 78, 107 et seqq., 232; Gromer 2006, p. 109 et seq.).

Automotive companies have had to adjust to accommodate the ever expanding market segments by expanding their own product selection. Since the 1970s, companies have been following one of two paths: either following the competition into a segment or being a step ahead of the competition by developing their own new vehicle category. While in the 1960s almost all vehicles could be assigned to one of two basic categories, passenger car or sports car, the number of types of cars multiplied over the next decade. From that point on, the customer could choose between hatchback, station wagon, economy car and coupé. The automobile market has developed two to four new market niches each decade, which split further the existing segments, whilst total market volume remains unchanged. Currently, the market has fragmented to such a level of detail that the automotive industry has resorted to creating new niches by combining existing vehicle segments. Examples of the so-called cross-over models of the past few years include sports utility vehicles (SUV) – a combination of the sports car and all-terrain vehicle, four-door coupés – which is the integration of a family passenger car and the sporty coupé, minivans – which are a combination of the characteristics of a large van with those of a compact economy car, or the luxury activity vehicle (LAV) – where the characteristics of a luxurious passenger vehicle are combined with a spacious station wagon (Becker 2007, p. 28 et seq., 78 et seq., 111).

To strengthen their own market position, many manufacturers pursued an acquisition strategy in addition to the current competition and concentration process. Model selection is expanded by acquiring competitors or individual brands. Over the past few years collaboration and strategic alliances have also been part of the manufacturers' repertoire for the same reasons. Following many years of experience with the independent subsidiaries Audi and Seat, the Volkswagen group was able to successfully integrate the brands Skoda, Bentley, Bugatti and Lamborghini into its group profile. For Renault, the acquisition of Samsung's car division and participation in the Romanian car manufacturer Dacia are also developing successfully. In contrast, the two premium sellers, BMW and Daimler-Benz have had negative experiences following their acquisitions of Rover and Chrysler respectively, as well as the participation of DaimlerChrysler in Mitsubishi. In each case they have retreated from these relations as far as possible, with the exception of the Rover Mini at BMW, which was developed into an independent brand. Until recently, it appeared that continuation of this consolidation process was among car manufacturers. However, mediocre profits and dissolutions of failed consolidations indicate that car companies have lost their immediate appetite for this problematic solution (Becker 2007, p. 75 et seq., 110; Gromer 2006, p. 123 et seq.; Rosengarten and Stürmer 2005, p. 146).

The advancement into market segments not addressed by a company's own product selection can take place without the addition of third-party brands by expanding into new model variants or by creating new brands. For example: Volkswagen began to enter the premium segment with the VW brand using the Phaeton model, even though their subsidiary Audi is already present in this segment, while Toyota created the brand Lexus to enter the American and the European premium market. By introducing premium models, volume dealers are able to improve their image and as a result, they are able to position their entire product selection at a higher level. However, when selecting such strategies OEMs must bear in mind that premium brands and mass produced brands are not only different with regard to brand maintenance, but are based on entirely different business models. The success pattern of a premium brand is based on above-average pricing with low sales volumes and resultant higher costs. Large sales volumes and associated scale effects are the deciding factors for volume brands. The risk of "trading up" is that the company must first invest to meet buyer preferences, whilst the new or repositioned brand is not expected to generate a rapid return, as initial growth is unlikely to be any more than that of the overall market (Diez 2005, p. 130, 134, 137 et seq.).

A strategy to expand a company's own vehicle selection tends to focus upon the consolidation of competitive positions by expanding to a vehicle category that has not yet been tapped. Volume manufacturers strive for the premium segment, which for many years has been more buoyant as it is less dependent on economic trends and is marked by a higher level of brand loyalty than lower priced segments. However, they are not converting their entire product line-ups, like Volvo and Audi did in the 1980s, but are instead combining volume and premium strategies (Diez 2005, p.133; Gromer 2006, p. 109 et seq.). In response, premium sellers have recently expanded their product selection with car models that are traditionally part of the volume segment. Ten years ago, automobiles in the premium class mainly included passenger cars, sports cars, cabriolets or station wagons of the main brands. However, when the former Daimler-Benz AG advanced into the compact car segment with the Mercedes-Benz A-Class and experienced a level of success, admittedly following initial start-up difficulties that came close to brand overextension, compact cars were no longer taboo for the premium class. Other sellers of premium automobiles followed, including BMW with the 1 series and Audi with its A3. An established premium brand with a robust image can transfer this image for the expansion of its product selection to new brand products in the lower market regions as long as the new models are not too distant from the traditional market core and avoid brand overextension. Such image transfer to expand the product selection to market niches is considerably easier when they are positioned close to the current models, such as the introduction of the Cayenne SUV by Porsche (Becker 2007, p. 11, 128; Diez 2005, p. 128 et seq.; Rosengarten and Stürmer 2005, p. 29, 146, 201 et seq.). An example of an expansion of a premium manufacturer to the volume segment that is not yet a complete success story is Daimler-Benz founding the Smart brand, only one year after the beginning of sales of the A-Class.

Down-trading strategies bear far less risk than the up-trading strategies for mass manufacturers described above. The successful acquisition of luxury brands into the group structure of premium or volume suppliers is more often observed. Fiat comes to mind with Ferrari and Maserati or Lamborghini by the VW group. There have also been numerous failures, but due to the low volumes in these markets, we will not address the luxury segment any further (Diez 2005, p. 134, 137 et seqq.; Rosengarten and Stürmer 2005, p. 29, 201).

Automotive manufacturing groups show a clear tendency to provide the full product range because they aim to attain a higher ability to respond to changes in demand and to reduce their risk of being directly threatened by the collapse of an individual market segment. Innovations from the premium segment flow down to a company's internal mass produced brands, after a time delay and patenting prevent or delay copying by the competition. The Western European automotive OEMs can steel themselves in the long-term against loss of premium model market shares to producers from cheap labour countries, as the emerging markets themselves will soon demand large volumes of premium vehicles as their wealth increases. Customers there will be ready to pay a fair price for premium innovations with premium quality. In the future, European companies will also advance into the low-cost car market to secure shares there as well (Becker 2007, p. 28, 35, 75 et seqq., 108, 136, 232; Diez 2005, p. 138 et seq.; Gromer 2006, p. 124 et seq.).

4.3.2 Modularisation

The increasing variety of brands and models created by the automotive groups induces vast numbers of variants, vehicle parts and components. This is added to by the companies' presence in more and more foreign markets, for which both regulatory issues and those of local taste must also be considered. As discrete part or component quantities drop, quantity discounts disappear whilst the costs for warehousing and commissioning skyrocket. The main goal of the automotive industry brand acquisitions was, besides an increase in revenue through scale effects, primarily the expansion of the product portfolio, leading to subsequent competitive advantage. Accordingly, the variety of variants at product level was a desired effect, but it has undesired side effects. Over the past few years manufacturers have attempted to meet this problem through modularisation and common parts strategies. Such approaches have been known and partially practiced for decades, but not with the necessary intensity that is required by today's market development and company structures (Becker 2007, p. 68; Sanz et al. 2007, p. 252; Wolters et al. 1999, p. 10).

The automotive industry must master this balancing act, not only to maintain product differentiation for the customer, but also, as far as possible, to standardise parts required for their entire model range. The old-fashioned approach of the platform strategy was usually limited to the standardisation of vehicle components that are rarely noticed by the customer, such as the use of the same chassis for two or more models of the same size class. In contrast, current modularisation concepts strive to build more complex modules or entire systems, which can be used in their basic forms in many vehicles. Individual specific components may vary within a module depending on the car model. Innovative modularisation concepts address the design of a standard base frame module that can be used in a large variety of vehicle derivatives of a size class and are enhanced with additional frame modules depending on the type, such as passenger car, station wagon, or coupé. As an example, HBPO, a joint venture of Behr, Hella and Plastic Omnium, and also including Faurecia, currently supply complete front ends. These consist of cooling components, air lines, lighting, hood lock with cable, air horns, bumpers and crash protection systems, and are provided for the BMW Mini and Audi A4 (Piller and Waringer 1999, p. 68 et seqq.; Sanz et al. 2007, p. 253; Wolters et al. 1999, p. 65).

There is now specific automotive terminology for this area. A module, such as a cockpit or exhaust module, is a spatially limited, ready-to-install unit with physically connected components. A system is a functioning unit of multiple modules that do not necessarily have to be physically connected, such as in an air conditioning system (Piller and Waringer 1999, p. 39 et seq.; Wolters et al. 1999, p. 62). A product platform generally consists of a combination of related parts or components that build a common structure and serves as the basic module for the production of a number of different products or an entire product family. A platform may represent the basis of a complex module and also of an entire vehicle, so that it is often understood to be the base module from the floor group, drive train, carriage and transmission system (Piller and Waringer 1999, p. 64 et seqq.). Such platforms are shared by the BMW Z4 and the BMW 3 series, Audi A3, Audi TT and Golf, and the Porsche Boxster and 911 (Rosengarten and Stürmer 2005, p. 142 et seq., 193).

Great savings in assembly costs are potentially available to the automotive manufacturers when they can limit the majority of the final assembly work to preassembled connected modules or systems. In recent years more and more production and development work has been assigned to the suppliers. This initially moved costs and risks to the weaker, mid-sized suppliers and it seemed to worsen their relative market position. However, the end effect was a core of suppliers who mastered these difficult circumstances and were able to establish themselves as a direct system or module supplier. The assignment of customer-specific orders to develop and produce systems or modules almost reversed the power ratio, because although the smaller suppliers often only served one major customer, this purchaser now completely depends on the quality of the development, production and on-time delivery from these first-tier suppliers (Becker 2007, p. 39 et seq., 111 et seq., 137, 170 et seqq., 181, 194 et seqq., 207; Gromer 2006, p. 127, 160 et seqq.; Sanz et al. 2007, p. 29 et seq.; Weigand 1999, p. 1; Wolters et al. 1999, p. 61 et seqq., 73).

Due to the changes over the past few years, the distinctive pyramid structure of the automobile market added-value chain has changed. In the beginning of the manufacturers' outsourcing wave, logistics services and then production volumes were affected. Today, development tasks are often completely outsourced to automotive suppliers.

In the future, first-tier suppliers will almost exclusively supply the automotive manufacturers with modules and systems as part of the preferred modular sourcing acquisition strategy. They can be assigned to two different categories based on the structure of their abilities. The most complex tasks fall into the system integrator's area of responsibility. These companies have taken on parts of the automotive manufacturers' core business and development, have extensive technological competence and ensure flawless integration of components into system modules. In contrast, the module suppliers limit their core competences to assembly services and logistics and have transferred development tasks to sub-suppliers. They obtain product development and innovations from system specialists (second-tier) who have almost completely halted assembly activities. The lowest level of the pyramid is the parts and component supplier (third-tier) who merely provide minor assembly and development services and mainly produce and deliver standardised parts as ordered (Becker 2007, p. 168 et seq., 181 et seq.; Gehr and Hellingrath 2007, p. 12 et seq.; Gromer 2006, p. 136 et seqq.; Sanz et al. 2007, p. 13; Weigand 1999, p. 13; Wolters et al. 1999, p. 69 et seqq.).

However, so-called development partnerships are no longer limited to vertical cooperation between customers and suppliers; rather, they are also found between different automotive manufacturers. Not only can scale effects be achieved in the area of development and construction, but also through central production of like parts and platforms. Examples of what is now called "badge engineering" are the almost identical models, the Ford Galaxy, VW Sharan and Seat Alhambra or Peugeot 104, Citroën C 1 and Toyota Aygo. The successful marketing of the almost identical SUVs, Touareg by Volkswagen and Cayenne by Porsche, shows that this strategy can be successful for products of different image and price levels in the market. Horizontal cooperation is currently limited to the mutual development and production of individual modules, such as the BMW 4-cylinder motor in cooperation between BMW and PSA (Becker 2007, p. 35, 75 et seqq., 129 et seq., 200; Gromer 2006, p. 179 et seqq.; Piller and Waringer 1999, p. 94, 105; Sanz et al. 2007, p. 204, 254).

While reducing their depth of value-added activity, with increased competition and a shift of power to the customer, automotive manufacturers recognised that they had to redefine their core competences. The focus of the core competences was subsequently shifted from development and production competences to brand management. Not too long ago, company divisions, including vehicle development, raw construction, painting, assembly and the in-house production of selected modules, such as motors, transmissions and exhaust systems, were off limits to the outsourcing strategists. Today, all value adding services that the customer will not directly connect with the manufacturer's brand, are open for disposition. Services, such as leasing, new car sales, used car sales, fleet management, financing or insurance are moving into the automotive manufacturer's focus. Only the design departments can be sure that they will remain in the hands of the manufacturers, while now, even image-bearing modules such as motors are not only produced by suppliers, but also purchased from competitors. Even if the final assembly is mainly located at the automotive OEM's plants, there are already examples of vehicles being completely produced by large first-tier suppliers or assigned manufacturers. For example, BMW assigned the production of the X3 and Mercedes-Benz the production of the G-Class to Magna Steyr, while Karmann produces the Audi A4 Cabriolet and Mercedes CLK Cabriolet (Becker 2007, p. 75, 85 et seqq., 108 et seq., 123 et seqq., 136, 175 et seq., 192 et seqq., 240; Gehr and Hellingrath

2007, p. 1 et seqq.; Gromer 2006, p. 136; Rosengarten and Stürmer 2005, p. 190 et seq.; Sanz et al. 2007, p. 27, 110, 253; Weigand 1999, p. 11 et seq.).

The choice between efficiency and flexibility regarding production capacities is now often made for the higher cost flexible production choice. If different car models can be produced at one location with the same assembly lines, such flexibility contributes to increased efficiency in production. This is in spite of higher initial investment, because the variance in demand for the individual segments can be balanced out, without incurring large overcapacities. Capacity is partially reserved for increases in demand and because these are not utilised most of the time, they represent dead capital. The investment behaviour of most manufacturers, with the expansion of the model selections, does not usually balance beyond this economic, planned under-utilisation of product capacities. Exogenous specified market volumes must currently lead to structural overcapacities worldwide. In Western Europe, the capacity utilisation over the past few years averages about 80%. Even in "booming" regions, the increased demand is generally addressed too far ahead of time and an expensive over-capacity "buffer" created. The automotive industry will have to address this problem in the near future; otherwise, some weaker manufacturers will have to withdraw unless overall market capacities are reduced (Becker 2007, p. 2, 21 et seqq., 83 et seqq., 106; Rosengarten and Stürmer 2005, p. 192).

Modularisation concepts can also be applied at the process level. It makes sense that with the product-orientated direction of the overall company, product-related development, production and manufacturing processes will be modularised. Although the creation of clear customer-orientated process modules within flat hierarchies increases the number of organisational interfaces, it contributes to the transparency of value-adding activity and reduces the overall need for coordination through the implementation of stable processes (Piller and Waringer 1999, p. 43 et seq.).

In addition to cost savings and reduced complexity with administrative tasks, modularity also delivers a direct customer benefit because the buyer of a new car can quickly and easily configure his car, make late changes due to flexible production and receive his car within a few days thanks to the short processing times. In the future, it may even be possible for customers to replace or exchange individual new modules, maybe because they offer different functions or because they have been updated (Piller and Waringer 1999, p. 50 et seqq.; Wolters et al. 1999, p. 88).

For further information on the issues of the sub-chapter modularisation see Krampf (2000) or Tietze (2003). For aspects of modularisation see Piller and Waringer (1999).

4.3.3 Construction of a Flexible Production Network

The established pyramidal market structure and tight vertical and horizontal business relationships in the automotive industry have made the company network the preferred form of organisation. The current trend for automotive suppliers and manufacturers to concentrate on their own brand-specific or customer-orientated core competences has resulted in more intensive mutual dependencies. One company alone can no longer maintain the competences required to service the market and meet competitive demands from across a broad spectrum of value-added levels. Therefore, the automotive industry has tended to form core competence-related segments and cooperate with partners who have complementing capabilities (Becker 2007, p. 67, 125, 137, 199; Piller and Waringer 1999, p. 93 et seqq.; Sanz et al. 2007, p. 111; Weigand 1999, p. 1 et seq.).

Market participants need to reduce uncertainty and increase their use of external capability to respond to the cost of their internal flexibility. One consideration is to decide to join an alliance or a network. Network organisations form where the automotive manufacturer represents the focal company within the overall network, while the first-tier and other suppliers take over the coordination of each sub-network (Sanz et al. 2007, p. 111, 255). The advanced development level of automotive networks requires demand-orientated planning and logistic methods to design value-added networks that are flexible, adaptive and cost-efficient. At the same time, the key to success is the overall view of networks as a unit of development, acquisition, production and distribution processes. The network participants have a common goal of becoming a customer-orientated, value-added chain, simultaneously optimising cost potential (Sanz et al. 2007, p. 397). With the forward displacement of work packets, many smaller suppliers are now asked to perform tasks that were only performed by the large supply companies. Currently, an ability to integrate into the network is the deciding factor for the success of the customer-supplier relationship. The relationships between first-tier suppliers and the automotive manufacturers must be maintained much more intensively today than ever before. The inclusion of the suppliers in development activities in the automotive network requires communication across different company levels (Sanz et al. 2007, p. 404).

Network competence, which is the successful planning, control, integration and surveillance of cross-company cooperation and efficient information exchange in global networks, is becoming a core competence in the automotive industry (Sanz et al. 2007, p. V et seq., 124).

A great deal of potential can be released in strategic networks. Outsourcing is transforming the industry and not only spreads the costs and risks of innovations to multiple partners by the integration of resources and competences, but speeds up the product development and marketing processes and brings growth. Companies that can contribute complementary resources and competences to networks can increase their ability to respond to market uncertainty. One of the central requirements for successful cooperation is a dependable basis of trust. This is promoted by mutual dependency and reduces behavioural uncertainties in inter-company communication (Sanz et al. 2007, p. 112 et seqq., 255). However, network engagements also conceal a number of risks, presented in the form of deceitful opportunistic partner behaviour, stronger or one-sided inter-dependencies, complexity-related additional costs, or an excessive need for coordination, which will end in inefficient decision-making processes (Sanz et al. 2007, p. 116 et seq., 124).

The current dominant prevailing planning procedure in networks still displays obvious deficits. Since optimised cross-network planning methods are currently rarely applied, each member of the network plans for his own individual security. Cascading these planning processes through a network results in delays from one supply level to another because needs forecasts, delivery forecasts and product requests are often not transferred to the next level in a timely manner. Many companies only determine net demands once each week using their own MRP system and then forward the information to the suppliers. The delay creates safety buffers at each subsequent level and results in emergency actions due to inaccurate planning and coordination lacking across the entire network (Gehr and Hellingrath 2007, p. VIII et seq., 14 et seqq., 25, 50).

The consequences are seen in the poor levels of flexibility when reviewing possible changes in demands or requested quantities. Cross-network troubleshooting management is almost impossible to maintain under these conditions, not to mention proactive troubleshooting management whereby responses to faults are possible even before the fault has actually occurred. In addition, there is no synchronisation of planning processes. Shipment planning is performed without optimising the shipment and transporting processes, and insufficient tracking and tracing functionalities do not permit exact reviews of the shipment status of goods and thereby their timeliness (Gehr and Hellingrath 2007, p. 7, 26 et seq., 49 et seq.).

Demand and capacity planning is currently not efficient. On the operational horizon it is based on customer orders, but on further horizons it relies on companies' sales budgets. Static parameters, such as fixed delivery times, are used for demand planning. These are usually not updated frequently enough and cannot reflect the current status of shipments dispatched or the supplier's current inventory. The effects of this lack of a planning method are even more significant when a company in a multi-level supply chain is distant from the actual individual generating the demand. Information lost when demands are forwarded worsens the forecasted values from one supplier level to another. This means data for dependable capacity and investment planning are not available. Furthermore, the recipients do not get final and current information about their supplier's capacity utilisation and therefore cannot estimate the supplier's ability to respond to changes in consumer demand (Gehr and Hellingrath 2007, p. VIII et seq., 49 et seqq.).

Complexity is increasing within supplier networks driven by increased dependability and interaction levels between companies and the variety and flexibility in production that is demanded by car buyers. New cross-network collaborative planning methods are needed for demand, capacities and inventory. Implementation requires all companies involved in the value-added network to have standardised planning software or at least software interfaces where all relevant information can be forwarded to the other participants. Standardisation of processes for cross-company interaction is, therefore, another major contribution (Gehr and Hellingrath 2007, p. VII et seqq., 9 et seq., 18, 51; Sanz et al. 2007, p. 396).

Automotive manufacturers generally hold control dominance over the supplier network and must therefore be as familiar as possible with the structures and processes of the network. As the central consumers, they want to be able to recognise any potential delivery bottlenecks early on and they want to be informed about current product movements using performance monitoring (Gehr and Hellingrath 2007, p. 5, 11). First-tier system and module suppliers are generally responsible for on-time deliveries of the correct quantities to the car dealers and therefore also require information about demand and capacity profiles of sub-suppliers. This information must be updated frequently to be able to proactively advise of problems and avoid short-term delivery bottlenecks. In addition, first-tier suppliers strive for optimal utilisation of their critical and expensive resources in terms of supply security and cost relevance and forward the bundled information of the planning system to their network partners (Gehr and Hellingrath 2007, p. 6, 15). The n-tier suppliers, which are often small and mid-sized companies, who often do not have their own high-performance planning systems, must be equipped with affordable standard compatible planning software that provides them information about the partners' planning activities, optimises the utilisation of their resources and ensures their ability to deliver (Gehr and Hellingrath 2007, p. 6 et seq., 15).

The collaborative demand and capacity planning process integrates mid- and long-term planning and optimises forwarding of demand forecasts. It also handles shipment planning and processing in the short term for a supplier up to receipt by the customer. Simulations are used to play through different planning scenarios, to generate demand forecasts and to determine how the network partners must respond to different, potentially critical, situations. If the actors want to lead their network successfully, they must orientate themselves around the three basic principles of collaborative actions – common benefit has priority over the benefit to an individual supply chain partner, the optimisation of the network has priority over the optimisation of the individual, and the potential benefit of rationalisation through collaborative actions must be divided in partnership (Gehr and Hellingrath 2007, p. 11, 18 et seq., 23, 27, 41 et seqq.).

In addition to the individual equipment options, car buyers now demand a high degree of flexibility in making changes to their vehicles' configuration, up until just before production begins or even during assembly. At the same time, each customer should receive their customised car on a binding delivery date. A software system for collaborative planning must be able to handle these rapid changes and new logistics requirements. In addition, they require the entire supplier network to be build-to-order-orientated. Already automotive manufacturers have modules and components directly delivered just-in-time or, better, just-in-sequence to the production line. The marketing to production philosophy is shown as a customer order-orientated one-piece flow. This means that in what has been termed the "pearl necklace", individual parts or module systems reach the assembly line at a defined point in time together with the higher-ranking building block, whether that is a module or the entire vehicle. Such exactly synchronised processes lead to increased complexity of planning and processing and are extremely prone to faults, including delivery delays caused by the road traffic situation. It is important to the automotive manufacturer that his suppliers can guarantee supply. Due to cost considerations, building to stock is not an option for suppliers, unless they happen to produce inexpensive standardised small parts. In addition, suppliers must also deliver to their customers just-in-sequence, so that the order in which their parts arrive matches the production sequence of the individual vehicles being built. Depending on the technology and module the sequence is implemented all the way to the forefront of the delivery chain (Becker 2007, p. 28, 67; Gehr and Hellingrath 2007, p. 9 et seqq., 13 et seq.; Gromer 2006, p. 161 et seq.; Sanz et al. 2007, p. 254 et seq., 397; Wolters et al. 1999, p. 64, 69).

Components delivered to the manufacturer for final assembly are generally complex, pre-assembled modules with equally complex and hard to transfer geometries. To aid efficiency, supply companies frequently relocate to industrial parks near their customer and pre-assemble their components there. First-tier suppliers in particular are inclined to follow their customers, placing assembly halls close to their locations. This occurs both nationally and internationally, because the majority of automotive companies generally acquire a system module from one proven supplier worldwide (Becker 2007, p. 236; Sanz et al. 2007, p. 254 et seq.; Wolters et al. 1999, p. 16, 69 et seq.).

The reasons for opening a new production location abroad differ for manufacturers and suppliers. Successful automotive manufacturers follow their demand abroad to learn about the markets directly and be able to respond more quickly. They are frequently followed by their first-tier. OEMs may want to leverage local advantages in the form of lower labour costs, and many n-tier companies move because they manufacture simple, labour-intensive parts, not because of their customers. Another motive for manufacturers' movement is the avoidance of high duty-related expenses for completely assembled cars. A product that requires final assembly and/or finishing is referred to as "completely knocked down" or "semi knocked down" (CKD or SKD). Plants that produce SKD or CKD may also be used to develop the market of a country if the expected quality output level there does not currently meet the requirements for a complete production location. Some automotive manufacturers move their production facilities away from their homeland to be able to win new customers located in foreign countries more easily. Last, but not least, engagement in other important markets reduces the currency risks because components are purchased with the same currency for which the finished product is sold. The history of automotive manufacturers' location changes began several decades ago in Latin America, continued in Eastern Europe, and is currently taking place in China and India (Becker 2007, p. 100 et seq., 189 et seqq., 202 et seq., 235 et seqq.; Sanz et al. 2007, p. 29, 42, 254 et seq., 400; Wolters et al. 1999, p. 17).

Other aspects should be taken into consideration before pursuing a relocation strategy. A volume manufacturer generally does not suffer negative consequences if it produces its cars in a cheap labour country as long as certain quality standards are met. A premium supplier may certainly follow its customers to an important foreign market with new production locations. However, the high quality and tradition of the home market are essential brand images of a premium brand. The production of premium cars in new boom regions, such as China or India, could damage the image of the vehicles, even if the customer base for premium cars is rapidly growing in such regions. Suppliers should always weigh the benefit of cheap labour against the cost resulting from higher logistic and transport expenses and brand implications before they decide on a new location (Becker 2007, p. 188).

As a result of the global structure of automotive networks and intensified cooperation of network companies on different levels, competition in the global automotive industry will become more and more a competition among these networks. In the future, the complete performance and quality of the network will be the deciding factor in the ability of an automobile brand to compete, not only the success of an automotive group at the top of a supplier pyramid[1].

4.4 Conclusion

Strategies are in place for the European automotive industry to confront the consequences of advanced core market saturation in Europe, North America and Japan and increasing competition from the Far East. Amongst these strategies BTO plays a key role. Modular product design is intended to make the advancing variety of options manageable for companies and affordable for customers. New collaborative planning methods are being developed to deal with the complexity of the multistage supply chains of the industry and to maintain its capacity to act. These will both be developed in more detail later in this book. History will be the judge of our success.

References

Becker H (2007) Auf Crashkurs. Automobilindustrie im globalen Verdrängungswettbewerb, 2nd edn. Springer, Berlin
Da Ke R (2007) Steiniger Weg nach Europa. Auto Prod 8:42
Diez W (2005) Strategiewahl – Premium- oder Massenmarkt? In: Gottschalk B, Kalmbach R et al. (eds) Markenmanagement in der Automobilindustrie. Die Erfolgsstrategien internationaler Top-Manager, 2nd edn. Gabler, Wiesbaden
Gehr F, Hellingrath B (2007) Logistik in der Automobilindustrie. Innovatives Supply Chain Management für wettbewerbsfähige Zulieferstrukturen. Springer, Berlin
Gromer S (2006) Die Automobilindustrie in Deutschland. Eine Untersuchung auf Basis des Konzepts der Koordinationsmängeldiagnose. Kovac, Hamburg
Knauer M (2007) Mehr als nur vier Räder. Automobilwoche 13:12
Krampf P (2000) Strategisches Beschaffungsmanagement in industriellen Großunternehmen. Ein hierarchisches Konzept am Beispiel der Automobilindustrie. Eul, Lohmar
Piller FT, Waringer D (1999) Modularisierung in der Automobilindustrie – neue Formen und Prinzipien. Modular Sourcing, Plattformkonzept und Fertigungssegmentierung als Mittel des Komplexitätsmanagements. Shaker, Aachen
Rosengarten PG, Stürmer CB (2005) Premium power. Das Geheimnis des Erfolgs von Mercedes-Benz, BMW, Porsche und Audi, 2nd edn. Wiley-VCH, Weinheim

---

[1] For additional information on successful supply networks, see Gehr (2007).

Sanz FJG, Semmler K et al (eds) (2007) Die Automobilindustrie auf dem Weg zur globalen Netz-
    werkkompetenz. Effiziente und flexible Supply Chains erfolgreich gestalten. Springer, Berlin
Schlott S (2007) Elektronisch die Sinne schärfen. Auto Prod 8:86–88
Scholz G (2007) Innovation muss bezahlbar bleiben. Auto Prod 5:54–55
Tietze O (2003) Strategische Positionierung in der Automobilbranche. Der Einsatz von virtuel-
    ler Produktentwicklung und Wertschöpfungsnetzwerken. Deutscher Universitäts-Verlag,
    Wiesbaden
Weigand A (1999) Integrierte Qualitäts- und Kostenplanung am Beispiel der Konzeptphase in
    der Automobilindustrie. Europäischer Verlag der Wissenschaften, Frankfurt am Main
Wolters H, Landmann R et al (eds) (1999) Die Zukunft der Automobilindustrie. Herausforderun-
    gen und Lösungsansätze für das 21. Jahrhundert. Gabler, Wiesbaden

Chapter 5
Outsourcing: Management and Practice Within the Automotive Industry

Jens K. Roehrich

Centre for Research in Strategic Purchasing and Supply (CRiSPS),
School of Management, University of Bath, Bath, UK

Abstract. With the continuing increase in competitive pressures in the automobile industry, the acceleration of cost and price increases and the omnipresent need for improvement of engineering productivity, managers have to constantly ensure the company's survival in the market. This perspective emphasises the need for managers to consider outsourcing in order to sustain a company's competitive advantage. The use of external assembly service providers in the automotive industry is widespread and is embraced in the build-to-order concept. Outsourcing capacity brings with it the risk of outsourcing competency. Managers in the customer-conscious automotive market have to thoroughly understand the concept and the associated risks in order to benefit from the outsourcing practice. Therefore, this chapter critically evaluates the benefits and risks associated with outsourcing core and supporting activities in the automotive industry. Moreover, the study draws on different theoretical positions delivering a rigorous description and comparison of the theoretical outsourcing standpoints. The research study is underpinned by empirical evidence from the automotive industry and concludes with a set of managerial implications to facilitate well-grounded outsourcing decisions within the automotive industry.

5.1 Introduction

The management practice of outsourcing a company's core and supporting activities in a variety of business functions has been a focus of attention, not only recently, but also during the last few decades. The build-to-order (BTO) paradigm presented in this book makes use of the outsourcing concept to deliver the flexible capacity necessary to deliver a car in 5 days. Apart from the growing volume of literature describing various facets of outsourcing, the consultancy business welcomes outsourcing as a weapon to fight within a competitive marketplace as well as to focus business strategies. This chapter presents a concise analysis of outsourcing core and supporting activities, with an emphasis on how to manage and evaluate outsourcing practice as the automotive industry moves towards BTO.

Outsourcing is prominent for many organisations thinking of survival in global competition. With the emergence of global, fast-changing markets, shorter product life-cycles combined with a whole-life costing approach, companies across industries are facing strong pressures with regard to their competitiveness and profitability. Customers are nowadays paying more attention to the innovativeness and the customisation of a product and service, taking high quality and delivery performance linked to low prices as a must. Consequently, companies have to constantly evaluate and improve organisational adaptability and competitiveness. This perspective emphasises the need for managers to consider outsourcing, among several other managerial practices, in order to sustain a company's competitive advantage over its rivals.

Outsourcing involves offloading a company's activities to external providers who possess expertise, innovative technologies and resources in its specialisation to carry out activities highly efficiently. The concept of outsourcing may bring about benefits to the outsourcing company that are vital for its competitive market position. But caution tempers the enthusiasm for outsourcing. Outsourcing has progressed from involving only peripheral business activities towards embracing critical core activities that contribute to a company's competitive advantage. As a result, outsourcing has become an increasingly complex and vital issue for many organisations.

Managers in the customer-conscious automotive industry have to thoroughly understand the concept and the associated risks of outsourcing in order to benefit from the practice. While it may be true that some companies base their out sourcing decisions purely on cost savings, companies must also evaluate decisions from several other beneficial perspectives. Yet, the risk of an inappropriate outsourcing decision may prove fatal and evidence is omnipresent across various industries.

Therefore, this chapter critically evaluates the benefits and risks associated with outsourcing core and supporting activities in the automotive industry. Furthermore, the study draws on different theoretical positions delivering a rigorous description and comparison of the theoretical outsourcing standpoints. The benefits of successful outsourcing and the risks and consequences of outsourcing failure are outlined. The research study is underpinned by empirical evidence from the automotive industry and concludes with a set of managerial implications to facilitate well-grounded outsourcing decisions. This chapter is beneficial to middle and senior managers in the automotive industry who are considering outsourcing or who already have outsourcing programmes in place.

5.2 Defining the Concept of Outsourcing

Before discussing the different approaches and their impact on outsourcing decisions, it is vital to rigorously define the concept of outsourcing. Therefore, the following section paves the way for a discussion on outsourcing core and supporting activities by collating the various outsourcing definitions. Outsourcing has been defined in various ways by practitioners and academics. What these ways have in common is the acknowledgement of risks and benefits that come along with outsourcing decisions. Definitions of outsourcing recognise the practice as "the procurement of products or services from sources external to the organization" (Lankford and Parsa 1999, p. 310) or as "the transfer of previously in-house activities to a third party" (Lonsdale 1999, p. 176).

Seen from the view of strategic flexibility, outsourcing is just one way in which the boundary of an organisation can be adjusted in response to changing global markets. It may concern either a firm's primary supply chains or its supporting activities (Lonsdale and Cox 2000). With regard to the growing interdependence of companies, a strategic flexibility perspective becomes increasingly important for a single firm. Recognising the controversial discussion about core competences in an organisation, Sharpe (1997, p. 538) defines outsourcing as "turning over a part or all of those functions that fall outside the organisation's chosen core competencies to an external supplier whose core competencies are the functions being outsourced". This definition is closely associated with the notion of the organisation's core activities, which raises the problematic issue of identification of what is core and what is supporting.

5.3 Approaches to Outsourcing

The following theoretical approaches have to be taken into consideration in order to derive well-grounded outsourcing decisions that are beneficial to the company. Among the most common theoretical approaches are the resource-based view, the resource-dependency theory, the transaction cost theory and the industrial network approach, which will be outlined in the following sections. The section on outsourcing approaches concludes with a summary contrasting the different approaches and their impact on outsourcing decisions.

5.3.1 The Resource-Based View – How Can a Company Sustain Its Competitive Advantage in a Dynamic Market Environment?

The resource-based view (RBV) understands the firm's resources as the foundation for its strategy. Focussing on internal resources and competencies to develop organisational capabilities in order to achieve a distinctive competence are central to this theory. A firm's competence can be described as "a competitively valuable activity that a company performs better than its rivals [and which] represents a competitively superior resource strength" (Thompson et al. 2005, p. 91). The RBV states that a firm will earn abnormal profits due to their lower costs gained by superior productive resources (Lonsdale 1999). Furthermore, companies can also earn abnormal profits through superior productive capabilities or through product differentiation, enabling the company to provide valued uniqueness and functionality for its customers and charge a premium. However, in order to sustain a competitive advantage with the RBV, the company should install protection around its resource position to avoid substitutions and imitations. Rumelt (1987) cited a list of "isolating mechanisms" including, for instance, buyer switching costs, reputation, producer learning and economies of scale for firms, that are needed to protect a company's favoured market position.

From an outsourcing perspective, another condition required to gain abnormal profit is to retain the valuable resources within the firm's boundaries. Williamson (1985) stated that it would be easier for a firm to do so if the resource is "firm-specific", i.e. that its value diminished outside the firm. Teece (1987) added the notion of "co-specialised", i.e. it is only valuable if it is used in combination with other resources within a firm, as another condition that keeps valuable resources inside the firm's boundaries. Therefore, the RBV proposed that if the firm outsourced certain resources, it will lose control over its strategic core. In other words, the company's competitive position depends on its ability to gain and defend advantageous positions concerning resources important to production and distribution (Wernerfelt 1984; Barney 1991).

In addition to deploying existing resources and capabilities, Grant (1991) cited that a company should also develop its internal resources and capabilities. In order to fully exploit a firm's existing stock of resources and capabilities, and to develop a competitive advantage, the external acquisition, i.e. outsourcing, of complementary resources and capabilities may be necessary. Grant (1991) argued that the outsourcing strategy not only secures the firm's existing resources, but also increases resources and capabilities as well as a company's strategic opportunities. Managers in the automotive industry have to understand that it is seldom enough to just possess a competitive advantage in the marketplace. The competitive market environment pronounces the need for timely responsiveness, rapid and flexible product innovation, coupled with the managerial ability to effectively coordinate and redeploy internal and external competences. The question to ask now is how do firms achieve and sustain a competitive advantage over their rivals?

The competitive advantage of firms is considered "as resting on distinctive processes (ways of coordinating and combining), shaped by the firm's specific asset positions (such as the firm's portfolio of difficult-to-trade knowledge assets and complementary assets), and the evolution path(s) it has adopted or inherited" (Teece et al. 1997, p. 509). Keeping in mind the definition above and the ever-changing market environment, a rather static viewpoint is less appropriate to answer the question of how a company sustains its competitive advantage.

A concept drawing on more dynamic rather than static capabilities should be deployed. Dynamic capabilities are defined "as the firm's ability to integrate, build, and reconfigure internal and external competences to address a rapidly changing environment" (Teece et al. 1997, p. 516). In markets where the competitive landscape is rapidly shifting, dynamic capabilities become the source of sustained competitive advantage (Eisenhardt and Martin 2000), giving organisations the ability to gain new and innovative forms of competitive advantage. Dynamic capabilities and the firm's adaptability are some of the keys to sustainable competitive advantage in the fast-moving automotive market. Companies in the automotive market have to constantly adjust and align their dynamic capabilities consisting of firm-specific strategic and organisational processes like product development. These alignments should be supported by strategic decision-making and senior management support in order to create value for companies working within dynamic markets.

5.3.2 Resource-Dependency Theory – Why Is a Company Dependent on Its External Environment?

The resource-dependency theory (RDT) differs from the RBV inasmuch as the former focuses on the external environment and not on the internal. Academics in the field argue that an organisation finds itself dependent on some elements in its external environment (Thompson 1967). The external dependency is caused by the firm's limited control over needed resources, such as land, labour, capital, information and/or specific products or services (Kotter 1979). An internal lack of resources can be overcome by the firm when it enters into exchange relationships with other firms, emphasising the dependency of the firm on its external environment. Therefore, the resource dependency theory emphasises the need for a company to adapt to environmental uncertainty, to handle problematic interdependences and to oversee resource flows (Pfeffer and Salancik 1978). Cheon et al. (1995) stated that the dependency of a firm on any other organisation, as occurs in outsourcing relationships, is determined by the importance of the resource to the organisation, the number of potential suppliers and the cost of switching supplier. In conclusion, a firm's outsourcing strategy consists of different degrees of dependency of one organisation on another in order to acquire critical resources that are not cost-effectively available internally.

A different perspective of outsourcing decisions is made by the transaction cost theory, which is not based on resources, but on costs that occur during transactions.

5.3.3 Transaction Cost Theory – Which Costs to Consider When Making Outsourcing Decisions?

The transaction cost theory (TCT) of a firm was introduced by Coase in the 1930s and further developed by Williamson in the mid-1970s until the mid-1980s. This approach states that the firm's economic activity depends on the balance of production economics against transaction costs. Transactions are seen as the exchanges of goods or services between economic actors, who are technologically separate units, inside and/or outside the organisation (Williamson 1985). The success of a firm in the view of this theory depends upon effective management of transactions.

The transaction cost approach provides, among other benefits, a generic framework for analysing outsourcing options. According to this theory, the choice lies between using an outsourcing provider and providing services in-house (Lacity and Hirschheim 1993). Two different types of costs have to be considered when making the decision between in-house production and outsourcing provider: production and transaction costs. On one hand, smaller production costs can be achieved by the decision to outsource, primarily due to the economies of scale that a provider might have (Cheon et al. 1995). On the other hand, Cheon et al. (1995) pointed out that outsourcing leads to higher transaction costs arising from negotiating, monitoring and enforcing contracts. In order to evaluate the decision whether to outsource or not, factors that influence the magnitude of transaction costs should be assessed. Williamson (1985) suggested three influencing factors: asset specificity, or the degree to which the transaction will produce an asset that is dedicated to a special purpose; the degree of uncertainty in the environment, such as unpredictable markets, technological and economic trends; and infrequency of contracting. Figure 5.1 illustrates the relationships between the influencing factors of transaction costs by Williamson, the trade-off between transaction and production costs and its consequences. Each of the three influencing factors raises the effort and cost of structuring an agreement between outsourcer and outsourcing provider that will assume the successful completion of the contract (Cheon et al. 1995).

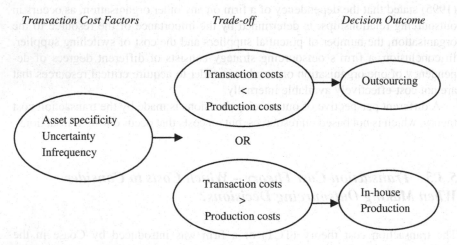

Fig. 5.1 A transaction costs perspective of outsourcing (Adapted from Cheon et al. [1995], p. 214)

The transaction cost theory is solely based on a single firm, neglecting the existence of other firms and their influence on the firm's behaviour. The industrial network approach takes the firm's wider network and its influence into account.

5.3.4   Industrial Network Approach – Why Should a Company Consider Its Wider Network?

The network approach focuses on entire relationships among buying and selling firms and has been influenced by the theory of social exchange. The industrial network approach provides, like the resource dependency approach, a perspective on inter-organisational relationships (Pfeffer and Salancik 1978). "[I]t is concerned with a focal organisation [and] attempts to describe the multiplicity of relationships of any industrial or commercial organisation" (Easton 1997, p. 103). The resource dependency view mainly concentrates on the individual relationship between firms, whereas the industrial network approach sees the network and its various relationships that exist among individuals, groups and organisations within a network.

Relationships are described in the literature as one of the most important assets of a company. "[A] relationship offers access to third parties who may have resources that are either valuable or essential to survival" (Easton 1997, p. 107). The relationship consists of several important elements, which should all be met in order to reap the maximum benefit. The elements are mutual orientation, dependence, the bond between firms, investments seen as the commitment of resources towards the mutual orientation, as well as atmosphere, described as the tension between conflict and cooperation (Ford et al. 1986). Consequently, "relationships form the context in which transactions take place" (Easton 1997, p. 111).

Nowadays, companies have a portfolio of different types of supply relationships with a number of suppliers of various capabilities and resources. Classifications of buyer–seller relationships can be based upon several variables such as involvement of the supplier, the strategic importance of the outsourced product/service or the complexity of the supply market. However, companies should always view relationships to their various suppliers in the network context, as the "total supplier network determines the efficiency and the effectiveness of the operations on the supply side of the companies" (Gadde and Håkansson 2001, p. 174). The network approach emphasises the combined and integrated capabilities and the exchange of information between the actors involved. Companies should also be aware that "any action undertaken by one actor impacts the others in several different ways" (Gadde and Håkansson 2001, p. 176). Therefore, companies need to include other network actors in their thinking and the company's own activities and resources should be seen as assets of the network rather than of a single company.

5.3.5   Summary of Outsourcing Approaches

Table 5.1 collates the four outsourcing approaches that should be taken into consideration. Each approach addresses different vital aspects of the company and its surrounding network. The amalgam of these approaches will prove useful in facilitating managers to arrive at well-grounded outsourcing decisions. The degrees to which the various approaches should be taken into consideration depend strongly on organisational factors and environmental conditions.

Table 5.1 A comparison of the four outsourcing approaches

	Resource-based view (RBV)	Resource-dependency theory (RDT)	Transaction cost theory (TCT)	Industrial network approach
Possibilities of outsourcing	Resources that miss some or several attributes such as rare, non-imitable, and non-substitutable	Resources that miss some or several attributes such as rare, non-imitable, and non-substitutable	Activities that do not require asset-specific investment	Issue of what the company can mobilise through the network Activities and resources that can be produced more efficiently within the network
Main variables and fundamental units of analysis	Resources (internal)	Resources in the external environment	Asset specificity, frequency, uncertainty (transactions)	Network, relationships and interactions (activities/actors/resources) Overall industrial network structure (How to mobilise resources from other actors?)
Limits to outsourcing	Core competences and activities that do not have these attributes, but cannot be sourced from the market	(Critical) core resources and valued resources	Activities that require specific investment (high specificity)	Core competences that would make the company redundant
Risks	Firm may become "hollow", loss of control over strategic core, loss of access to assets	Loss of advantageous position (resources), loss of control over strategic resources, become dependent on too many firms	Transaction costs more than production costs, outsourcing activities with high specificity	Firm does not manage the relationships properly and fails to mobilise resources, activities and actors

Understanding the concept of core and supporting activities is crucial for companies in the automotive sector when considering the outsourcing option. The following sections will explore the controversial debate over core and supporting activities and identify different models that seek to distinguish between the two.

5.4   The Quest for Competitiveness – Identifying Core and Supporting Activities

"In principle any part of the value chain can be externally sourced, though for that to occur for a particular firm would imply that the company has no unique value-adding capabilities, and that its processes are available for replication by competitors and possibly by customers" (Jennings 2002, p. 26). Since the 1980s there has been a trend to reduce the degree to which companies are vertically integrated. Which activities should be considered in an outsourcing decision is a topic of an ongoing controversial discussion among practitioners and academics. Outsourcing can concern either a firm's primary supply chain(s), and consequently its core activities, or it can concern its supporting activities. Figure 5.2 depicts the distinction proposed by Arnold (2000).

Arnold's outsourcing model is build upon four elements of outsourcing: the subject, the object, the partner and the design. The subject of outsourcing can be considered as the company that wants to outsource processes. The objects are the processes that might be outsourced. Arnold (2000) distinguished between four different activities of a firm. The core activities of a company are defined as "all activities which are necessarily connected with a company's existence" (Arnold 2000, p. 24). The other three activities range from core-close to disposable activities. A disposable activity is an activity that is readily available on the market. Outsourcing partners can be considered as all partners who might be considered as possible suppliers, even an in-house supplier such as an independent business unit of the company (Arnold 2000). Moreover, the outsourcing design is strongly influenced by the market and company structure and will give a guideline for up-coming outsourcing decisions.

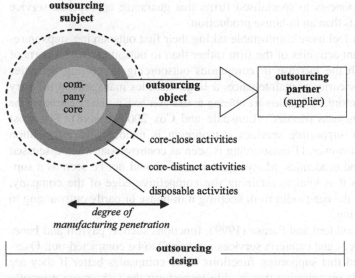

Fig. 5.2 Outsourcing model by Arnold (from Arnold, 2000, p. 24)

"One of the primary driving forces for companies to rely increasingly on outsourcing is gaining access to the resource collections of specialised actors" (Gadde and Håkansson 2001, p. 125). Authors who support this argument note that the outsourcing company can concentrate on its own resources in order to improve and exploit its capabilities. According to Quinn and Hilmer (1994), a company's focus on core activities has two main strategic advantages. First, a company that concentrates solely on its own resources and capabilities will maximise its return on these internal resources, and second, the further developed capabilities may function as an entry barrier.

The author's empirical data, derived from interviews conducted with middle and senior management in the automotive industry, illustrates that activities concerned with product development, including design, international marketing, innovation and technology, and the final assembling of vehicles are considered core. Activities concerned with the IT function are seen by most of the managers interviewed as core-close, whereas supporting activities are found in areas such as assembling parts or systems, design modelling, human resources, finance and accounting, manufacturing and customer services.

5.4.1 Outsourcing Supporting Activities in the Automotive Sector

The majority of outsourcing activities are concerned with outsourcing supporting activities (also called non-core activities) as shown in several surveys such those as conducted by PA Consulting (1996) and Cox and Lonsdale (1997). Companies that are well-positioned in their markets have already understood and embraced the concept of outsourcing non-core activities (Quinn and Hilmer 1994). Baden-Fuller et al. (2000) stated the example of manufacturing firms outsourcing standardised sub-components to specialised firms that guarantee appropriate service levels at lower costs than an in-house production.

Managers often feel more comfortable taking their first outsourcing steps in relatively unimportant activities of the firm rather than to outsource crucial parts of the business. With the increase in companies outsourcing supporting activities such as catering, security or maintenance, a large facilities management industry has evolved. Therefore, "[s]uppliers are offering to take over both the management and the operation of whole business premises" (Lonsdale and Cox 2000, p. 448). The most prominent part of supporting services outsourcing is undoubtedly information technology (IT). However, IT outsourcing is seen as controversial when discussed by practitioners and academics. Many argue that IT should not be seen as a supporting activity as it is vital to retaining the competitive edge of the company. Suggestions cover the bandwidth from keeping it in-house or partly outsourcing to complete outsourcing.

According to Lankford and Parasa (1999), functions such as payroll and benefits, human resources and cafeteria services are "ripe" to be contracted out. Overall, it can be noted that supporting functions serve companies better if they are given to an outside organisation that is able to perform the tasks more promptly and cost-efficiently. Managers in the automotive industry who were interviewed state that activities such as catering, security or cleaning and maintenance of buildings were completely outsourced to external providers as automotive companies have no competences and no benefits in carrying out these activities. Despite outsourcing supporting activities, the pressure on firms to further improve efficiency and their competitive advantage still remains present (Baden-Fuller et al. 2000). Many practitioners and academics are asking the question: are there some aspects of the primary supply chain (the company's core business) that could be or even should be outsourced?

5.4.2 The Core Competence Concept in the Automotive Sector

The opinion of traditional approaches to strategy that state that outsourcing aspects of the core business is risky is supported by many practitioners and academics. Companies may lose their competencies and become hollow (Prahalad and Hamel 1990). Furthermore, negative outsourcing effects can be experienced when competitors are able to steal key aspects of the firm's knowledge base (Bleeke and Ernst 1991). As core activities mostly lie in the area of core competences that are important for the survival of the company, an understanding of the core competence concept is vital. The approach of Prahalad and Hamel (1990) suggests that only goods and services that are considered core competences should be produced internally. Alexander and Young (1996) highlight the different ways in which the term core is defined by managers. The four meanings of "core activities" according to Alexander and Young (1996) are:

1. Activities traditionally performed internally with long-standing precedent.
2. Activities critical to business performance.
3. Activities creating current or potential competitive advantage.
4. Activities that will drive the future growth, innovation or rejuvenation of the enterprise.

Considering the fast-moving automotive market, Granstrand et al. (1997) add a new term to the concept of "core" or "distinctive" competences as suggested by Prahalad and Hamel. Granstrand et al. (1997) recommend that the "management in large firms needs to sustain a broader (if less deep) set of technological competences in order to coordinate continuous improvement and innovation in the corporate production system and supply chain" (p. 18). Moreover, they see the necessity for the management to do this "in order to explore and exploit new opportunities emerging from scientific and technological breakthroughs" (Granstrand et al. 1997, p. 18). Granstrand et al. (1997) recommend that large companies should become multi-technology companies and further introduce us to a more accurate description of large multi-technology firms' competences, which they call distributed technological competences. The so-called distributed technological core competences can be found in a number of technological fields, in different parts of the organisation and among several distinctive strategic objectives of the corporation.

5.4.3   Outsourcing Core Activities

Practitioners and academics, who challenge the traditional point of view of outsourcing, argue that outsourcing core activities could dramatically improve the shape of the company, combined with leanness and agility. This perspective draws on the industrial network perspective where external suppliers are able to provide benefits for the outsourcing company and gives access to the exploitation of network competences. However, for most firms, but not for all, the protection of core activities is a necessity.

Academics have theoretically and empirically researched four different circumstances under which outsourcing what seems to be core gives the company a competitive advantage over its rivals. The four circumstances proposed by Baden-Fuller et al. (2000) are as follows:

- *Catch-up:* a company has fallen behind its rivals despite a slow moving environment.
- *Changing value chains:* changing customer needs calls for a firm to respond.
- *Technology:* new technology made the firm's core obsolete.
- *Emerging markets:* rapid changes in customer demand and technology make new markets available to the firm.

Baden-Fuller et al. (2000) developed a 2×2 matrix (Table 5.2) that plots the four different circumstances depending on the two dimensions, which are technology environment and customer needs.

Managers whose company find itself in one or more of the circumstances described in Table 5.2, should consider outsourcing part of the company's core activities in order to gain back the firm's competitive advantage. Moreover, Baden-Fuller et al. (2000) argue that when the core competencies that a company currently possesses fade away, outsourcing could bring back its competitive advantage. This

Table 5.2 Four categories of outsourcing core activities (from Baden-Fuller et al., 2000, p. 287)

	Technology is evolutionary	Technology has become revolutionary
Customer needs are evolving	**Catch-up** What should be a core competence is in fact of little value, because of firm failure. The firm must catch up with rivals who have evolved faster. **Key issue:** Building new competence.	**Technology shifts** The key technologies required to fulfil customer needs have changed. The company needs to buy in new skills to stay in the competitive race. **Key issue:** Accessing new competence.
Customer needs are changing rapidly	**Changing value chain economics** The source of profit is shifting in the value chain. What was critical is now peripheral, and is now outsourced. **Key issue:** Cost.	**Emerging markets** Typically the firm does not possess either technology or customers to fully exploit the market. Rivals are frequently in similar positions. **Key issue:** Fast track innovations to get to the market first.

view is supported by Alexander and Young (1996), urging managers not to stop outsourcing when it comes to core activities. According to their research, problems occur mostly at the start of outsourcing as managers lack a consistent definition of the notion of "core activities".

Additionally, managers in the automotive industry should be aware of the shifting nature of core activities, prescribing the concept of "core" in a dynamic and evolving rather than static manner. Various academics (e.g. Lonsdale and Cox 1998) urge managers to focus not only on what is core today, but on what will be core to the company in the future. The disadvantages of this interpretation are most obvious in fast-moving technology markets, where companies may lose their competitive advantage "overnight". Therefore, companies should adopt a more strategic view on outsourcing, which may include the transfer of an entire product, a product line, or even an entire plant to achieve strategic value.

The author's empirical research in the automotive industry illustrates that certain activities in a company's core functions are partly or completely outsourced to external providers. For instance, the production of body modules, chassis, seats and interiors is commonly outsourced. However, automobile manufacturers state that the production of drive trains and their components still remain mostly in-house, as it is seen as one of the core activities. In contrast, almost no automobile manufacturer invests in stamping and pressing machinery anymore. Another point of interest for the automobile industry and its supplier is the growth of a few mega suppliers who will be given major pieces, such as sub-assembling complete systems.

Furthermore, interviewees cite that logistics activities including warehouse and inventory management are partly outsourced. Managers interviewed expect a significant rise of outsourcing to external logistics providers within the next few years due to their efficiency and cost savings. While considering IT activities as close to core, companies in the automobile industry have outsourced most activities concerned with hardware and application maintenance, documentation development and application support to external providers. In addition, repetitive and generic activities in the accounting and finance as well as human resources functions, such as bill procedures, revenue accounting and payroll processing, have been partly or completely outsourced. However, even more specific activities, for instance, those associated with the credit and collection function, have been partly outsourced as companies expect to leverage the best-in-class credit and collection capabilities of its outsourcing providers.

In conclusion, a prerequisite for successful outsourcing is that the company is able to identify which are the core and which are the supporting activities, adapting a dynamic and forward-looking management style. Additionally, problems may occur if companies outsource today's supportive activities, which become important in the future. In-sourcing of these activities in the future is likely to be a costly, time-consuming and problematic process. Once outsourcing decisions are agreed upon, the careful selection of the appropriate partner is vital to outsourcing success. Outsourcing does not finish with signing the contract and should rather be seen as a relationship that includes support for the partner with expertise and managerial resources.

5.5 A Snapshot of Outsourcing Benefits and Risks

The following sections critically review and discuss benefits and risks closely linked to the managerial practice of outsourcing. This should by no means be considered either a complete list of benefits and risks of outsourcing or a judgement for one of the sides, but instead facilitate managers' well-grounded outsourcing decisions. Practitioners and academics alike argue in favour of both benefits and risks of outsourcing. Whereas, one side argues that outsourcing helps to reduce costs, provides capacity on demand and that it might give access to advanced technology, the opposite side describes outsourcing management practice as risky. Outsourcing risks include the loss of control, the loss of flexibility, the loss of qualified personnel and the loss of competitive advantage. It can be stated that most potential benefits are mirrored by potential risks.

5.5.1 Outsourcing Benefits

The perceived benefits of outsourcing have changed over the years. At the beginning of the outsourcing trend, a couple of decades ago, the cost factor was perceived as the ultimate reason to outsource. Nowadays, companies see additional benefits in the opportunity to concentrate on their core business, the increased service predictability and the ability to effectively manage costs. In order to reap the best possible advantages for the outsourcing company, it is suggested that automotive companies develop flexible outsourcing partnerships and programmes that are designed to meet their unique needs and cultures.

5.5.1.1 Cost Efficiency and Economies of Scale

As mentioned above, cost reduction has been the predominant motive for outsourcing (Lankford and Parsa 1999). Managers often find that outside firms produce more cost-effectively than a company could in-house. A typical example is the acquisition of capital-intensive investment in computer hard- and software, which could be avoided or at least minimised by the company if data processing services were contracted out.

In order to achieve cost reduction and maintain required standards, the suppliers need to have access to superior cost drivers, such as economies of scale, learning and low cost locations (Jennings 2002). As discussed in this chapter, the outsourcing company should include the transaction costs, the sum of negotiation, search and contract enforcement costs in order to identify the overall outsourcing expenses and to arrive at a realistic cost picture. In addition, the benefit from supplier's investment and innovation (Lonsdale and Cox 2000), and access to skills (Akomode et al. 1998; Embleton and Wright 1998) can improve cost performance as well as productivity and speed. However, managers have to acknowledge that a trade-off between cost and time can sometimes lead companies to outsource activities that could have been done cheaper internally, because the activities cannot be completed in-house within a certain timeframe.

A quantification of cost savings is highly dependent on the specific company and the activities outsourced. However, the following figures provide managers with a rough estimate of possible cost reductions. Blumberg (1998) proposes that outsourcing can lead to a 20–40% reduction in costs, while Lankford and Parsa (1999) state more moderate figures, proposing "at least 15%, and sometimes 20–25%" savings. However, many academics and practitioners warn not to expect savings at all as less than half of the outsourcing companies achieve a reduction in overall expenditure.

5.5.1.2 Focus on Core

"Outsourcing can help the management of a firm redirect its attention to its core competencies instead of having to possess and keep updated with a wide range of competencies" (Kotabe and Mol 2005, p. 120). One of the most controversial issues of outsourcing is the debate over core activities. Despite the problematic definition of a company's core, the question of whether to outsource core or not is discussed at length by practitioners and academics. An organisation's resource availability is mostly limited and resource expansions are made upon the value of the area, which further leads to a necessary distinction between core (valuable) and supporting (less valuable) activities. Most authors also state that outsourcing is a way to reduce tied up investments and scarce resources, especially when needed for core business activities (Johnson 1997). This view is further underlined by the author's empirical findings from the automotive industry. Interviewees state that outsourcing helped to relieve managers' responsibility for repetitive or generic business tasks, allowing them to concentrate on high-level management and other value-adding activities.

5.5.1.3 Product and Service Quality Improvements

The specialised knowledge of the contractor is one of the major advantages of outsourcing. The specialised knowledge can be seen as the main requisite to deliver high quality products and services. Quality improvements are cited by many authors as a motivation for outsourcing (McFarlan and Nolan 1995; Embleton and Wright 1998). The author's empirical findings showed that the better the skill set of an external provider, the higher the quality standard that may result. Interviewees emphasised that external suppliers may bring about higher quality due to their expertise and experience in an activity.

According to Akomode et al. (1998), the outsourcing company enjoys various benefits regarding quality improvements. Empirical findings based on the author's work prove that suppliers are eager to please the outsourcing companies due to several aspects. First, external suppliers always attempt to avoid cost and time consumption. Second, in order to meet the contractual standards, suppliers favour higher quality to avoid rework. Third, suppliers like to build long-term relationships with the outsourcer and will see the delivered good as an advertisement.

5.5.1.4 Time-to-Market Reduction and Reduction of Capacity Constraints

Speed improvements are recognised by most authors as benefits of outsourcing, leading to improved time-to-market (Lonsdale and Cox 2000), improved performance (McFarlan and Nolan 1995) and greater productivity and time savings (Embleton and Wright 1998). Outsourcing can overcome the company's own capacity constraints in meeting volume of sales. Companies that produce for markets with seasonal or cyclical patterns (demand fluctuation), resulting in under-used in-house capacity, may consider outsourcing as beneficial (Fill and Visser 2000). The potential for improved flexibility may apply not only to the volume of output, but also to the ability of the organisation to change the product range in response to market conditions. Many of the flexibility benefits are associated with long-term strategic rather than short-term operational flexibility (Brown 1997). In order to reap the benefits of flexibility, outsourcing should be considered as a long-term, strategic decision.

5.5.1.5 Resources, Innovation Availability and Risk Allocation

One of the most obvious reasons to outsource is non-availability of the required resources within one's own company (Johnson 1997). Companies in the automotive sector are undergoing constant reorganisations that lead to divested resources. Moreover, outsourcing relationships can deliver innovative product and/or service solutions that emerge from the combined knowledge and expertise existing within different organisations. Companies may profit from involving suppliers in product and process developments. Along the same lines, working closely with the outsourcing provider may help to share risks associated with outsourcing that are related to changing market conditions, new government regulations, shifting financial conditions and evolving technologies. Empirical data show that companies in the automotive industry may also benefit from external providers' innovative ideas and the extended market access through providers' distinctive networks.

5.5.2   Outsourcing Risks

Despite the extensive list of benefits for a company provided by outsourcing, several reports (e.g. PA Consulting Group 1996) state that only a small percentage of managers felt that outsourcing had met its objectives, with the majority reporting a mediocre outcome. Criteria to measure outsourcing success and failure are as diverse as opinions about outsourcing. Nevertheless, companies that wish to outsource need to consider both sides and evaluate their own position in terms of benefits and risks involved.

5.5.2.1 Missed Cost Targets

The failure to achieve anticipated cost improvements frequently occurs during outsourcing (Cross 1999). Outsourcing contracts usually aim for cost savings of about 25%, as discussed in Sect. 5.5.1.1 (Lankford and Parsa 1999). However, empirical results show that the average saving only accounts for less than 10% and a large portion of outsourcing clients may only break even or worse, find their costs increased (e.g. Embleton and Wright 1998; Ketler and Walstrom 1993). Moreover, interviewees reported that on some occasions, outsourcing costs exceeded the expected level, while quality standards were not fulfilled.

The author's empirical evidence illustrates that large companies in the automotive industry found prospective outsourcing providers unable to match their own internal economies of scale and many specialist suppliers did not have an effective scale greater than that of their customers. In addition, the costs invested to support the ongoing relationship and the supplier's premium often made outsourcing more expensive (Embleton and Wright 1998; Akomode et al. 1998). Rising transaction and coordination costs make it necessary to prioritise outsourcing providers. Furthermore, companies should be aware of the risk of being leveraged by their suppliers, particularly when outsourcing into a limited supply market (Lonsdale and Cox 2000). They state that having assured the contractual business, suppliers may pursue additional services, which lead to cost escalations. Additional services may include consultancy or business process re-engineering. In order to avoid such practices, the relationship should be protected by contractual safeguards and the formulation of exact specifications that guide both partners.

5.5.2.2 Low Quality Standards and Increased Time-to-Market

The risk of interruptions of supply or capacity limitations are problems that may occur when companies outsource activities (Embleton and Wright 1998). Several suppliers will attempt to reduce costs, not through cost-effective management, but rather by using cheaper resources and materials. Furthermore, empirical data collected by the author show that the anticipated speed, frequently based on an assumption of supplier scale of operations, may not be achieved. Reasons are manifold and may include the augmenting complexity of transactions between the outsourcing company and the supplier or simply the over-optimistic estimates of individuals. The use of external suppliers can also imply a reduction in the opportunities with which to achieve differentiation through the use of more widely available activities and components (Alexander and Young 1996). Additionally, the author's empirical findings illustrate that increased requirements for time-consuming and complex quality controls can diminish the cost reduction of the outsourcing arrangement.

5.5.2.3 Loss of Competences

If tasks are not too specialised, the remaining work force can be moved to a different project, which helps to maximise workers' productivity. In addition, transferable skills of workers can be leveraged. For instance, companies who completely outsource their IT staff are forced to use external expertise when minor IT problems occur. For companies pursuing outsourcing, inter-functional competences may get lost, as resources, tangible or human, will be contracted out to a supplier.

Loss of intellectual property rights and confidentiality leaks are another outsourcing risk, particularly if the supplier provides services to the company's competitors (Lonsdale and Cox 2000). Empirical findings show that managers in the automotive industry are aware of the risk that valuable data might fall into competitors' hands. Managers reported that tight data guidelines had to be incorporated into the outsourcing contract to avoid data misuse.

Additionally, outsourcing may lead to a loss of internal coherence, which means that the gap left by the outsourced activity negatively influences other internal departments or groups.

5.5.2.4 Danger of Dependency and Declining Brand Share

Another main risk of outsourcing is concerned with the danger of becoming dependent on suppliers, thus leading to reduced control over activities. Therefore, managers should be aware of how firms can become dependent on suppliers. According to Lonsdale (1999), outsourcing into a limited supply market, poor internal alignment and contractual incompetence in the face of different degrees of asset specificity are the main ways in which dependency emerges. Interviewees in the automotive industry underline the potential risk of outsourcing companies losing control and consequently becoming dependent on suppliers.

Closely linked to the loss of an organisation's core competences and the danger of dependency is the risk of a declining brand share. Quinn and Hilmer (1994) illustrate that companies outsourced the manufacturing of parts, which seemed at that time to be minor components, and taught suppliers how to achieve the required quality. Later these companies found that their suppliers were unable or not willing to supply them as required. By that stage the company had already lost the ability to produce the parts in-house and could not stop its suppliers from supporting competitors or entering downstream markets on their own. Therefore, managers have to take into consideration the dynamic nature of core activities when making outsourcing decisions.

5.5.2.5 Problems of Reversibility and Flexibility

Outsourcing a specific responsibility and adjusting or even erasing the internal resources accordingly is often linked with the loss of strategic flexibility and reversibility (Embleton and Wright 1998). Managers should consider the risks associated with the difficulties of rebuilding or acquiring internal resources and the knowledge needed to bring the outsourced activity back into the company. With any outsourcing agreement the company loses some control to its outsourcing provider. Lonsdale and Cox (2000) argue that the time of contract renegotiation, when the balance of power has changed, is most critical for outsourcing companies. Company's options may be limited and the decision to cancel the outsourcing contract may no longer be practical as the outsourcing provider has acquired all the necessary skills. Companies may be "forced" to renew the contract as no other supplier will be able to fulfil the required standards.

5.6   Managerial Implications for (Beneficial) Outsourcing Decisions

The amalgam of theoretical and empirical findings presented in this chapter provides several managerial implications. The following recommendations should support middle and senior management in the automotive sector in deriving benefits and eliminating (or at least mitigating) various risks from outsourcing activities.

5.6.1   More Attention Should Be Drawn to the Selection of Outsourced Activities

Outsourcing core activities is highly risky as it puts the company on the fast track to the "hollow organisation". Companies should carefully assess each activity to ensure that an appropriate decision-making process led to the outsourcing decision. Consequently, managers have to fully understand what is involved in each activity and how they are linked to each other before they can be outsourced. Many of the outsourcing risks experienced by companies in the automotive sector could have been avoided by proper activity assessment. If management is inexperienced in outsourcing activities, they should start with more repetitive activities that are not core to the company.

5.6.2   Assessing Future Core Activities Properly

Core activities should be evaluated in both a static perspective of what is core at present and also what could become core in the future. Companies positioned in the fast-moving, innovative, technology-based automotive market, should embrace the concept of dynamic capabilities. This perspective will help a company to move and stay ahead of its competitors in the future.

5.6.3   Be Careful with Customer-Centric Areas

Attention should also be drawn to outsourcing consumer-centric areas such as call centres. Outsourcing the interaction between the company and its customers is risky. Quality guidelines must be in place and the outsourcing provider must convey the "culture and message" of the outsourcing company.

5.6.4   Setting out Clear (Contractual) Guidelines

Prior to outsourcing activities, companies should have a clear understanding of how to manage and measure the provider's performance. The outsourcing company should set clear guidelines and requirements for the outsourcing provider in areas such as product quality and service level standards. This will help to avoid future misunderstandings and conflicts about requirements and standards that have to be met by the provider. Obligations for each side should be made clear and contractually enforceable safeguards should be installed to protect the ongoing relationship.

5.6.5   Outsourcing Providers Should Be Selected Carefully

Once management has decided on what to outsource, a careful selection of the outsourcing provider is vital for business success. The author's empirical data illustrate that financial stability, expertise and the "cultural fit" are the major assessment criteria to be used for outsourcing providers. Outsourcing companies should have a thorough assessment scheme in place that follows a rigorous, structured process. In return, several outsourcing risks can be eliminated or at least minimised. Managers should be aware that once activities are outsourced, bringing activities back into the company or switching outsourcing providers is a time- and cost-consuming process.

5.6.6   Outsourcing Relationships Have to Be Managed

In order to minimise the risks inherent in outsourcing, the relationship between the two parties has to be managed carefully. The provider should be supervised and assessed to ensure that contractual requirements are met. Management should not neglect the communication and cultural barriers between the outsourcing company and the provider as they often cause problems and lead to underperforming providers. The barriers are often more pronounced when the outsourcing company and the provider are geographically and culturally distant.

5.7 Conclusion

In the face of rising profit pressures, automotive companies are increasingly pursuing outsourcing practice as part of their BTO strategies. This also allows them to gain access to suppliers' innovative research and technology expertise, to focus resource allocation towards core activities and to cut expenses.

Outsourcing activities are concerned with both core and supporting activities. In addition to outsourcing repetitive activities in several areas, more analytic or strategic activities in areas such as finance and accounting or human resources are contracted out to external providers. The automotive industry even partly outsources core activities where resources and capital are not sufficient to carry out these activities in-house.

This chapter considered the benefits and risks of outsourcing for BTO. It examined core and supportive activities of business functions such as manufacturing, human resources, information technology, logistics, accounting and finance. Apart from the cost and quality aspects and the possibility of focussing on core activities, companies may enjoy several other outsourcing benefits. According to the author's empirical data, outsourcing offers access to specialist knowledge and expertise, to innovative technologies and processes and to economies of scale to perform activities cheaper and of better quality.

But caution must temper enthusiasm for outsourcing. In the highly risky and customer-conscious automotive market management should be aware of several outsourcing risks. Outsourcing raises fears among managers across sectors about losing power and control over activities and proprietary knowledge. Furthermore, poor performance delivery of external providers is feared by automobile companies as they cannot tolerate mistakes, especially where regulations have to be met and vehicles delivered more quickly to the final customer. Interviewees stated that a thorough provider assessment as well as integrated quality tests for all outsourced activities helped companies to achieve beneficial results.

Consequently, empirical findings indicated that outsourcing presents a significant challenge in managing relationships with external providers and underlined the need for managers to be fully aware of the risks and benefits involved in outsourcing activities. An understanding of the concept of core and supporting activities, as presented in this chapter, is vital to arrive at well-grounded, beneficial outsourcing decisions that will enable companies to transition to BTO production.

References

Akomode OJ, Lees B, Irgens C (1998) Constructing customised models and providing information to support IT outsourcing decisions. Logist Inform Manag 11(2):114–127

Alexander M., Young D (1996) Strategic outsourcing. Long Range Planning 29(1):116–119

Arnold U (2000) New dimensions of outsourcing: a combination of transaction cost economics and the core competencies concept. Eur J Purch Supply Manag 6:23–29

Baden-Fuller C, Targett D, Hunt B (2000) Outsourcing to outmaneuver: outsourcing re-defines competitive strategy and structure. Eur Manag J 18(3):285–295

Barney JB (1991) Firm resource and sustained competitive advantage. J Manag 17(1):99–120

Bleeke J, Ernst D (1991) The way to win in cross-border alliances. Harv Bus Rev 69(6):127–135

Blumberg DF (1998) Strategic assessment of outsourcing and downsizing in the service market. Manag Serv Qual 8(1):5–18

Brown M (1997) Outsourcery. Manag Today January:56–60

Cheon MJ, Grover V, Teng JTC (1995) Theoretical perspectives on the outsourcing of information systems. J Inform Tech 10(4):209–220

Cox A, Lonsdale C (1997) Strategic outsourcing methodologies in UK companies. CBSP working paper, University of Birmingham

Cross M (1999) Getting a grip on outsourcing. Computing 13 May:48–50

Easton G (1997) Industrial networks: a review. In: Ford D (ed) Understanding business markets, 2nd edn. Dryden, London, pp 102–128

Eisenhardt KM, Martin JA (2000) Dynamic capabilities: what are they? Strat Manag J 21:1105–1121

Embleton PR, Wright PC (1998) A practical guide to successful outsourcing. Empowerment in Organizations 6(3):94–106

Fill C, Visser E (2000) The outsourcing dilemma: a composite approach to the make or buy decision. Manag Decis 38(1):43–50

Ford D, Hakansson H, Johanson J (1986) How do companies interact? Ind Market Purch 1(1)26–41

Gadde LE, Håkansson H (2001) Supply network strategies. Wiley, Chichester

Granstrand O, Patel P, Pavitt K (1997) Multi-technology corporations: why the have 'distributed' rather than distinctive core 'competencies'. Calif Manag Rev 39(4):8–25

Grant RM (1991) The resource-based theory of competitive advantage: implications for strategy formulation. Calif Manag Rev 33(3):114–135

Jennings D (2002) Strategic sourcing: benefits, problems and a contextual model. Manag Decis 40(1):26–34

Johnson M (1997) Outsourcing in brief. Butterworth-Heinemann, Oxford

Ketler K, Walstrom J (1993) The outsourcing decision. Int J Manag Inform 13:449–459

Kotabe M., Mol MJ (2005) Competitive (dis)advantage of outsourcing strategy. In: Ancarani A, Raffa M (eds) Sourcing decision management. Edizioni Scientifiche Italiane S.p.A., Rome, pp 115–126

Kotter JP (1979) Power in management. Amacom, New York

Lacity M, Hirschheim R (1993) Implementing information systems outsourcing: key issues and experiences of an early adopter. J Gen Manag 19(1):17–31

Lankford WM, Parsa F (1999) Outsourcing: a primer. Manag Decis 37(4):310–316

Lonsdale C (1999) Effectively managing vertical supply relationships: a risk management model for outsourcing. Supply Chain Manag 4(4):176–183

Lonsdale C, Cox A (1998) Outsourcing: a business guide to risk management tools and techniques. Earlsgate, Boston

Lonsdale C, Cox A (2000) The historical development of outsourcing: the latest fad? Ind Manag Data Syst 100(9):444–450

McFarlan FW, Nolan RL (1995) How to manage an IT outsourcing alliance. Sloan Manag Rev 36(2):9-23

PA Consulting Group (1996) Strategic sourcing: international survey. PA Consulting, London

Pfeffer J, Salancik G (1978) The external control of organisations. Harper & Row, New York
Prahalad CK, Hamel G (1990) The core competence of the corporation. Harv Bus Rev 68(3):79–91
Quinn JB, Hilmer FG (1994) Strategic outsourcing. Sloan Manag Rev 35(4):43–55
Rumelt RP (1987) Theory, strategy and entrepreneurship. In: Teece JD (ed) The competitive challenge: strategies for industrial innovation and renewal. Ballinger, Cambridge, pp 137–158
Sharpe M (1997) Outsourcing, organisational competitiveness, and work. J Labor Res 18(4):535–549
Teece DJ (1987) The competitive challenge: strategies for industrial innovation and renewal. Harper and Row, New York
Teece DJ, Pisano G, Shuen A (1997) Dynamic capabilities and strategic management. Strat Manag J 18(7):509–533
Thompson JD (1967) Organisations in action. McGraw-Hill, New York
Thompson AA Jr, Strickland AJ III, Gamble JE (2005) Crafting and executing strategy: the quest for competitive advantage; concepts and cases, 14th edn. McGraw-Hill, New York
Wernerfelt B (1984) A resource-based view of the firm. Strat Manag J 5(2):171–180
Williamson OE (1985) The economic institutions of capitalism. Free Press, New York

Part II
Modularity

Chapter 6
An Overview of Modular Car Architecture: the OEMS Perspective on Why and How

Philipp Gneiting and Thomas Sommer-Dittrich

Daimler AG; Group Research & Advanced Engineering;
Materials, Manufacturing and Concepts

Abstract. The diversification of an OEM's model range serves to address as many customers as possible and thus cover as much of the market as possible. Here, the recognisable external difference between the cars therefore plays an important role in market positioning. However, the large number of variants also generates high cost; all part of mastering this complexity. One of the solutions to overcoming this problem is the use of common parts and modules across different models. The goal is to keep the internal complexity to a minimum whilst maintaining greater external variance, i.e. that which is visible to the customer. This chapter gives a brief overview from the OEM's perspective and serves as an introduction to the other chapters in this section.

6.1 Introduction

The objective of this chapter is to first shed more light on the subject of modularisation from the OEMs' perspective and present it as a product structuring strategy. We will also address the reasons for the introduction of a modularisation strategy for automobile manufacturers.

With modularisation, the product is divided into partial systems so that complexity is manageable and can be differentiated from the environment as a system. Here the sub-systems are arranged hierarchically. Ideally, a module consists of sub-modules of the next hierarchy level (Hruschka 1986). The key driver is for these partial modules to fulfil one or more partial functions within the overall system and that they are interchangeable. Modules can therefore also be understood to be components that are connected with each other through interfaces. New product variants can be created by combining different modules. If the products of a product family consist of such a set of modules, then this is a modular product family (Junge 2005). Another aspect is that a partial module can be replaced in a product in the event of modifications or new developments without having to adjust the entire

101

product. The modules thus not only make complexity more manageable, but also reduce the further work required when making changes to the product over time.

The design of a modular product results in the question: how will the individual modules be defined? Piller and Waringer's definition is that a module is a unit that can be defined according to its characteristic function of installation in a larger unit and comprises multiple units (Piller and Waringer 1999).

6.2 Internal and External Reasons for Modular Car Architecture

Next, we will examine in more detail the drivers for the introduction of the modularisation strategy to automobile manufacturers. There is a differentiation between internal and external drivers. External drivers refer to the customers' needs and wants. These drivers include:

- Individualisation and shorter product life cycles
- Lower price
- Shorter delivery times

There are also intra-corporate drivers that arise due to the manufacturer's interests. These include:

- Reduction of complexity in production and logistics
- Balanced utilisation of capacities
- Reduction of development costs

Thus, an OEM wants to keep the number of car model variants as low as possible to realise cost-effective product development and production. In contrast, the customer wants to be offered a broad range of vehicle options, or in other words, he wants to drive a highly individualised car. Automobile manufacturers must take the customer-orientated path when searching for market advantage by providing the variety of options demanded.

In addition to options, the diversification of the model selection serves to cover as many of the market niches as possible. The key is maximising the externally recognisable differences between the cars to facilitate market positioning. One major success factor for OEMs in the future will be to offer attractively designed, functional products and make them in a high number of variants. In spite of strategies such as packaging special equipment, as frequently done by the Japanese manufacturers, in order to keep production variety low, European premium manufacturers continue to be forced to offer a higher degree of individualisation to maintain their competitive advantage. The high number of variants means that significant effort is employed in managing the increased complexity, though this is made easier through the use of modules.

Another benefit of the use of modules is the ability to offer the customer the option to further customise his vehicle in the after-market. The customer may exchange individual modules to gain additional functions to suit a need. It would be conceivable that even the vehicle type could be changed by replacing a chassis module, such as changing a sedan into a station wagon. Even updates for certain functions could be of interest to a customer who does not want to acquire an entire new car. This type of functionality places a new demand on manufacturers to provide for compatibility of modules over time.

As was stated, wide variety comes with higher fixed costs in production, logistics and development that must be distributed to a low number of products. However, it is difficult for the manufacturers to pass these costs on to the customer. Customers can quickly obtain comprehensive information regarding offers by the competition and can easily compare the price difference between a personalised product and the less expensive products made by competitors with a predetermined special equipment package. It is therefore clear that the customer will only accept a limited amount of additional costs for additional equipment.

Even when the manufacturer meets the customers' individual requests, they will not accept delivery times of several months. The goal must be shorter delivery times. The use of standardised and customised modules reduces the processing time because the standard modules can be kept in stock and are readily available when the customer's order is received. The processing time for the order is thus reduced to the time that is required to install the individual modules.

Another aspect is the option to increase the utilisation of capacity at the module plants. Car manufacturer's success remains significantly dependent upon the utilisation of plant capacities. Because of globalisation and worldwide competition, investments in production are made to improve competitive position through adding variety whilst maintaining competitive pricing. However, based on current estimations, utilisation of capacity is 75% worldwide. This overall situation results in the need to develop cars that can be produced cost-efficiently, even those produced in small quantities. Vehicle variants that can be manufactured flexibly at many locations must be offered. The module strategy supports this as it facilitates balanced capacity utilisation of module production and the use of different common combinations in the production of finished goods. The later the variant is established during the production process the more scale potential can be utilised during preceding phases. The use of a module in different finished products weakens the effects of volatile sales variances and reduces the risk of insufficient capacity utilisation at the production line for the module. Capacity is also better utilised through flexibility of the production processes. Shifting the creation of variants to the final production process also offers the advantage that the order penetration point can be delayed. The order penetration point is the point at which a common product is customised according to the requirements of a specific customer order and assigned to that order.

Due to competition, manufacturers in the automobile industry are forced to continuously offer new car models to their customers. This leads to shorter product life cycles and thereby increases development efforts. Many variant-specific assemblies must be developed. Market demands for new car model variants, which must be developed and produced quickly and cost-effectively, cannot be met using the conventional approach to vehicle production. The cost of development and changes to production logistics often exceed the acceptable level for economic returns. This issue can also be controlled through modularisation. The division of a car into modules provides the option of revising or replacing existing modules to create a new or updated vehicle, instead of the more costly option of new vehicle development. This means that, in principle at least, modules can be developed independently so that only individual modules are affected by the development of variants. The module concept enables the engineers to successively introduce new technologies to improve weight, stability and quality of individual partial modules without having to redefine the overall layout of the vehicle. New partial technologies will therefore flow into the series much more quickly than is common today because they can be integrated in an individual series regardless of the overall life cycle. Modularisation is also advantageous with the development of completely new car models. One can utilise this kit, further expand proven modules and develop any necessary new modules. Subsequently, development times are shorter and shorter product life cycles can be achieved.

The structure of vehicles must change to support BTO, along with the processes to support the complex delivery chain ranging from tier n suppliers to dealers accordingly. Modularisation is a potential response to this demand as well, and may be the only way to convert the entire delivery chain from a push system to a pull system.

6.3 Requirements for the Production and Supply Chain

Now that the reasons for modularisation have been described, we will address the question of how to implement modularisation. Modularisation means, in most cases, using a common platform and adding customised or non-customised components (modules), which may differ with regard to performance, characteristics and quantity. In some cases even the common platform approach may be omitted. Automobile manufacturers have made enormous efforts trying to reduce the complexity due to product planning and preparation by utilising suitable modularisation concepts. A car might be divided into seven main modules (Fig. 6.1):

- Front end
- Engine compartment
- Passenger compartment front
- Passenger compartment back
- Rear end module
- Outer panels
- Exhaust system

All of these elements should be defined according to the customer's order and then connected to each other during final assembly.

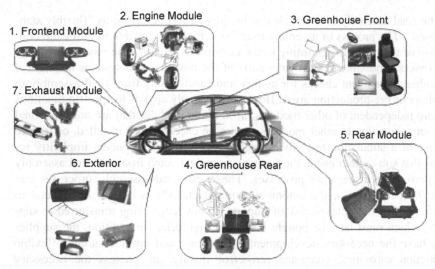

Fig. 6.1 Possible modules of a car

Individual modules must have standardised interfaces, be accessible and they must be connected "kinematically". This last point deserves greater clarity. Some connection techniques cannot be applied because pre-installed modules may contain integrated components such as cables, plugs, plastic covers or similar, that cannot tolerate the high temperatures that are generated in some bonding processes and radiate through the modules. Therefore, there will be a need to develop and apply special types of welding and glued, screwed or riveted connections.

In addition, the current production sequence used in continuous automotive production systems can no longer be followed. Finished chassis are currently brought to the paint shop as a whole piece, but with the indicated approach of modularisation of a vehicle into seven components, the various parts of the vehicle must be painted independently. They will not be joined until the vehicle modules are assembled. This raises the challenge of colour matching.

Overall, the modular structure of the car must match the manufacturing concept of the automobile and the approach must not diminish model and brand identity. The use of modules from economy cars in higher priced luxury models is self-prohibiting if the customer notices them in any way. Even if the buyer of a luxury model is not initially aware of a reused module, if they find out later through a trade publication for example, this may damage a brand's value in the eyes of a customer. Therefore, such installation of modules in different vehicle classes must be managed carefully, seen as brand-specific and carried out with the greatest consideration paid to the automobile manufacturer's customer structure.

Basic principles of logistics bring their own limits for modularisation. Modules that are very difficult to handle and transport will result in an over-proportional increase in logistics costs.

The final assembly of a vehicle can be viewed and described as "flexibly standardised". The process of assembly must be sufficiently flexible to allow the connection of the modules according to the customer's request. The modular strategy also makes it possible to shift large parts of the modules' sub-systems testing and individual component checks for quality and functionality from final assembly to suppliers or pre-production areas. However, this only applies to specific functions that are independent of other modules. Production modules that are not functional without connection to other modules cannot be checked until installed, or require the use of a simulation test environment. The modularity concept implicitly requires that sub-system installation processes are extracted from the final assembly and performed in separate processes. These extracted assembly processes can either be performed at the automobile manufacturer's facility or outsourced to suppliers. This presents the risk of important knowledge being transferred to suppliers, which must later be bought back at a high price. In addition, the supplier must have the necessary development competence and provide suitable flexible production capacities, guarantee zero-error quality and possess the necessary competence to coordinate their sub-suppliers.

6.4 A Possible Approach to the Design of Modular Automobiles

First of all, an OEM has to decide which functional or structural components of a vehicle should be modularised. Here the expected module dimensions, weight, testing abilities or logistic factors play an important role. Afterwards the components of the module itself have to be defined. Depending on these decisions it becomes clear what and how many interfaces, both internally and externally, have to be standardised and which modules are customer-independent. Customer-independent modules may be built-to-stock or built-to-order, depending upon the results of cost–benefit analysis. In contrast, individual customised modules will not be produced until the corresponding order for such modules has been received.

Once the modules for a vehicle have been determined, the modules' life cycle must be managed. If it becomes necessary to modify a module due to new statutory regulations, the impact of such modification must be determined within a short period of time. This impact analysis would include any other modules beyond existing interfaces and interactions that would be affected. A comprehensive change in the management process must be initiated that integrates all parties involved.

Another major question is how to control modularised product manufacturing. One of the main reasons for an OEM to switch over to a modularised product is to reduce complexity. One approach is to hand over some of the planning, coordinating and sourcing processes to reliable first-tier suppliers, which can provide the needed know-how, flexibility and integrity. Following this strategy would mean that the OEM would no longer deal with all the single components of a car and could therefore expect efficiency savings.

6.5   Conclusion

This chapter provides a brief overview of the numerous challenges and benefits of modularised architectures from an OEM's perspective. The following chapters will further address the detail of the practical application of modularisation in automobile construction. More detailed explanations on the topic of planning and control will be provided later in this book.

References

Hruschka P (1986) Principles of modularization (Prinzipien der Modularisierung). HMD 23(130):67–75

Junge M (2005) Controlling modular product families in the automobile industry. Development and application of the modularization balanced scorecard (Controlling modularer Produktfamilien in der Automobilindustrie – Entwicklung und Anwendung der Modularisierungs-Balanced-Scorecard). Wiesbaden

Piller F, Waringer D (1999) Modularity in the automobile industry (Modularität in der Automobilindustrie). Shaker, Aachen, p. 39

Chapter 7
The Modular Body

Andreas Untiedt

ThyssenKrupp Steel AG

Abstract. Modularity has been a commonly used design approach for some time in automotive design and production. Modularity in design aims to separate a system into smaller parts that can be autonomously produced and then utilised in a range of differing products to allow various functionalities to be achieved. The requirement to produce vehicles of increasing variety to meet niche market requirements, together with a need to maximise capacity utilisation leads a drive for greater flexibility in production systems as well as for designs that allow late vehicle configuration. The ability to create a full range of body variants that comprises ostensibly common parts and that requires low tool investment provides an efficient solution. The modular body contributes through the presentation of the development of the ModCar. The body consists of four modules and different combinations of modules result in a variety of vehicles. There are many parts and beams even within these modules which have the same shape and hence can be manufactured on the same machine. This overall approach will not only reduce the manufacturing costs and complexity, but also the development time of new vehicles.

7.1 The Idea of Modularity

Modularity has been a commonly used design approach for some time in automotive design and production (Automobil Produktion 2004; Dudenhöffer 1999; Koeth 2004; McAlinden et al. 1999; Mercer Management Consulting 2001; Mutegi et al. 2000; Neumann 1998). Modularity in design aims to separate a system into smaller parts that can be autonomously produced and then utilised in a range of differing products to allow various functionalities to be achieved. An automotive example of modular design is the ability for some models, such as the Mini, to allow "snap in" upgrades. Upgrades may be related to higher performance, a more powerful engine, for example, or perhaps to allow further vehicle variants by using a new body module. The essential characteristic is that they do not require any change to other units of the car, such as the chassis or seating compartment. Whilst the concept of modularity has been in common usage in the automotive industry, here we will describe the requirements for a modular body, something that to date has not been achieved on a commercial scale. Besides the potential for reduced cost by taking advantage of economies of scope and flexibility in the design process, modularity provides a number of other benefits such as augmentation, where a new solution may be obtained by utilising a new module. Our approach to modular design, then, is an attempt to combine the advantages of standardisation (high volumes and lower manufacturing costs) with those of customisation.

High external variance is driven by customer demand and describes the number of features that can be offered by the car manufacturer, like air conditioning or different levels of torque and power output.

Internal variance describes the number of different parts that are needed to reach the external variance for the customer. The general intention of a car manufacturer is to have a high external variance whilst shrinking down the internal variance to reduce complexity and storage costs for BTS systems and BTS parts. This means simplification, such as offering different power outputs using few engine components. Using one type of engine block, one type of cylinder head with electrically driven valves (i.e. EMVT by FEV) and a belt-driven variable gear box radically reduces the complexity and the internal variance without affecting the external variance. The different power output is created by different control units delivering different timing signals to the valves.

In addition to aiming for high numbers of variants at low complexity there are more factors driving the automotive industry:

- Environmental regulations – increasing requirements to improve recyclability.
- Customer choice – customers requiring not only quality, performance and reliability, which are now taken as a given, but increasingly putting an emphasis on style, novelty and brand values.
- Globalisation of trade, supply and competition affecting both vehicles and components.
- Strong pressures to reduce costs from vehicle manufacturers affecting the whole supply chain.
- Shorter product life cycles as customers expect better performance, style and features.
- Fragmentation of markets with growth in specialist niche market vehicles that meet particular life style and other requirements.
- Much shorter order lead times with competition to be able to deliver a custom-built vehicle within a very short time.

The customer's increasing emphasis on style and outer appearance has a great impact on the vehicle body (Fig. 7.1).

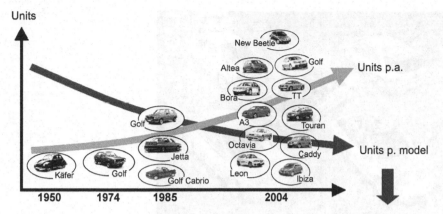

Fig. 7.1 Development of vehicle variant numbers

The profitability of the automotive sector is highly dependent on the level of capacity utilisation. However, the increasingly global nature of the industry has led to greater investment in manufacturing capabilities in many areas of the world and increasing competition in terms of price and features. The automotive industry is typified by over-capacity in many markets, with estimates of 20% overcapacity world-wide. There are now serious concerns that any down-turn in demand will lead to plant closures.

The requirement to produce vehicles of increasing variety to meet niche market requirements, together with a need to maximise capacity utilisation is leading to a drive for greater flexibility in production systems as well as for product designs that allow greater potential for late configuration of vehicles. The ability to produce highly individualised vehicles by offering a great number of variants has an impact on the body structure as well as on other components such as the cockpit and engine. In order to remain competitive, production and development costs need to be controlled. As the number of vehicle variants rises, so does the requirement for differing parts and components. As a consequence, fixed costs, such as those for tooling, increase significantly in high-variant production. This is because costs need to be allocated to lower quantities of parts, thus increasing the piece price. The following examples show some concepts of how companies are adapting to meet these demands.

7.2 Modularity Concepts

7.2.1 DaimlerChrysler Vario Research Car

In 1995 Mercedes-Benz revealed a completely new research vehicle, consisting of a basic section with two doors and an exchangeable module giving the standard vehicle its different appearance as a hatchback, notchback, sedan or pick-up

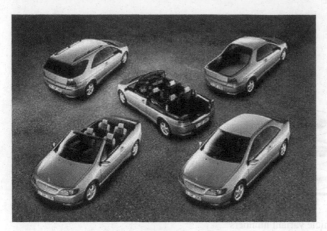

Fig. 7.2 Variants of the DC Vario Research Car

(Daimler 2004). It only took a few steps to exchange these modules and provide a specific function and shape and it could be done by the customer himself. For example, the sedan could be changed into a pick-up or the hatchback into a notch-back (Fig. 7.2).

In short, the Vario Research Car succeeded in combining four different vehicle concepts into a single car, giving customers the opportunity to change their car according to their current requirements. The aim of the engineers in Stuttgart was to create a vehicle that consists of a base structure that houses all basic components like front windscreen, doors, cockpit and seats. This independent unit remained unchanged and could be fitted with any of the four optional modules, making a "four-in-one-vehicle" out of it. The module joints were positioned beneath the waistline along the sides, along the side windows and the top edge of the wind-screen frame and at the rear. The metamorphosis could take place within a couple of minutes, requiring little technical know-how from the operator. Electromagnets and special locking systems helped to keep the modules securely fastened to the basic body. Electrical power and control was provided by a standardised connector at the rear. Electrical components were detected automatically. Due to the use of lightweight material like carbon fibre-reinforced plastic the weight of the ex-changeable modules varied between 30 and 50 kg. According to DaimlerChrysler, this lightweight design provided high levels of strength and crash performance.

7.2.2 DaimlerChrysler MoCar

In 2001 DaimlerChrysler released the MoCar study regarding a new modularity concept: future cars will not be assembled from many separate parts, but will con-sist of only four completely pre-assembled, fully equipped and painted modules (Elbl-Weiser 2003; Truckenbrodt 2001). Accordingly, the study is also known as

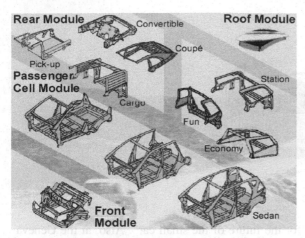

Fig. 7.3 Modules of the DC MoCar

"quartering the car". Like every OEM, DaimlerChrysler faces the challenge of finding a cost-effective way of offering an increasing number of model variants and more individualised cars. And many OEMs focus on the platform strategy to adapt to the changing demands, as described earlier. The Modularised Car (MoCar) concept offers an alternative solution for the inexpensive development and production of individual vehicles within one brand and car line, based on a highly modular structure of the vehicles and the decentralised, modularised production.

Primarily, the current modularity concepts are based on the supply of modules and parts to a central assembly plant, where the modules and parts are fitted into a painted body. One disadvantage of this concept is the transfer of know-how towards the suppliers. Moreover, the concept offers a high degree of flexibility (engines, equipment, trim, roof), but only as long as the modules are compatible with the body. The concept reaches its limits when more basic bodies are involved. This is where MoCar claims its merits. According to the MoCar study the four vehicle modules, front end, safety cell, rear end and roof, are produced separately, including all integrated components ranging from electrical harness to trim. Each module is produced according to the customer's requirements. Only during the final production step there are four modules that are joined together through defined interfaces, resulting in a finished vehicle. In this way DaimlerChrysler expects to produce more vehicles with more model variants at less expense. The standardised interfaces enable the combination of a multitude of module variants, generating an enormous number of different vehicles (Fig. 7.3).

7.2.3 Fiat Ecobasic

In 2000, the Fiat Ecobasic won the Environment Award at the automotive world awards ceremony (Maier 2003; Perini 2007; Rossbach 2001). The concept car was

Fig. 7.4 Fiat Ecobasic

judged as being "a blueprint for the future of the small car". Also, at the Geneva Motor Show of the same year this concept was presented to the public, focussing on a new production approach to recyclable and low-cost automobiles. This prototype was the most environment-friendly Fiat ever.

A test conducted by the German TÜV showed that the Ecobasic produces 76 g/km of CO_2, 0.025 g/km of particulates and needs less than 3 l/100 km of fuel. The car has a top speed of 160 km/h, which is electronically limited and needs 13 s for acceleration from 0 to 100 km/h. The maximum torque of 160 Nm is provided by the four cylinder diesel engine at 1,800 rpm. Maximum power output of 45 kW is received at 3,500 rpm. The price of this vehicle will lie roughly at €5,000, provided the overall production volume is at least 200,000 units per year. The intention of Fiat was to have a show car presenting a catalogue of solutions for components and systems in the future rather than building it precisely as it is shown (Fig. 7.4).

As well as the vehicle, the production method is also innovative. The new vehicle structure and assembly concept allocates new roles to suppliers, plants, marketing techniques and dealers. The spaceframe-like body consists of parts manufactured by conventional press technology. Here the main parts and components are fixed. According to Fiat, the body in white weighs only 150 kg, half of the weight of a conventional body shell. Compared with an aluminium structure the production costs are estimated to be two-thirds less.

The non-load-bearing parts whose surface needs to look good have been made out of thermoplastics. To be able to create the complex shape of the rear window, polycarbonates are used, which are treated with anti-scratch solution. The front end is made out of polypropylene to improve low-speed impact absorption. The benefit that comes out of this is that a paint shop is no longer required for this kind of vehicle production because the plastics themselves are coloured. Only one vehicle variant can be built up at the assembly line. Other variants like four or five doors have to be configured by the dealer by changing the version coming from the vehicle plant. Also, logistics play a great role. New ways of storing, distributing and delivering have been chosen and it is estimated that the internet will be of great importance here.

7.2.4 General Motors Autonomy

The technical lay-out of the General Motors Autonomy study clearly differs from what today is considered as standard (Automobile 2005; GM 2007; Rossbach 2001). First, the vehicle focuses on the use of fuel cell propulsion. Second, with electrical motors in the wheel hubs, the propulsion system is inspired by a concept originating in the early days of automotive production and which is now commonly used for vehicles with hydraulic motors. Third, the "x-by-wire" technology replaces the mechanical operation of steering, brakes and all other vehicle systems. The result is a completely new and unusual vehicle architecture, based on an 18-cm high chassis that contains all the technical components required for propulsion, including the fuel cell stacks, hydrogen storage tank and suspension system. Based on its appearance, it is generally known as the "skateboard chassis".

Due to the x-by-wire technology and the absence of a conventional powertrain, the position of the driver can be anywhere on the chassis. There are virtually no boundaries for the body, except for the attachment points to the chassis, and it can thus be shaped individually. Within this concept, a customer might even own or lease a set of body variants that he can mount onto the chassis depending on his requirements. Through the separation of the chassis and the vehicle body the universal chassis hugely simplifies production and maintenance and allows the development of many different vehicles within a reduced time frame. At the 2002 Detroit Motorshow General Motors unveiled the two-seater Autonomy, but the same chassis could easily accommodate a ten-seater van as well. The chassis can be designed to be scalable, enabling different wheelbases and increasing the range of possible applications. Safety, stiffness and driving properties should be largely generic issues.

The core of the electrical system is the universal "docking port" in the centre of the chassis. Depending on the body that is currently plugged onto the chassis, the driving characteristics can be adjusted by software making the chassis adaptive to the body. From a driving characteristics point of view, the skateboard design offers a low centre of gravity combined with acceptable ground clearance and good weight distribution. This is an excellent combination for both handling characteristics and stability, even with high body variants.

Crash performance benefits from the rigid structure of the skateboard and the absence of "hard" mechanical components that may cause injury to passengers, such as pedals, the steering column or the engine. Thanks to the unrestricted position of the driver and the plain floor, the interior can be designed to offer maximum protection for all occupants by building a safety concept around them.

All in all, the Autonomy is a ground-breaking concept that has the potential to bring about the renewal of the automotive industry. In 2005, the Autonomy concept made a step further towards reality: at the international auto show the Sequel was unveiled, covering technologies like fuel cell propulsion, x-by-wire and wheel hub motors. Here GM was able to double the range and to halve the time of acceleration from 0 to 60 mph compared with current fuel cell vehicles in less than 3 years, according to Larry Burns, GM vice president of research, development and planning.

7.2.5 Ford Skala

To stay in line with the market development, Ford, along with other OEMs, rely not only on a platform strategy, but are also developing further strategies that better allow the creation of many vehicle variants within a shorter life-span (Figs. 7.5, 7.6) (Baumann et al. 2002).

Platform strategies use as many common parts over different vehicle variants as possible, but still have to handle high capital investment in presses, tooling and automation processes, which can only be assigned to one variant and hence drive

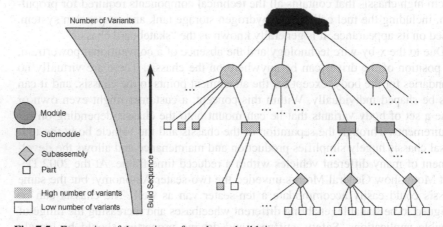

Fig. 7.5 Expansion of the variance funnel over build time

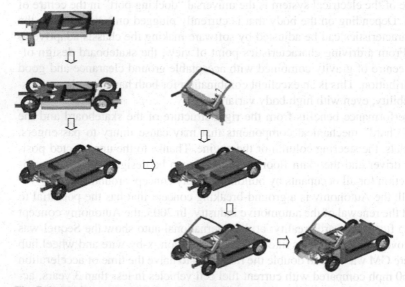

Fig. 7.6 Modular assembly of the Ford Skala

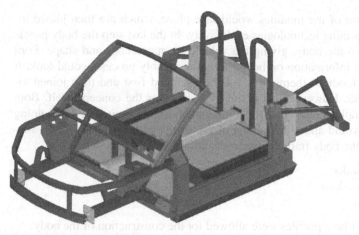

Fig. 7.7   Ford Skala structure

parts costs up. Long machine changeover times for panels require the manufacturing of large batches. The result is higher stock volume and hence cost. In addition, paint shops are frequently unreliable and colour is a major variant, a combination that poses a challenge in correct sequencing on the final assembly line.

Great importance has to be attached to managing the increasing number of variants. First of all, it is essential to focus on what the customer really requires and avoid the creation of variants that are not in demand. Deleting obsolete functions and options is welcome since it is the first step to reducing complexity. Additionally, variants should be built up as late in the assembly process as possible. A prerequisite for this would be a modular vehicle structure. Figure 7.7 shows an example of modular differentiation. The so-called mass customisation describes the development and production of vehicles at a high level of customisation going in line with a high overall production volume.

To reduce the number of variants several functions can be incorporated into one part or assembly, which leads to less complexity as well as designing the same interfaces for different module variants. Standardisation of interfaces prevents unnecessary tooling change during final assembly. The main principle is to generate high outer complexity and variance for the customer, whilst keeping the inner complexity lower by using as many common parts as possible.

In addition to this, which basically depicts the platform strategy, Ford took a direction that can be summed up by the following three points:

- Separation of the frame structure that bears the mechanical load of the body and panels that give the vehicle its typical shape and design.
- Separation of the vehicle into modules, namely front end, rear end, underbody and greenhouse to ease the creation of new body variants
- Scalability. Body parts have a simple structure and hence can easily be made longer or wider to change the main dimensions of the vehicle for the development of new derivates.

First, the production of the modules would take place, which are then joined together using cold joining technologies exclusively. In the last step the body panels are then attached to the body, giving the vehicle its appearance and shape. Ford has not been given information on how the final assembly process would look. It is not clear if the modules themselves will be equipped first and then joined together or vice versa. However, this is not of interest for the concept itself. Both scenarios can be implemented since the joining methods are limited to cold joining technologies and would allow the joining of pre-equipped body modules. To support scalability of the body frame only three kinds of components were used:

- Cold formed blanks
- Aluminium extrusions
- Sandwich plates

Only straight or 2D-bent profiles were allowed for the construction of the body.

7.2.6 Fiat Dual Frame

Fiat developed the Dual Concept in 2000 (Fig. 7.8) (Fiat 2006, 2007). Its aim is to show a new structure that offers several advantages, especially for niche car applications. Niche cars typically have low production volumes and hence limit the manufacturer's investment.

The Lancia Lambda first used the dual frame concept as far back as 1922 and was probably the first car with a load-bearing body. Up to this point in time, cars consisted of a ground frame where the components were fixed and the body was bolted on. The monocoque solution is still used for high production volumes today.

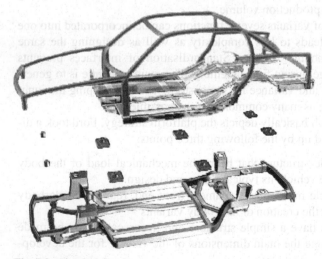

Fig. 7.8 Fiat dual frame

Monocoque structures require significant tooling investment and therefore require high production volumes just to break even. For this reason Fiat returned to the concept of building a frame for niche vehicles; this time using modern materials and technologies. The body was separated into two elements where the lower part carried main components like the engine, gear box and suspension whilst the upper element supported the surface body panels giving the model its individual shape. The stiffer the lower frame was designed the more freedom was given to the upper frame in terms of dimensions, material reduction in parts of the upper frame and styling of the vehicle. Both frames are fixed together using flexible blocks. These blocks can be adjusted hydraulically to adapt to different road load conditions and vibrations, helping to reduce noise levels within the passenger compartment.

7.3   Basic Conditions for Body Design

7.3.1   Logistics Requirements

Since logistics play a major role within the concept of delivering a customer-ordered car within a short lead time, the BTO strategy has to be taken into account when choosing the modularity concept. The intent is to have a reduction in production time, a simplification of the order and delivery network and a reduction of fixed capital within the whole process.

The development of the ModCar and its variants has to consider production scenarios that utilise as many common parts as possible. Parts that are unique for each variant will require manufacturing technologies that provide low fixed costs and hence are in line with their planned quantities. Reducing complexity is a key success factor for stockless production and is explored further in other chapters.

7.3.2   Complete Vehicle Structure

For designing a vehicle structure some prerequisites have to be set, including requirements for stiffness and crash protection, a feature list in connection with package information and design surfaces. All this input is then used for the first draft of the body structure, which depicts the routing of the load paths without going into further detail.

With regard to BTO, it is particularly important to consider further special requirements originating from the planning of the vehicle production and logistical structure. The aim of creating a BTO vehicle that will be built within a short lead time has an important impact on the product's structure and the whole production process. The following points must be taken into account to deliver in accordance with the demands of the BTO strategy, a short production time and a simplification of the product structure for complexity reduction:

- Modular body structure
- Standardised interfaces are needed to enable assembly of different body modules to create variants
- Body frame design, separation of functional structure and styling surfaces, removal of conventional paint shop
- Scalability of body for variability
- Vehicle is split into seven main modules, the body has to follow this structure
- All body modules will be pre-assembled with vehicle components before they are joined together in final assembly; hence, cold joining technologies are required for this latter operation to prevent localised damage

The virtual structure of a complete vehicle has been defined, which meets the requirements laid out above, and is called the ModCar. It depicts features that will be offered to the customer, either standard equipment or optional features like white indicators, cruise control, keyless entry etc. These requirements then have to be translated to the parts that need to be offered. If for example air conditioning is ordered by the customer, a compressor, radiator and pipes are needed, as well as the control unit with switches within the cockpit. The impact of these secondary requirements from the customer perspective is relatively low, but is significant to all supplier partners involved and for the production process.

7.3.3 Assembly Order

To save production time it was decided to split the vehicle into seven main modules, namely front end, engine module, greenhouse front, greenhouse rear, rear end module, exhausts and covering. The intention is to build all modules separately before joining them together to get a complete vehicle. Individual modules are fitted with their corresponding equipment and features and are only joined together in final assembly. Then, any remaining modules, e.g. exhausts and coverings, are fixed to the vehicle, as well as some smaller module bridging parts. Fig. 7.9 shows a scheme of this process.

7.4 ModCar Body Design

7.4.1 Scope of Research

The R&D for this vehicle has been carried out on a conceptual level. We have used the small car segment for 2010–2015 as a primary focus. The data generated

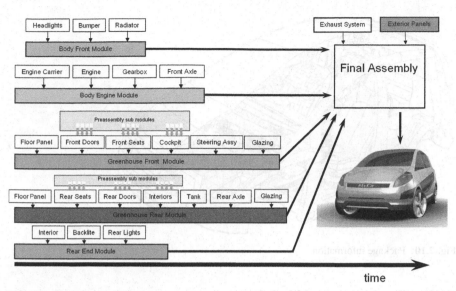

Fig. 7.9 Vehicle assembly process

provided the basis not only for the structure of the car, but also the basis for logistical calculations and capacity planning. The general frame conditions for the vehicle, like crash performance and packaging, were relevant for the design as well as requirements for this special case like vehicle styling, number of variants, separation of mechanical structure and panels, modular lay-out and choice of cold joining technologies.

The body design comprises a structure built of simple parts in terms of shape and manufacturing technology and provides the required stiffness and load routing. Materials have then been chosen to suit those technologies as well as the requirements that come from loads that would be applied to the vehicle. Lightweight design has also been an important issue so the additional challenge here was to have the ability to reduce material thickness and hence body weight by choosing novel materials.

Since we are describing a concept, the vehicle presented will not have details of components and their fixings to the body, sealing investigations for closures, quantitative tolerance investigations, a detailed BIW assembly fixing concept, cataphoretic painting outflow check-up and eye ellipse analysis. When arranging the package data the cockpit, engine block with gear box, styling surfaces, seats, dummies, wheels and different tank systems have been considered. Figure 7.10 gives an overview of the package amount.

CAE crash simulation has been carried out taking into account several standard Euro-Ncap cases. Details can be found in the corresponding chapter. Since no prototype has been built, physical tests have not taken place within this project.

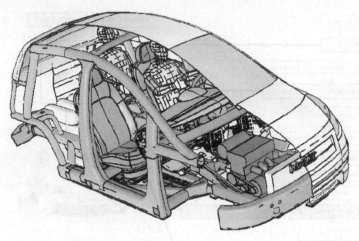

Fig. 7.10 Package information

7.4.2 Combination of Modules

Taking into account the variant types that have been chosen for this project, namely three-door, five-door, wagon and convertible, the modular structure has to fulfil the requirements of the assembly concept, the product structure and the use of as many common parts as possible over the whole range of vehicle variants. These requirements lead to the module variance depicted in Fig. 7.11.

The body consists of four modules; namely, the front end, engine module, greenhouse front and greenhouse rear. Different combinations of different modules result in a variety of vehicles. The interfaces are all oriented longitudinally to the vehicle direction of travel to ensure a feasible joining process. The distinction between the types within a module group will be illustrated on the following pages. The front end and engine module of the body frame are the same for all vehicle variants; the greenhouse front and rear each have three variants offering the possibility of the assembly of four different vehicle types. There are many parts and beams even within these modules that have the same shape and hence can be manufactured on the same machine.

7.4.2.1 Front End

The front end has three main functions (Fig. 7.12) (Allianz 2004). It stiffens the body and protects the vehicle in the case of a front-end crash. Additionally, it supports the front-end frame by carrying the headlights, radiator and outer skin. The two crash boxes have a waveform and provide energy absorption for the RCAR test (Research Council for Automobile Repairs) by being the first part to crumple. In this kind of test only the crash box should be damaged to make it simple and inexpensive to repair the vehicle and hence keep insurance premiums low.

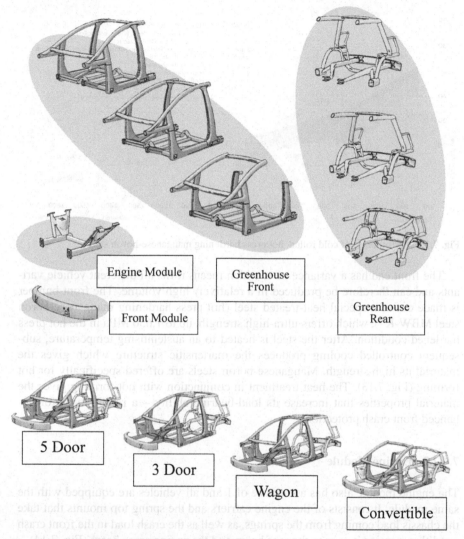

Engine Module

Greenhouse Front

Front Module

Greenhouse Rear

5 Door

3 Door

Wagon

Convertible

Fig. 7.11 Combination of different module variants

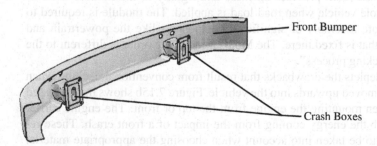

Front Bumper

Crash Boxes

Fig. 7.12 Body front end module

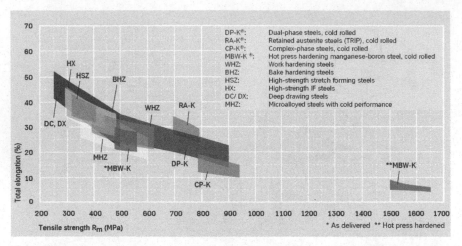

Fig. 7.13 Performance of cold rolled, hot press hardening manganese-boron steel

The front end has a variance of 1, which means it fits all different vehicle variants and can therefore be produced in a relatively high volume. The front bumper is made out of a special heat-treated steel (hot press hardening manganese-boron steel MBW-K®), which offers ultra-high strength, up to 1,650 MPa in the hot press hardened condition. After the steel is heated to an austenitising temperature, subsequent controlled cooling produces the martensitic structure which gives the material its high strength. Manganese-boron steels are offered specifically for hot forming (Fig. 7.13). The heat treatment in conjunction with hot forming gives the material properties that increase its load-bearing ability – a requirement for enhanced front crash protection.

7.4.2.2 Engine Module

The engine module also has a variance of 1 and all vehicles are equipped with the same module. It consists of the engine carriers and the spring top mounts that take the chassis load coming from the springs, as well as the crash load in the front crash condition, passing it over to the roof beam and the engine cross beam (Fig. 7.14).

The engine cross beam provides stiffness and rigidity to the module itself as well as to the whole vehicle when road load is applied. The module is required to be stiff for the safe assembly of significant components like the powertrain and other equipment that is fixed there. The ModCar mounting order is different to the conventional "decking process".

Figure 7.15a depicts the drawbacks that result from conventional decking when the powertrain is moved upwards into the vehicle. Figure 7.15b shows the increased package area when mounting the engine from the top or front. The engine carrier also has to absorb the energy coming from the impact of a front crash. These requirements have to be taken into account when choosing the appropriate material and manufacturing method and much work has gone into this area of research.

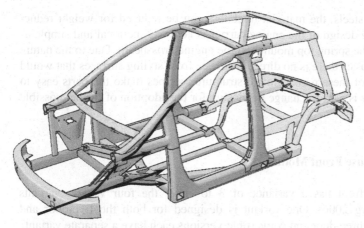

Fig. 7.14 Load path for the front crash condition

The retained austenite Steel RA-K® 40/70 that has been chosen for this application is an advanced, high-strength steel characterised during forming by the "TRIP" effect (transformation-induced plasticity steel) and has a tensile strength of 700 MPa and a yield strength of 400 MPa. With retained austenite steels, the retained austenite is transformed into martensite during forming – giving rise to TRIP. This also allows large material elongations to be achieved at high strengths. The material is characterised by high work hardening, even with large strains, and high bake hardening potential, particularly after prior forming. RA-K® steel displays high energy absorption capacity under dynamic loads. The micro-structure consists of retained austenite embedded in a ferrite/bainite matrix. The ferrite, including the bainitic ferrite, accounts for up to 90% of the structure, and an amount of martensite may be present. Compared with dual-phase steels these grades offer elevated strength as well as even better cold formability and work hardening. The result is a high ability of energy absorption. Since RA-K® can achieve higher strength at comparable elongation and higher yield points than cold

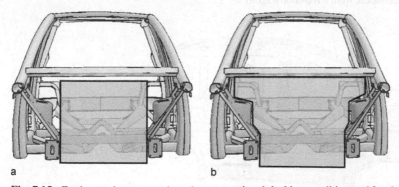

a   b

Fig. 7.15 Engine package space in **a** the conventional decking condition and **b** when mounting it from the top or front

rolled dual-phase steels, the material thickness can be reduced for weight reduction purposes. The design of the engine carrier is kept as functional and simple as possible, as with the spring top mount and the engine crossbeam. Due to the nature of the modular chassis there is no direct influence from styling surfaces that would dictate the shape of these parts. Simple structured shapes make the parts easy to manufacture and it is easy to change the design for the adoption of further possible vehicle variants.

7.4.2.3 Greenhouse Front Module

The Greenhouse front has a variance of 3 to cover the four vehicle variants (Fig. 7.16) (Osburg 2006). One variant is designed for both the five-door and wagon whilst the three-door and convertible versions each have a separate variant.

Since the only difference between the three-door and the five-door vehicle is the position of the B-pillar, the intention was to create a flexible design that consists of the same parts with just a variable location of the B-pillar. Figure 7.17 shows the design measures to fulfil those requirements.

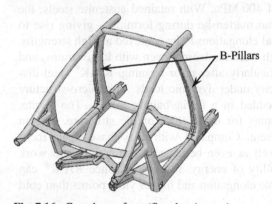

Fig. 7.16 Greenhouse front (five-door/wagon)

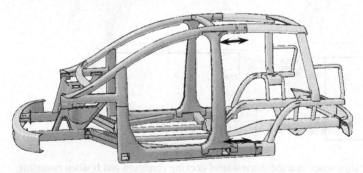

Fig. 7.17 Variation of the B-pillar position to suit to either three or five doors

For the interface areas where the B-pillar makes contact with the sill and with the roof beam, a constant shape in the x-direction is needed to ensure adaptability to the wagon/five-door and three-door designs. The roof beam itself has not got a straight routing in the x-direction, as featured on the sill, since it follows the requirements of the stylists' outer surfaces. The flexibility of the B-pillar position needs indentations in the upper area of the roof beam. This reduces the number of possible manufacturing methods that may be applied for the latter. Inner hydro forming was chosen to give this part the shape postulated.

Other parts, like the passenger frame, have to follow the shape of the outer design surfaces and hence need manufacturing methods that provide this, such as inner hydro forming and deep drawing.

A- and B-pillars as well as sills have special requirements concerning crash protection, be it the pole or the side impact crash standards. In contrast to the engine carrier, which absorbs energy by plastic deformation the area at the passenger frame does not have that much space for deformation. The space between the passenger and an intruding object is much smaller than in frontal crash condition so the objective for these parts would be intrusion prevention rather than energy absorption. This requires strong and stiff materials. These parts also have a complex shape, following the outer surface, so materials also need to be highly deformable and hence need great toughness. Conventional steels only offer a trade-off, having either high rigidity and low toughness, or vice versa (Fig. 7.18).

To combat this contradiction, research is ongoing within the steel industry towards developing a steel class that unites both material capabilities. There are some steel types that can be named here that additionally provide the potential of lightweight design through wall thickness reduction compared with conventional steel grades. Dual-Phase Steels as well as Retained Austenite Steels (TRIP) show

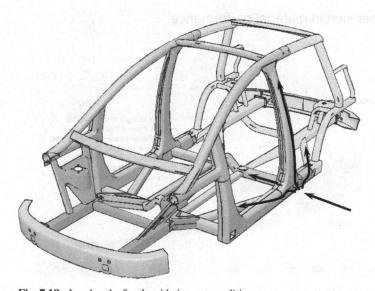

Fig. 7.18 Load paths for the side impact condition

the tendency to increase both toughness and rigidity. Dual-Phase Steels have a balanced ferrite and martensite content and offer a particularly attractive combination of high strength, low yield-to-tensile ratio, good cold formability and weldability. These steels are suitable for cold forming operations involving a high percentage of stretch forming in the production of complex structural components and body parts. Their good dynamic properties make dual-phase steels ideal for crash relevant components.

Retained austenite steels are advanced high strength steels characterised during forming by the "TRIP" effect. Compared with dual-phase steels, these grades offer elevated strength as well as even better cold formability and work hardening. These properties are achieved due to the presence of metastable retained austenite in a ferrite/bainite matrix. In comparison to the cold rolled dual-phase steels DP-K®, the retained austenite steels can achieve higher strength at a comparable elongation and higher yield points. The cold rolled, retained austenite steel RA-K® from ThyssenKrupp Steel is ideal for the manufacture of parts that are difficult to form and hence require high percentages of stretch forming and deep drawing, such as complex, strength-relevant structural components. See Sect. 7.4.2.2 above for further details of this type of steel.

Looking forward to potential new materials, a new steel grade, whose performance in terms of relation of elongation to strength is significantly improved, has emerged. X-IP® steel, which stands for "extreme strength and formability through induced plasticity", provides unusually good formability, comparable to lower strength conventional steels. Additional work-hardening potential facilitates significant increases in strength. This steel grade is still under development and

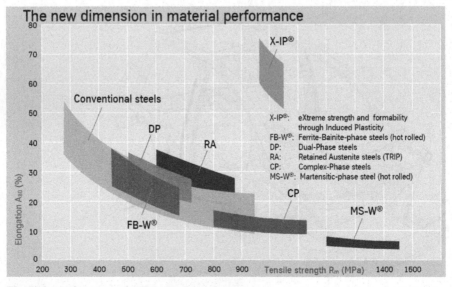

Fig. 7.19 Performance of different steel grades

promises great benefits in the area of lightweight engineering. Higher strength leads to the possibility of wall thickness reduction and hence to light part design.

Good formability is required where parts have a complex shape so engineers are limited to manufacturing methods like deep drawing and hydro forming. These methods require high tool investment and hence are only cost effective if the production volume is high enough. Parts that need no complex shape should have a simple structure and can be produced using methods like roll forming or bending. These production processes are cheaper and the constant cross-section of the part over its length makes it easy to modify and create new vehicle variants (Fig. 7.19).

7.4.2.4 Greenhouse Rear Module

Focussing on the Greenhouse rear, again, there are three different variants that support four different vehicles (Fig. 7.20). We again have one variant for both the three- and the five-door, one variant for the wagon and one for the convertible.

Scalability is, as in the other modules, provided by employing the simplest component parts to build the structure, with more complex parts and processing only used in areas that are affected directly by the outer surface. For further vehicle modifications concerning wheelbase extension the 6 beams that are connected to the greenhouse front would have to be modified, as there are two boot beams, two rear sills and two roof liners rear. The simple shape of those parts provided a good prerequisite for creating new vehicle variants in a short time span at low production volumes.

Similar to the front-end module the greenhouse rear is fitted with crash boxes to limit damage and repair cost in the case of a low speed crash. To maintain simplicity and low cost the parts above the level of the spring plate differ depending

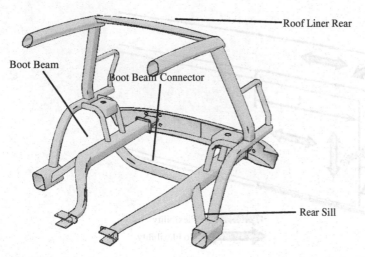

Fig. 7.20   Greenhouse rear body module

on variant type, whilst the area below consists of parts that are the same for all variants, apart from additional reinforcements for the convertible. The consequence of this coherent approach is a higher volume for each part and hence more numerous and affordable manufacturing possibilities from which to choose.

7.4.3   Tolerance Compensation

The body modules must be joined together and this raises a number of challenges. As the modules form the body frame they provide the car with stiffness. Several interfaces are needed between the main modules. As well as achieving an acceptable level of body stiffness, tolerance compensation has to be provided at the joints to ensure that the modules will go together (Fig. 7.21). As it is proposed that the body modules are only joined together after they are equipped with components and interior materials, we are also limited to cold joining technologies. The use of counter brackets is also not possible, so screw bonding has been identified as the most appropriate joining technology.

The edges of the front sill are cut out to generate flanges that are flexible only perpendicular to their surface, but they will still take load in the other direction, as Fig. 7.21 depicts. The flexibility allows tolerance compensation. The walls of both sills can be brought in line before being fixed together. These joining propositions are still at the conceptual level and parameters have to be defined to achieve a reliable system that fulfils all its requirements, but the current outlook is promising.

When joining the greenhouse rear to the greenhouse front there are six interfaces that need to be brought into line and joined together. Therefore, a way of compensating for geometric and dimensional tolerances had to be thought of. Both

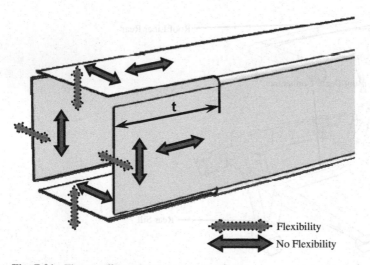

Fig. 7.21 Flanges allow tolerance compensation

modules have to be moved together until the boot beam connectors go on block with the greenhouse front in the x-direction. In this stage direction y and z are free. The flanges are then folded over while the centre connector of the greenhouse rear is used as a counter support. When folding is finished both parts are joined together. This procedure leads to complete tolerance compensation. The sills use a different system.

7.5 Conclusion

We recognise the OEMs' need to have capacity utilisation of their plants high on their "radar" at all times in order to maintain cost-optimised vehicle production plants. On the other hand we see customers demanding a highly individualised car.

Modularisation offers an opportunity to create a win-win situation. The ability to create a full range of body variants that comprises ostensibly common parts and that requires low tool investment provides a logical and efficient solution. It is also essential, for minimising the risk of low plant capacity utilisation, that all variants can be assembled on one assembly line. In the case that the demand for one vehicle variant is lower than expected, the probability is high that there is one other variant that compensates by having higher demand. It can be expected that the overall demand curve for all variants will be at a more constant level, which would then make vehicle production and plant capacity utilisation very effective. Effectiveness from a body design perspective means the assembly of a load-bearing body frame that is separated from design surfaces and panels. Panels have a high variance since they are offered in a wide range of colour types. It is therefore recommended that this complexity is moved to the end of the assembly process.

If it is possible to design the body frame in such a way that many simply shaped parts are involved, manufacturing methods can be used that require low investment. This would allow vehicle variant production at low volumes and low risk. Design changes for new variants could easily be implemented and even modifications to create cars belonging to other vehicle segments could be accomplished. This would reduce not only manufacturing costs and complexity, but also the development time of new vehicles.

References

Allianz (2004) Allianz Zentrum für Technik: Neuer RCAR-Crash. Available via:
    http://azt.allianz.de/azt.allianz.de/Kraftfahrzeugtechnik/Content/Seiten/Presse/
    Pressemeldungen/herstellerinfo_neuercrash.html. Cited 22 October 2007
Automobile (2005) GM sequel fuel cell concept, Automobile online –
    http://www.automobilemag.com/auto_shows/2005_chicago/0503_gm_sequel/.
    Cited 9 December 2007
Automobil Produktion (2004) Sonderausgabe: Studie: "Future automotive industry structure
    FAST 2015"

Baumann M, Hänschke, Sweeney (2002) Die modulare und skalierbare Karosserie. IIR Fachkonferenz Karosserie-Modularisierung, 9–11 July 2002, Karlsruhe

Daimler AG (2004) Vario research car – four cars in one. Available via: http://www.daimlerchrysler.com/dccom/0-5-7182-1-392471-1-0-0-348452-0-0-135-7165-0-0-0-0-0-0-0.html. Cited 5 December 2007

Dudenhöffer F (1999) Merger Puzzle in der Autowelt. Absatzwirtschaft 1:1–9. Available via: http://www.absatzwirtschaft.de/pdf/sf/merger.pdf. Cited 15 October 2007

Elbl-Weiser K (2003) Automobile coatings of the future – new solutions for modular automobile designs, BASF Coatings AG. Available via: www.corporate.basf.com/basfcorp/img/innovationen/felder/mobilitaet/e/PI_MrKreis.pdf. Cited 12 October 2007

Fiat (2006) Automotive space frame design: Tipologie Costruttive. Available via: http://www.marcotraverso.it/spaceframe/it/1-4-3.htm. Cited 17 October 2007

Fiat (2007) Innovation at Fiat: from design to prototype. Available via: http://www.madeinfiat.com/aug00/briefa.htm. Cited 17 October 2007

GM (2007) Available via: http://www.electricdrive.org/index.php?tg=articles&idx= Print&topics=47&article=584. Cited 17 October 2007

Koeth C-P (2004) Gefangen in alten Konzepten. Auto Ind 7–8:34–39. Available via: http://www.agiplan-gmbh.de/pdf/04_09_08_ver_auto_ind_7_8.pdf. Cited 15 October 2007

McAlinden SP, Smith BC, Swiecki BF (1999) The future of automotive systems. What are the efficiencies in the modular assembly concept? In: Michigan Automotive Partnership Research Memorandum, No.1. Available via: http://www.umtri.umich.edu/content.php?id=332. Cited 15 October 2007

Maier G (2003) Fiat Ecobasic: Tre litri arranciata. In: Focus online. Available via: http://focus.msn.de/D/DL/DLB/DLBO/DLBOP/dlbop.htm. Cited 17 October 2007

Mercer Management Consulting (2001) Automobiltechnologie 2010. Technologische Veränderungen im Automobil und ihre Konsequenzen für Hersteller, Zulieferer und Ausrüster. Available via: http://www.mercermc.de/mapper.php3?file=upload_material%2Ftuv%2F3.pdf &name=Technologische_Ver_nderungenim_Automobil_und_ihre_Konsequenzen_f_r_ Hersteller_Zulieferer_und_Ausr_ster.pdf&type=application%2Fpdf. Cited 15 October 2007

Mutegi C, Macheleidt S, Matalla M (2000) Umsetzung hybrider Wettbewerbsstrategien in der Automobilindustrie

Neumann M (1998) Neuere Trends der Modularisierung im Automobilbau. Dissertation, University of Göttingen, Germany

Osburg B (2006) ThyssenKrupp steel: the global specifications for Steel 2015+, integration of global market requirements in the development process and services of a steel manufacturer. World Automotive Steel Assembly 2006, Berlin

Perini G (2007) Fiat Ecobasic, CarDesign. Available via: http://www.cardesign.to/numero_000/fiat_ecobasic/fiat_ecobasic.htm. Cited 15 October 2007

Rossbach R (2001) Fiat Ecobasic: Kompakt in der Zukunft. In: Prova-Magazin für Automobile Avantgarde. Available via: http://www.prova.de/archiv/2001/00-artikel/0016-fiat-ecobasic/index.shtml. Cited 15 October 2007

Truckenbrodt A (2001) Modularized cars – die Quadratur des Autos. Verkehrswissenschaftliche Tage, Dresden

Chapter 8
Modular Concepts and the Design of the ModCar Body Shell

Maik Gude and Werner Hufenbach

Institut für Leichtbau und Kunststofftechnik, TU Dresden, Germany

Abstract. To achieve the aim of building a car to customer order within 5 days, an innovative modular concept car was conceived that supported the planned cost-effective built-to-order proposition and stockless production. In a multi-stage design process the automotive body shell, outer panels and selective outer structures have been developed to fit over the modular body frame, offering numerous advantages. The project had a twin aim of meeting not only the technical requirements, which included not only modularity, safety, low weight and a panoramic view, but also the emotional design aspects, including looking "sporty" and "agile". The appearance of the car body shell for the ModCar is of great importance if it is to be commercially viable and achieve success in the marketplace. Developments in lightweight materials and process technologies enabled the development of advanced function-integrated lightweight vehicle modules for series manufacture. The ModCar fulfils the criteria regarding passenger protection with respect to the Euro-NCAP standards. In addition, simulations show that the essential bending stiffness has been achieved. An optimised lightweight door module consisting of novel materials demonstrates good overall performance with regard to static load and crash behaviour testing. As such, the ModCar demonstrates that the novel concepts employed may be used to successfully produce a vehicle that is saleable in the European market.

8.1 Introduction

In order to build a car to customer order within 5 days an innovative modular concept car (ModCar) was conceived that supported the planned cost-effective built-to-order (BTO) proposition and stockless production. The proposal for a form of modularity that includes flexibility and low complexity, as well as the associated manufacture and assembly strategy, required novel modular concepts and associated design studies. Whilst there are clear advantages of modularity for production, this approach must also deliver a car with attractive styling and not result in additional vehicle weight. Therefore, in addition to the development of a modular lightweight car body, the appearance of the car body shell was of great importance for the ModCar if it was to be commercially viable and achieve success in the marketplace.

The Institute of Lightweight Structures and Polymer Technology (ILK) at TU Dresden, Germany, provided the design (styling) for the ModCar body shell as well as the technical engineering of selective outer panels and outer structures. Following a non-variable parts strategy, within an interactive design process of body shell and space frame, a vision for the outer appearance was produced for different vehicle variants (three-door hatchback, five-door hatchback, five-door notchback, convertible and wagon). The project had a twin aim of meeting not only the technical requirements, which included modularity, safety, low weight and a panoramic view, but also the emotional aspects, such as looking "sporty" and "agile".

Modern processes for styling and the technical engineering of vehicles are strongly interlinked and often dominated by the vehicles' appearance. The adaptation of the body shell and the space frame is done in an interactive process, with the help of virtual mock-ups that serve for collision and penetration analyses. The digital mock-ups are also used for subsequent technical engineering tests involving quasi-static and highly dynamic (crash) loading conditions.

Within the technical engineering phase the outer panels are designed, materials chosen and elaborations made for the novel fastening elements and joining technologies as well as colouring techniques to guarantee class-A surface standards. As an example, the structural design of the novel modular concept door is explicitly demonstrated. In order to meet the modularity requirements the modular door, with integrated plastic side impact protection, is characterised by its flexible design that enables the use of identical parts over a broad range of variants. Demands for a light weight, manufacturer requirements, several legal directives and regulations in terms of crash and impact behaviour are fulfilled using novel lightweight materials. In order to assess the suitability of the lightweight materials and compare their performance, accompanying simulations of the structural behaviour of the selective door frame are performed.

The body shell components are implemented into a bill of materials dependent on each individual ModCar variant. The bill of materials serves as a basis for the elaboration of adapted supply and assembly concepts, as well as for the development of a logistics chain for the ModCar.

8.2 Styling Process and Concepts of the ModCar

The first step during the development of the ModCar body was the styling of an aesthetically pleasing outer body shell. The design had to incorporate the modularity and light weight requirements, as well as the restrictions of the manufacturing of the outer panels and car body. The basic dimensions of the car wheelbase, height and width, are defined according to the envisaged market segment, in this case the "small car". The design of the outer shape depends on the different variants of the ModCar. These variants are hatchback, wagon, notchback (this style is also variously referred to as a "saloon", "limousine" or "sedan") and convertible. The hatchback can also be produced as a five-door or three-door variant. The appearance of the ModCar is strongly interlinked with the development of the modular body frame.

Within the styling process of the modular car a philosophy was defined as a guideline with regard to the identified target buyer group, and contains the three main themes: sportiness, agility and vision. Furthermore, following the latest trends in automotive design, the ideas of visible safety elements and of an advanced panoramic view using a multifunctional materials systems such as carbon fibre-reinforced polymers or function-integrated transparent thermoplastics should be focussed on.

One of the first styling drafts of the outer shape is shown for the five-door hatchback in Fig. 8.1, giving an idea of the surface curvatures and light reflections of the body shell. The design is strongly driven by a sporty appearance combined with visible side impact protection and a panoramic roof.

At an early stage the design was transmitted across all the different ModCar variants to support the modularity concept. This facilitates the reuse of as many body shell parts across the variants as possible and reduces the manufacturing complexity and costs. The shape at the front and the side area is therefore identical for all variants; only the rear differs in size and shape, depending on the variant (Fig. 8.2).

The visual design of the ModCar body shell was continually re-aligned with the evolving design of the body frame (see also Chap. 7). In a continuous process, the body shell and body frame designs were compared, with an increasing level of detail, in order to achieve product development conformity in terms of constructed space, connections, field of view and avoidance of intersections. Fig. 8.3 shows as an example the "collision" check of the body frame and body shell for the hatchback variant. This test ensures that the body frame fits within the proposed body shell design.

The body shell consists of different outer panels and outer structures respectively. Developments in lightweight materials and process technologies enable the development of advanced function-integrated lightweight modules for series manufacture. Developments in the area of innovative glazing technologies, for example,

Fig. 8.1 Draft side and front design of the hatchback

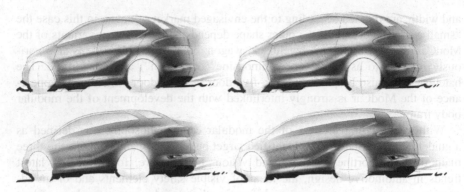

Fig. 8.2 Draft side and front design of the hatchback, notchback and wagon

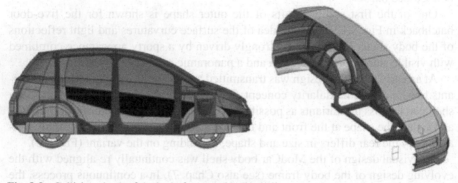

Fig. 8.3 Collision check of the body frame and body shell

enable the realisation of novel, strongly curved, function-integrated glazing modules. The use of transparent light thermoplastics such as polycarbonate (PC), polymethyl methacrylate (PMMA) or polystyrene (PS) and two-step injection moulding technologies presents a new approach for the development of fully equipped modules and thus supports the 5-day car concept of using preassembled function-integrated parts. The rear design provides an example of the application of advanced materials and processes within the ModCar. This consists of a broad, complex shaped rear plastic window that provides a panoramic view and has integrated within it a reinforcement frame with a hinge rail and lock attachment, a defogging system, rear lamps, a third stop lamp, an antenna and a spoiler (Fig. 8.4).

The modular design also enables flexibility in the rear module, as demonstrated with the customised opening mechanism designed for the hatchback and shown in Fig. 8.5. Different variants were elaborated – one-piece or two-piece hatchbacks. All opening mechanisms are applicable to the current design, so that the customer has a choice between different opening mechanisms. This choice may be exercised at order or even during the late assembly phase without changes to the surrounding outer panels and body frame being required.

Fig. 8.4 Hatchback design with function-integrated rear module

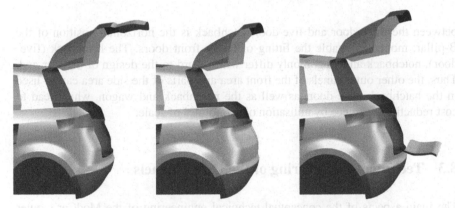

Fig. 8.5 Customised opening mechanisms for the hatchback

Fig. 8.6 Final design of the hatchback (five-door)

The final design of the ModCar five-door hatchback variant with its main design elements, the panoramic roof and rear window, visible side impact protection, powerful front and sporty rear, is shown in Fig. 8.6.

The derived variants of the ModCar final design are shown in Fig. 8.7, which also indicates the basic front module, which is kept visually the same in all derivatives, whereas only the marked rear ends differ for each variant. The main difference

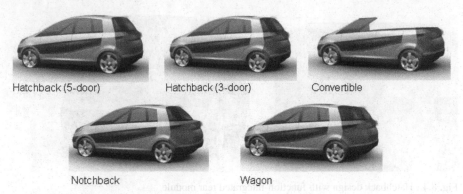

Hatchback (5-door)   Hatchback (3-door)   Convertible

Notchback   Wagon

Fig. 8.7 Variants of the ModCar

between the three-door and five-door hatchback is the horizontal position of the
B-pillar, moved to enable the fitting of larger front doors. The hatchback (five-
door), notchback and wagon only differ with regard to the design of the rear end.
Thus, the other outer panels of the front area and parts of the side area can be used
in the hatchback (five-door) as well as the notchback and wagon, which lead to
cost reduction per piece by utilisation of economies of scale.

8.3   Technical Engineering of the Outer Panels

The main aspects of the conceptual technical engineering of the ModCar's outer
panels are the choice of suitable materials, definition of partitioning lines in accordance with the modular design concept and styling, as well as the use of joining
strategies that are tailored to the ModCar's assembly strategy and consider the
challenges of a multi-materials design of the body frame and body shell.

8.3.1   Choice of Lightweight Materials

The outer panels have to fulfil different mechanical and visual requirements as
well as legal directives regarding crash and impact behaviour. As the ModCar
concept is of a self-carrying body frame, the outer panels and outer structures are
not designed to bear substantial mechanical loads, excepting wind loads, misuse
cases and loads that maintain the integrity of the components (e.g. the door intrusion test). Innovative developments in the area of polymer materials and associated manufacturing processes offer the potential for the design of lightweight outer
panels using different polymers such as reinforced and non-reinforced thermosets
as well as thermoplastics (Table 8.1) (see also Strümpel et al. 2006).

Table 8.1 Overview of potential polymer materials and associated manufacturing processes for use in the ModCar body panels and outer structures

	Material	Typical manufacturing process
Thermoplastic matrices	Non-reinforced thermoplastics, c.g. PA, PBT, PP and thermoplastic blends like PA-ABS	Injection moulding
	Glass mat-reinforced thermoplastics (GMT)	Compression moulding
	Long fibre-reinforced thermoplastics (LFT)	Compression moulding
	Endless fibre- or textile-reinforced thermoplastics based on hybrid yarns (GF-PP)	Compression moulding
Thermosetting matrices	Moulding masses and moulding batches, e.g. bulk moulding compound (BMC) and sheet moulding compound (SMC)	Compression moulding, injection moulding
	Fibre-reinforced polymers	Reinforced reaction injection moulding (R-RIM), structural reaction injection moulding (S-RIM), long fibre injection (LFI)
	Endless and textile fibre-reinforced thermosets	Resin transfer moulding (RTM), resin infusion (RI), resin film infusion (RFI)

A broad variety and many variants of the materials and processes mentioned can be found from different companies including BASF, Bayer, Dieffenbacher, Engel, Krauss-Maffei Kunststofftechnik, Peguform, Saint Gobaint or Visteon, each using their company-specific notations. Additional issues, besides the process and material properties themselves, are price, capability of large scale production, manufacturing process economics, recycling and questions regarding colouring methods. An overview of some of the materials currently available and their properties is shown in Table 8.2.

Table 8.2 Mechanical properties of selected materials (Herd 2006; Hufenbach et al. 2006; Kompetenznetz RIKO 2003; N.N. 2007)

Property	Material				
	Unit	SMC	Triax® 3157	Twintex® T PP 60	LFT PA66-GF40	LFI-PUR (20% GF)
Density	g/cm³	1.8	1.15	1.5	1.45	1.0
Tensile modulus	MPa	8,000	2,000	15,000	14,000	3,147
Tensile strength	MPa	80	50	350	130	33
Coefficient of expansion	$10^{-6}K^{-1}$	10–18	70	5.4	22	<35
CDC-capability	–	On-line	On-line	Off-line	Off-line	On-line

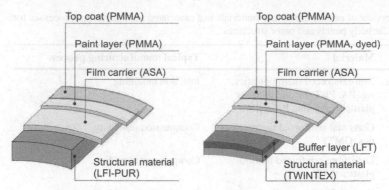

Fig. 8.8 Composition of foils for paintless film moulding (*left*) and usage of buffer layers to avoid print-through in the case of endless fibre-reinforced or textile-reinforced polymers (*right*)

The materials SMC or Triax® and Ultramid TOP (Feldmann 2007) as thermoplastic blends, and LFI-PUR, can pass through the cathodic dip coating (CDC) painting process typical of current automotive production lines without incurring any thermally induced damage. However, the concept of the ModCar includes the removal of the central paint shop for cost, environmental and late configuration reasons. This requires pre-coloured surface panels to be produced using novel technologies such as paintless film moulding (PFM®). This process uses coated or dyed foils to form the outer surface of a panel, consisting of several layers – a top coat, a paint layer and a film carrier (Fig. 8.8). PFM® skins are thermoformed and then back-moulded with a 20–30 % glass fibre-reinforced supporting panel material.

Parts made using fibre-reinforced or textile-reinforced polymers like Twintex® as a structural support generally need additional layers to guarantee a class-A surface finish. The main problem is "fibre print" through to the outer surface, where a fibre pattern is still visible. The "fibre print" is caused by the highly inhomogeneous material structure, which leads to non-uniform process-related shrinkage.

The non-uniform shrinkage of highly inhomogeneous materials is strongly dependent on the difference in the coefficients of thermal expansion (CTE) of the fibre and matrix of the single components. In extensive numerical and experimental investigations it was shown that the effect of non-uniform shrinkage at the outer surface can be substantially reduced by the integration of a buffer layer. The buffer layer, a short fibre- or mat-reinforced layer, is made of a material with a lower CTE and preferably a higher Young's modulus than the matrix and hence compensates for any shrinkage that may cause surface imperfections.

The more homogeneous material structure of SMC, Triax®, LFI-PUR and LFT all give a class-A-standard surface finish, which fulfils automotive customer requirements without the need for additional layers and is hence a lower cost alternative.

8.3.2 Partition of the Outer Panels

The partition of the body shell into single outer panels is key to the success of the ModCar common parts strategy and has an impact on a number of different additional aspects of the vehicle, e.g. the choice of suitable materials, associated processes including manufacture, joining technologies, tooling costs, functionality and flexibility, transportation and logistic issues and any repair costs. Styling is also a key driver, so partitioning has to be done using a complex decision process in which the pros and cons are weighed up in order to find an optimal compromise. The definition of partitioning lines of the ModCar and the resulting panel components are demonstrated for the hatchback variant. Figure 8.9 gives an overview of the split outer surface, where each outer panel is dyed with a different colour.

In order to reduce the number of components, the different panels should be compatible with as many variants of the ModCar as possible. The main panels of the body shell are the front panel, front fender, bonnet, roof panel, C-panel, door panels, back panel and rear wing (Table 8.3). For the ModCar the front panel, the front fender and bonnet, the door module and the roof and the C-panel are identical for the hatchback (three-door and five-door), notchback and wagon variants. The same front panel, front fender and bonnet can also be used in the convertible variant. The only modules that are variant-specific are the back panel and the rear wing. In order to reduce tooling costs the shortened rear door of the ModCar three-door variant remains an independent module.

As part of the design brief logistics costs were taken into account. The panels are designed to occupy the smallest area possible for transportation to the decentralised assembly shops within the ModCar assembly strategy. This is a key advantage over the present cars, where many different parts with large spatial volumes are used to build up the car bodies of the different car variants.

Fig. 8.9 Outer surface of the ModCar (hatchback) and the partition strategy

Table 8.3 Non-variable parts of the ModCar body shell

No.	Outer panel	Hatchback (five-door)	Hatchback (three-door)	Notchback	Wagon	Convertible
1	C-panel	X	X	X	X	
2	C-panel (convertible)					X
3	A-pillar panel					X
4	Front panel	X	X	X	X	X
5	Front fender	X	X	X	X	X
6	Bonnet	X	X	X	X	X
7	Roof	X	X	X	X	
8	Rear wing (hatchback)	X	X			
9	Rear wing (notchback)			X		
10	Rear wing (wagon)				X	
11	Rear wing (convertible)					X
12	Front door panel (five-door)	X		X	X	
13	Front door panel (three-door)		X			X
14	Back door panel (five-door)	X		X	X	
15	Back panel (three-door)		X			X

8.3.3 Joining Concepts for Outer Panels

The ModCar body frame is made up of colourless steel profiles with a constant cross-section, avoiding complex geometries and section sweeps. This provides low production costs and facilitates the modularity concept. In contrast to this, the design of ModCar aims to present the images of sportiness, agility and vision by providing well-shaped free-form surfaces and defined and complicated shapes. To bring those two structures together, special connections are necessary.

The connection between the outer panels and the steel body frame has to fulfil the requirements of great stiffness and strength, modularity, outer surfacing fit and finish, design and colour as well as compensation for manufacturing tolerances and differences in the thermal expansion of the steel body frame and plastic panels.

One method of connecting the two structures is to integrate the connection elements in the outer panels, hence increasing their complexity. The outer panels would therefore require structurally relevant functions (outer structure). These functions could be achieved by mounting bars, ribs, pins and other elements onto them, which would require localised increases in thickness and changes in material mass. The advantages of this approach are lower weight and a highly integrated product. However, the main disadvantage is that integrated elements like bars, ribs and pins can affect the surface finish of the outer panels. Another disadvantage of

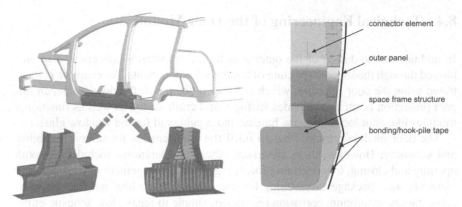

Fig. 8.10 Connector elements, B-pillar/door sill and joining concept

a highly integrated structural panel is caused by the different thermal expansion coefficients of the body frame and outer panels. A uniform clearance has to be guaranteed between body and panel, and the mounting system would have to provide a very wide tolerance band.

Alternatively, it is possible to mechanically decouple the outer panels from the body frame. To retain the class-A finished outer panels with great stiffness and strength a differential mode of construction can be used, whereby special multifunctional connector elements are used between the body frame and outer panels (Fig. 8.10). These fibre-reinforced thermoplastic structures are highly complex moulded parts that integrate the functions of sealing, covering and placing of cable trees, but are only occasionally visible to the vehicle owner, for example in the door entrance area.

The paint-free body frame structure will then be fully enclosed and thus covered by these connector elements. They can be screwed or snapped onto the body frame structure in the assembly shop. The snap pins and bolts can be welded onto the body frame modules by a welding robot before the application of an anti-corrosion coating.

The tolerances between connector element and body frame can be overcome, due to the relatively short distances involved, by the use of tolerance-friendly joining technologies on the outer panels and the connector elements (bonding, hook and loop etc.). To fulfil the requirements of modularity, stiffness and strength the connector elements will be fixed by screws in highly stressed areas and "snapped on" in secondary structural areas.

The connector elements, produced by injection moulding, provide a perfect surface onto which to bond the outer panels. The adhesive can be pre-applied during the production of the connector elements and activated before the marriage of the frame and the outer panels. The outer panels will primarily be joined to the connector elements by adhesives and hook/loop fastening. Self-tapping screws will be used in well-hidden areas where bonding or velcro fastening is not possible. Bonding technologies will be used in those areas where impact and misuse are most likely.

8.4 Technical Engineering of the Door Module

In addition to the design of the outer panels, the modularity concept is also employed through the support structure of body shell components. An example is illustrated using the door structure, which consists of the panel, the door frame, an impact protection beam that provides stiffness and crash safety, as well as functional modules like door locks, mounts, hinges, and a guide rail for the window glass.

The door module primarily has to fulfil the requirements for safety, reliability and economy. However, there are many other considerations, including smooth opening and closing, low operating force, ergonomic arrangement of the operating elements, low package dimensions for the opened door, low weight, stiffness, crash energy absorption, corrosion resistance, simple to repair, low acoustic emission and ability to mass produce.

Following the non-variable parts strategy, the door module can be conceived so that non-variable parts are used and the additional requirements are achieved. An opening mechanism based on a flexible telescopic lever was developed, which enables the easy wide opening of the door and gives low package dimensions for the opened door. The telescopic arm is fastened in the middle of the door structure using a stiff Y-beam. The Y-beam provides side impact protection in combination with a profiled fibre-reinforced polymer sheeting (Fig. 8.11).

To reduce logistics complexity the diversity of the components used was minimised using modular design, exemplified by the three- and five-door variants, which were designed so that they differ in their structural length only. The difference between the door length of the three- and five-door variant is about 80 mm. The door frame can be partitioned into two frame parts, which are connected by spacers (Fig. 8.12). The front and rear parts of the door frame consist of half shells made of textile-reinforced thermoplastics, manufactured by compression moulding. The half shells are bonded to complete the front- and rear-part of the door frame, which are identical for the five-door and three-door variants. These parts are joined together by three spacers, manufactured using a blow mould process, which are bonded to the front and rear sections. Using the same parts for the door frame of the variants clearly reduces the tooling costs. The mounting flanges at the door frame have a flexible design, which enables the fastening of the same Y-beam and profiled fibre-reinforced polymer sheeting in both the five-door and three-door variants.

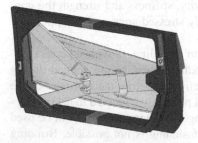

Fig. 8.11 Model of the modular door structure

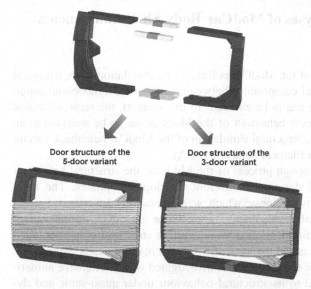

Door structure of the 5-door variant   **Door structure of the 3-door variant**

Fig. 8.12 Modular design of the front door for the five-door and three-door variants

Table 8.4 gives an overview of the resulting individual door parts and their use in the hatchback, notchback, wagon and convertible variants.

The design of the door module was successfully tested for its structural behaviour under loading. The tests included typical quasi-static load cases such as door lowering, torsional loading, over-opening and push/pull loading. Further structural simulations of the door module with regard to its crash performance have been performed in combination with the numerical crash tests of the ModCar body frame.

Table 8.4 Non-variable parts (door frame)

No.	Part	Hatchback (five-door)	Hatchback (three-door)	Notchback	Wagon	Convertible
1	Door frame – front part	X	X	X	X	X
2	Door frame – rear part	X	X	X	X	X
3	Three spacers for the five-door variant	X		X	X	
4	Three spacers for the three-door variant		X			
5	Profiled sheeting	X	X	X	X	X
6	Y-beam	X	X	X	X	X
7	Mount for door-opening mechanism	X	X	X	X	X
8	Door lock	X	X	X	X	X
9	Metallic inserts	X	X	X	X	X

8.5 Structural Analyses of ModCar Body Shell Components and Structure

The technical engineering of the ModCar is linked to a simulation of its structural behaviour. Here, an optimal compromise between modularity, lightweight design and structural performance has to be ensured. In this context, the static structural behaviour as well as the crash behaviour of the ModCar has to be analysed at an early stage of development. Structural simulation of the ModCar hatchback variant was performed using Finite Element Analysis (FEA).

As part of the iterative design process of the ModCar, the structural behaviour was continually optimised during the different development phases. The main focus was placed on crashworthiness, which was extensively investigated using crash simulations during all development stages. Once a ModCar design variant that fulfilled the crash requirements was found, further structural properties, such as the body-bending stiffness and torsional stiffness, were numerically determined. Besides the body frame, the modular door is investigated within extensive numerical simulations with regard to its structural behaviour under quasi-static and dynamic loading in order to find the best compromise among weight, manufacturer requirements (stiffness and deformation limits) and several legal directives. The following illustrations refer to the final development stage of the conceptual ModCar variant.

8.5.1 Deformation Behaviour of the Modular Door

For the developed ModCar door with its new opening mechanism the quasi-static load cases of "door lowering" (a vertical loading case in which somebody climbs on the door when it is open), "door over-opening" and "door torsion" known from conventional doors were adapted, associated simulations were performed and the fulfilment of the technical requirements was proven. In addition, to crash-test behaviour virtual door intrusion tests, known from industrial standards, were successfully performed for the novel modular door. During these tests the door is placed in a reduced space frame structure to take the connection stiffness into account. A pole is placed in the middle of the door. The pole is assumed to be perfectly rigid. The pole is slowly displaced by 300 mm after contact with the outer surface of the door. A range of performance factors (contact force versus the pole displacement, the steady rising of the force displacement curve, no final failure of load-carrying parts, maximisation of the absorbed energy and minimisation of the peak force) are then used to make an assessment of the door's structural performance with regard to its crash behaviour. The effects of the door intrusion test are illustrated in Fig. 8.13 for the novel modular door, which has been proven to fulfil the required criteria.

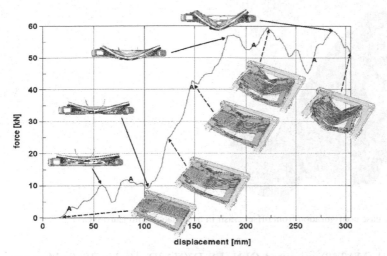

Fig. 8.13 Deformation behaviour of the modular door in the door intrusion test

8.5.2 Crash Behaviour of the ModCar

In an advanced design stage, the virtual model of the ModCar body frame was completed by the door module, the other outer panels and structures and full scale crash simulations were conducted. Since European countries are the target market of the ModCar, the crash investigations were performed according to the definitions of the "European New Car Assessment Programme" (Euro-NCAP). The consumer-oriented Euro-NCAP definitions were chosen instead of the ECE R95 European legal directive due to the more stringent Euro-NCAP requirements. The most relevant crash cases, front offset impact, side impact and pole test were considered. For the front offset impact test the vehicle hits a deformable barrier with a speed of 64 km/h and an offset of 40%. The side impact and pole test are characterised respectively by a deformable barrier of 950 kg, which hits the car at 48 km/h in the side area, and by a rigid pole, which is hit by the vehicle at 30 km/h.

The crash behaviour simulations were performed using LS-DYNA software. Data to model the undercarriage and the motor package were adopted directly from vehicles in the same market segment. For the front offset and the side impact test simulation-approved deformable barrier models created by CADFEM were used. The parts of the vehicle structure that were not modelled by finite elements were considered as additional mass points giving a total structural weight of 1,320 kg, including the crash test dummy. All material models used in the crash simulations consider the strain rate dependency of the material's properties. In order to properly model the energy absorption capabilities of the composite materials used in the outer panels and structures, an elastic-orthotropic material model in combination with Hashin's stress failure criterion as well as with the Matzenmiller

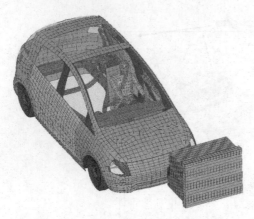

Fig. 8.14 Load case configuration front offset crash in accordance with Euro-NCAP

damage model (MAT58) was used (N.N. LS-DYNA3D; Hashin 1980; Matzenmiller et al. 1995).

The configuration of the front offset crash with the certified crash barrier in accordance with Euro-NCAP is shown for the ModCar in Fig. 8.14.

For the assessment of success or failure in crash behaviour, the loading of the passengers, e.g. in terms of head injury criterion (HIC) values is usually considered. This requires a completed vehicle design structure that includes interior components and safety systems such as seat belts, seat belt fasteners and airbags, which were not available for the ModCar as a concept car. Therefore, the assessment of the ModCar was performed on the basis of the deformation behaviour of the car body in terms of intrusions, reaction forces, courses of energy and decelerations or accelerations respectively. For the front offset crash, qualitative and quantitative assessment criteria included the guaranteed stability of the roof frame, a maximum front wall intrusion of less than 100 mm, deceleration measured at the B-pillar of less than 80 g and a relative deformation between the A-pillar and B-pillar of below 40 mm.

Figure 8.15 shows the general deformation state of the ModCar for the front offset crash in the state of the maximum deformation at about 100 ms as well as characteristic displacement time courses. Instabilities of the roof frame have not been observed and the front wall intrusion of 21 mm and a calculated relative deformation between the A-pillar and B-pillar of 16 mm are far below the required limits. Furthermore, the maximum deceleration at the lower B-pillar shows values around 40 g. In combination with suitable safety systems, low passenger loadings can be expected in a real front offset crash.

The side crash has been simulated using a deformable barrier with a mass of 950 kg that hits the car at 48 km/h in the area of the front door and the B-pillar (Fig. 8.16).

The assessment of the crash behaviour was again based on characteristic deformations of the body frame and door components. Figure 8.17 shows the global deformation of the ModCar and the course of the maximum displacement in the front door at a maximum value of 162 mm. Neither unacceptable intrusions into the survival space nor instabilities of the B-pillar occur. The maximum acceleration measured at the lower B-pillar can be considered safe.

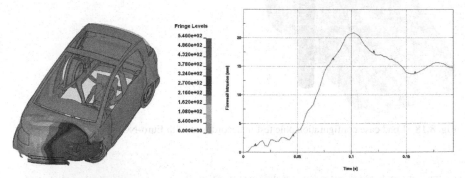

Fig. 8.15 Deformation behaviour of the ModCar due to front offset crash loading (values in mm)

Fig. 8.16 Load case configuration side impact in accordance with Euro-NCAP

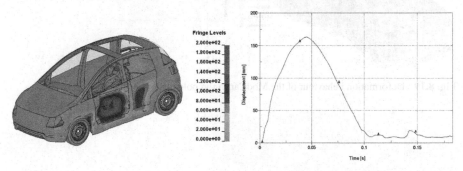

Fig. 8.17 Deformation behaviour of the ModCar due to side impact loading (values in mm)

Fig. 8.18   Load case configuration pole test in accordance with Euro-NCAP

The third investigation, the pole test, depicts the most severe impact scenario, with a rigid pole impacting the body frame near the passenger (Fig. 8.18). In order to withstand this massive localised load, the door structure and the passenger cell must redistribute the load to the surrounding structures and minimise the intrusion into the passenger safety zone.

Figure 8.19 illustrates the global deformation of the ModCar during the pole test. The deformation is characterised by large deformations in the sill and the front door. The maximum intrusion is calculated as 185 mm, such that a distance of 20 mm between the door structure and the dummy remains and no contact is made. The deceleration at the B-pillar is calculated as 31 g and thus lies within the acceptable range if adequate safety systems such as side airbags are used.

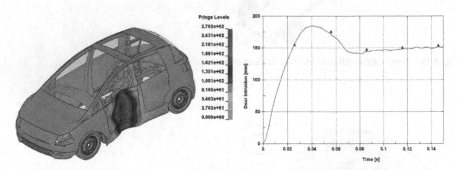

Fig. 8.19   Deformation behaviour of the ModCar in the pole test (values in mm)

8.5.3 Stiffness Behaviour of the ModCar

The torsional and bending stiffness of the final ModCar body frame was numerically determined after successful completion of the crash simulations. The corresponding load cases are illustrated in Fig. 8.20 and Fig. 8.21 together with selected deformation results. To determine bending stiffness forces are applied with a simulated load representative of two passengers in the front seats. The body structure is fixed against vertical displacement at all undercarriage connections. The torsional load case is simulated with the help of a pair of forces applied at the front suspension strut; the rear undercarriage connections are fixed.

The bending and torsional stiffness obtained are 11,800 N/mm and 7,406 Nm/° respectively. In comparison, production cars of the same class have a bending stiffness of about 12,200 N/mm, so the bending stiffness of the ModCar body frame is within the required range.

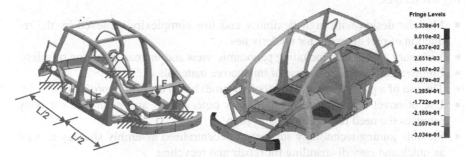

Fig. 8.20 Determination of bending stiffness: boundary conditions and deformation results (mm)

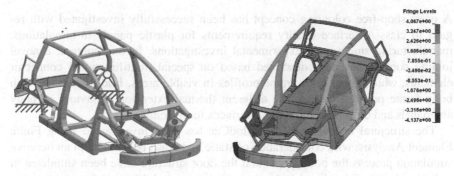

Fig. 8.21 Determination of torsional stiffness: boundary conditions and deformation results (mm)

The ModCar torsional stiffness of 7,406 Nm/° is below the average in modern vehicles that have a torsional stiffness of 10,000 to 14,000 Nm/° and needs further optimization.

8.6 Conclusions

The success of a car built to order depends strongly on both the novel modular technical solutions and appealing styling. The ModCar is demonstrative of this innovative concept, showing the potential of modularity, which enables production flexibility combined with low complexity and a stockless process, in a package that should appeal to end customers.

In a multi-stage design process the body frame, an automotive body shell, outer panels and selective outer structures have been elaborated, which offer the following advantages:

- Modular design with high flexibility and low complexity achieved by the reduction of part diversity for all derivates
- Visionary appearance including panoramic view and increased feeling of safety
- Lightweight design by the use of innovative materials
- The use of a paint-free car body (body frame) based on simple profiles
- Use of novel colouring technologies for outer panels and paint-free car body removes the need for a central paint shop
- Flexible joining technology supports a decentralised assembly strategy as well as quick and easy dismantling for repair and recycling
- Increased passenger comfort achieved using innovative space-saving door-opening mechanism.

A paint shop-free colouring concept has been successfully investigated with regard to class-A surface quality requirements for plastic panels in simulations, manufacturing studies and experimental investigations. In this context, a novel joining concept has been described based on special multifunctional connector elements, which cover body frame profiles in visible areas, bridge gaps between body frame profiles, balance the different thermal extension behaviour of body and panels and offer smooth bonding surfaces for the outer panels.

The structural behaviour of the ModCar has been investigated using Finite Element Analysis, with consideration of static and crash behaviour. In an iterative simulation process the body frame and the door structure have been simulated in order to identify the optimal compromise between modularity, lightweight design and structural performance. The ModCar fulfils the criteria regarding passenger's protection with respect to the essential load cases in accordance with Euro-NCAP standards. Moreover, the required bending stiffness as a essential assessment criterion of the ModCar body frame has been verified, whereas the torsional stiffness needs to be increased. An optimised door module consisting of novel materials such as textile-reinforced thermoplastics shows very good overall performance with regard to the static and crash behaviour as well as the lightweight design. As such, the ModCar demonstrates that the novel concepts employed may be used to successfully produce a vehicle that is saleable in the European market.

References

Feldmann J (2007) Konzentration auf Produkte mit Optimierungspotenzial. In: Konstruktion. Springer VDI-Verlag, Berlin

Hashin Z (1980) Failure criteria for unidirectional fibre composites. J Appl Mech 47:329–334

Herd O (2006) Langglasfaser verstärkte Kunststoffe und ihre Einsatzgebiete im Automobil. Kunststoffkolloquium FH, Deggendorf

Hufenbach W et al (2006) Textile-reinforced composite components for function-integrating multi-material design in complex lightweight applications. Colloquium of DFG-Collaborative Research Centre SFB 639, Dresden

Kompetenznetz RIKO (2003) Realisierung innovativer Konstruktionswerkstoffe aus nachwachsenden Rohstoffen; Vortrag auf der AGRITECHNICA 2003 von W. Frehsdorf, Krauss-Maffei, Munich: "Verarbeitung von n.R. im LFI-Verfahren". Available at http://www.riko.net/download/kwst2003_frehsdorf.pdf

Matzenmiller A, Lubliner J, Taylor RL (1995) A constitutive model for anisotropic damage in fiber-composites. Mech Mater 20:125–152

N.N. (2007) LS-DYNA3D users manual, V971. Livermore Software Technology Corporation Available at http://www.twintex.com. Accessed 15.02.2007

Strümpel F, Palik M, Söchtig W (2006) From reaction processes to injection moulding material and process engineering that's right for your application. 9th International AVK Conference, Essen, September 2006

Chapter 9
Complexity Cost Management

J. Schaffer and H. Schleich

Leuphana University, Department of Automation and Production Technology, Lueneburg

Abstract. Complexity cost management refers to costs in industrial production processes that are directly or indirectly related to the handling, management or creation of different variants of a product. The current understanding and state of awareness of the influence of complexity on cost structures is reflected in the results of a survey carried out by the authors. The methodology used to analyse the complexity costs of existing processes is described and a case study from the automotive industry, which is highly affected by variant-driven complexity costs, is introduced. Furthermore, a model to calculate potential complexity costs, when changing product variety based on the number of variants, variant drivers and the characteristics of variant drivers, is shown. Finally, an easy-to-handle tool, which allows more detailed determination of complexity costs based on changing product variety parameters, is presented.

9.1 Introduction

The level of complexity that characterises a manufacturing system is often determined by the overall complexity of the manufactured product itself. However, the amount of complexity that results from creating variants of a particular product sometimes has a far greater effect on the structures and processes used in production. This aspect of complexity is referred to as "variant-driven complexity", the effects of which impact heavily on total production costs.

In well-developed economies, cars are not only expected to meet customers' personal needs, but also need to make a social or lifestyle statement. The number of individual choices and amount of diversity have therefore become important issues in customer decisions, and respectively, in automotive marketing during the last few decades.

Previously, vehicle manufacturers offered the market unique models that had a small variety of attributes and long life cycles. Today, vehicle manufacturers are required to provide a high product variety to remain competitive as they are facing increasingly sophisticated customers and fast-paced technological developments in their industry (Chakravarty and Balakrishnan 2001).

As a result of this wider variety, organisations can maintain or increase their market share, increase their prices, and serve consumer needs better by closely matching customer preferences and offered products (Benjaafar et al. 2004; Klapper 2005). But with increased diversity of products offered and with option lists lengthening, the number of variants to be manufactured rises exponentially. In addition to the number of extra variants caused by the need to deliver customer choice (Build Combinations), there are the variants created by technical needs the customer does not determine (e.g. different injection and exhaust systems to fulfil environmental standards in different countries or set-ups for different climate conditions) and boundary conditions (e.g. creating different badges or brands for cars that are technically the same for marketing and distribution reasons). This results in organisations having variant-driven manufacturing systems that bear the risks of higher costs, lower efficiency and limited flexibility. The compromise adopted by all major OEMs up to the beginning of the 21st century is called mass customisation. This process still allows production of large quantities of most components, but also offers the chance of individualisation of the product at several points in the supply chain.

Some major OEMs in the automotive sector, which explicitly herald individualism as an outstanding characteristic of their products, do not have sufficient systems or tools to control the costs of creating and handling the amount of variant-driven complexity that it generates (Schleich et al. 2005). In this situation, it is difficult to accurately assess the cost–benefit ratio of introducing the additional variants needed to develop platform or option-bundling strategies. Economically important operational and strategic decisions regarding investments, running costs, market share and revenues are routinely estimated with no regard to the true related production costs.

In order to provide a succinct overview of the theory behind manufacturing complexity as a result of product variety and the main implications for the automotive industry supply chain, the next section gives a definition and some clear examples of costs caused by variant-driven complexity. We will then show an efficient method of analysing the resulting complexity costs in existing industrial production processes. This overview is based on a survey showing the current situation regarding the awareness and availability of information relevant to complexity costs in the manufacturing industry. Methodology to derive complexity cost data is also provided and the correlation of the different types of complexity cost fractions is explained. We also introduce a linear model we have successfully used to estimate changes in the costs of variant-driven complexity when changing the overall number of variants produced, the number of variant drivers, or the number of characteristics of those variant drivers. Finally, we discuss a related concept for a more detailed complexity cost model that considers non-linear influences.

9.2 Product Variety

9.2.1 Product Variety in Lean Manufacturing Environments

The rapid increase in product variety is a key trend in many industries worldwide. The Japanese approach to manufacturing, pioneered by Toyota and frequently called lean manufacturing, has become increasingly essential in achieving the required efficiencies. In the changing business environment there is an increasing requirement for a lean, flexible and highly responsive, yet stockless, supply chain. Lean manufacturing therefore is an important issue change management approach that has to be implemented. Without serious attention to the cost impact of variety and complexity, costs can only be expected to rise further, and moreover, uncontrollably. This means there are two important trends and issues that greatly affect each other. The closer an industry moves towards lean and ideal manufacturing and production processes, the more important awareness of the cost of the increased variety becomes. While this increased variety, offered over time, may provide a competitive edge to companies on the market (Kahn 1998), the central strategic question with regard to product variety concerns the "optimal" or "appropriate" level of variety. On the one hand, offering variety increases costs; on the other hand, it can provide product differentiation in the market, thus leading to a higher market share and sales volume (Lancaster 1990).

As product variety increases, firms can experience a reduction in supply chain performance due to the effects of dis-economies of scale, with potential negative impacts on component prices, lead times and component inventory levels. There is also a risk that direct manufacturing costs, overheads, lead times and inventory levels in the company's internal operations might increase if batch sizes remain unchanged. This may, in turn, lead to longer supply lead times and, consequently, to higher inventory and back order levels. Finally, the number of parts in stock increases, which induces higher inventory costs. This is something that lean manufacturing environments aim to avoid (Thonemann and Bradley 2002; Fisher and Ittner 1999; Salvador et al. 2002). The authors believe this ultimately leads to higher variant-driven complexity costs. A simple management tool could aid many companies throughout the supply network to drive costs out of, rather than down, the supply chain. This will act as a strong catalyst for truly lean manufacturing.

9.2.2 Product Variety in the Automotive Industry

A key trend in the automotive industry over the last decade has been the expansion of traditional vehicle model ranges with the so-called "cross-over" and niche vehicles (Holweg and Greenwood 2001).

The traditional segments of small cars, sub-compacts, compact cars, family cars and executive class have been joined by SUVs (sport-utility vehicles), MPVs (multi-purpose vehicles), minivans and others. Thus, the number of models offered in Europe is increasing rapidly; in 1990 a total of 187 models were offered, which increased to a total of 315 models in 2003 (Midler 2005). Along with the number of models produced the number of body types has also increased. Those produced by the top eight vehicle manufacturers in Europe doubled between 1990 and 2002, rising from 88 to 179 (Pil and Holweg 2004), confirming the trend towards diversification and segmentation in the automotive market.

In their study of product variety, Pil and Holweg (2004) also analysed the product variants of the best-selling 20 vehicles in the European market in 2002. The results showed that there is no consistent picture of the overall level of variants offered in Europe. Their research identified no discernable pattern in the overall variety offered by model, either in relation to sales (i.e. low sales equal low variety), or in relation to model segment (i.e. higher level cars offer more variety). The analysis also identified a group formed by BMW and Mercedes whose total possible combination of different vehicle options and variations surpassed the order of 10^{16}, reaching the order of 10^{24} for Mercedes' E Class model.

9.2.3 Variant-Driven Complexity

The following two figures give a simple example of the complexity arising from the introduction of one new variant to the standard variant of a car. The variant driver shown as an example is the sunroof. The driver has two characteristics, which means that the car can be produced in two versions "with sunroof" or "without sunroof".

In the no-variant scenario, the sunroof is a standard feature on each car. There is an operator who fits the sunroof to every car on the assembly line. The customers pay for the value of the work done by this employee. Thus, the employee is adding a specific amount of value to the car, which is finally paid for by the end-customer (see Fig. 9.1).

In the second simple scenario, the sunroof is an option. The customer has the choice to order the car with or without, a sunroof, resulting in two variants. Assuming, that the operator in scenario one is able to fit a sunroof to each car on the line, this employee will be put in a situation where he or she is out of work when a car without a sunroof is on the line. Usually, certain labour-intensive options are subject to an overall line balance in order to level the volatility of option-specific

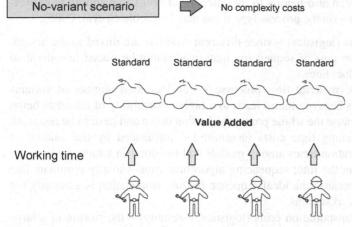

Fig. 9.1 There is no complexity cost in a no-variant scenario

work content, which results in line balance losses of up to 10% in automotive production. If the considered option is not included in the list of line balance criteria, waiting times will occur as soon as the build sequence does not comprise a perfect option mix. This waiting time is not adding value to the car and translates into a complexity-related cost, as the employee usually still gets compensation for unavoidable idle time (see Fig. 9.2).

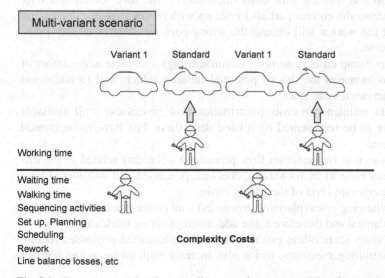

Fig. 9.2 The existence of variants implies a set of non-value-adding times and costs

Other activities can also belong to the category of non-value-adding (complexity) costs. Depending on the process type these may be (Schleich et al. 2005).

- Sequencing costs (logistics) – since different products are mixed in the assembly sequence, the cost of sequencing materials cannot be traced to individual products or product lines.
- Downtime costs (material flow processes) – an increased number of variants (and the changes between them) leads to a higher probability of mistakes being made that may cause the whole process to be shut down and have to be restarted.
- Line balance/waiting time costs (assembly) – influenced by the amount of variation in operation times among models and the order in which model types are sequenced on the line; sequencing algorithms cannot totally eliminate line balance losses because the ideal sequence for one workstation is generally not ideal for other workstations.
- Walking time/transportation costs (logistics, assembly) – the storing of a large number of different variants will lead to an increased requirement in transportation, since they cannot all be picked up at one particular point.
- Set-up costs (manufacturing) – when certain lot sizes are manufactured, the change from the specification of one lot to another will cause set-up costs (including down time)
- Storing costs (logistics) – product variety has a significant impact on inventory level; more parts and lower volume per part also increase the coefficient of variation in demand for a particular part, requiring greater safety stocks.
- Costs of stock-outs (material flow processes) – if the above-mentioned safety stocks do not last as calculated, an increased risk of stock-outs occurs; the costs of potential stock-outs include additional labour to expedite parts, as well as the costs of line stoppages, rework, and quality problems due to actual stock-outs.
- Part selection and walking time costs (assembly) – the time for an assembly worker to access the correct part also goes up with product variety, increasing the risk that the worker will choose the wrong part, resulting in quality problems and rework.
- Rework costs, scrap costs (assembly, manufacturing) – an increased number of variants also increases the risk of potential failures, which lead to additional rework or increased scrap rates.
- (Master) data maintenance costs (information flow processes) – all available variants have to be represented by related data; these data have to be created and maintained.
- Data handling costs (information flow processes) – the data related to the different variants have to be forwarded, checked, processed and considered with regard to the relevant level of the supply chain.
- Production planning costs (planning processes) – all above-mentioned processes have to be planned and therefore cause additional planning work and costs.
- Controlling costs (controlling processes) – all implemented processes require adequate controlling measures, which also increase with an increased number of variants.

Complexity costs, which are not necessarily related to the number of variants, but to their existence, can be found in the field of investments. When only one version of a product is being built, the operator knows which process steps to take; the equipment can be run with one single, non-changing setting. As soon as the first variant that cannot be distinguished by the operator by its appearance is introduced onto the same production line, additional equipment (e.g. RFID readers or barcode scanners) have to be procured to inform operators and/or equipment which different process steps have to be carried out for the different variants. For example, the torque required to tighten cylinder head bolts may change when the cylinder head is made of a different material due to the desired power level or purpose of an engine. This material difference may not be visually apparent.

When the number of variants continues to increase, fixed step costs occur. These include, for example, certain storage and picking systems that support the manufacturer in the creation of the variants. Increasing numbers of parts mean increasing storage space and picking bins, which cannot be procured in a single quantity, but necessitate the acquisition of storage racks that might hold 6, 10, 100 or 1,000 new parts.

Most of the complexity-related activities suggested above are relevant to several processes within the supply chain and can be analysed at the highest process level for the complete car or broken down to the lowest level of the supply chain and to suppliers on any level from tier one down to tier n.

Product variant-driven complexity, from this point of view, is dependent on the number and combinations of variants. Thus, if there are no variants at all (just one standard car type with no choice for the customer), there is no variant-driven complexity.

In the academic literature, complexity is often equated with variety. This view does not capture the reality that there are extremely complex processes in which no product variants at all are created. This is a part of non-variant-related complexity, which is not referred to in this chapter. However, a major part of complexity is caused by variety, which is clearly related to the number of variants created in a process. As there is a stringent correlation between these parameters, variant-driven complexity may be represented by the number of product variants and/or build combinations (Schleich and Schaffer 2006).

9.3 The Complexity Cost Model

9.3.1 Basic Survey

At the beginning of the development of the complexity cost model a survey of OEMs and suppliers in the automotive industry was conducted. The aim was to specify requirements for, and to evaluate the possible impact of, a tool that enables companies to evaluate the variant-driven aspects of cost, whilst efficiently avoiding detailed calculation of the complete variety range for that product. The survey comprised several aspects related to variant-driven complexity. Amongst others, these were:

- Trends in product variety within manufacturing companies
- Evaluation of complexity as a cost driver
- Correlation between variants and complexity costs
- Effects of variant-driven complexity on lead times
- Trend in product variety within manufacturing companies

When considering product variety, not all of the companies asked were aware of the actual variety they offer and produce with their products. These companies could only give a rough estimate of the number of variants they produce. Some companies' product variety is much bigger than the product range produced and sold annually. Driven by their customer demand, they only sold 50–80% of the total product diversity they offered. This was especially true of the premium car segment. The choice for the customer in terms of possible build combinations is significantly higher than the number of cars produced. Among the suppliers questioned, 85% of the companies produce and sell all of their offered product variants during 1 year.

Having followed the trend towards individualisation and mass customisation in the past, most companies stated that offering product variety has led to significant increases in the numbers of variants. Extrapolating from the historical trend, about half of the interviewees anticipate further significant growth in variety (Fig. 9.3). Thirty per cent believe that variety will remain at today's level; others even expect decreasing figures.

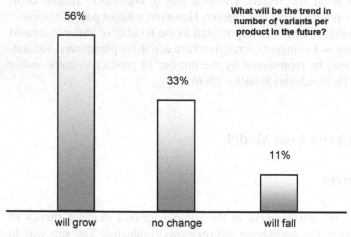

Fig. 9.3 Most managers expect the product variety to grow further in the future

9.3.1.1 Evaluation of Complexity as a Cost Driver

The additional complexity resulting from creating and handling variety was identified as an important cost driver in production by 64% of the respondents (Fig. 9.4) (Schleich et al. 2005).

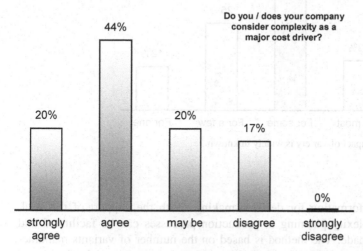

Fig. 9.4 Managers consider complexity a major cost driver

9.3.1.2 Correlation Between Variants and Complexity Costs

Our research shows that in most companies detailed analysis is required to determine this type of cost. However, most managers and engineers do not have the resources or expertise to carry out this kind of financial analysis (Martin and Ishii 1997). As a consequence, few managers have sufficient cost–benefit information about their product variants (Fig. 9.5).

A common practice is to allocate costs that are related to variant-driven complexity (e.g. storage costs for an increased number of parts, or administrative costs to handle and maintain an increased number of part numbers) as overhead costs to all manufactured products regardless of the impact of their variant-driven complexity. This usually results in the calculation of production costs that may not be fully related to reality.

These practices can lead to a situation in which it is difficult to measure the cost–benefit ratio of introducing additional variants, or indeed discontinuing existing ones. Financially important decisions in terms of investments, running costs, market share and revenues are based on rough estimates regarding the related production costs, with no real application to what is really happening in the business. The model presented here has been tested with observable real life data and includes complexity costs; it can also be easily applied and provides a comprehensive

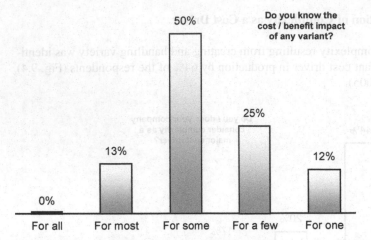

Fig. 9.5 The cost impact of variety is widely unknown

set of relevant information for decision-making. With the support of this tool, variety and complexity planning in production processes can be facilitated and efficiency gains made. The method is based on the number of variants manufactured and the characteristics of variant drivers. In subsequent models, further influences from the percentage of option penetration or take rate (describing the percentage of vehicles or other products that are equipped with a particular option) and the point of product differentiation will be included.

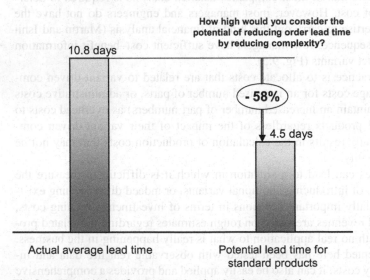

Fig. 9.6 The majority of order lead time is related to complexity

9.3.1.3 Effects of Variant-Driven Complexity on Lead Times

Another important aspect to focus upon on the journey towards lean and stockless production is the order lead time, which is also a cost issue due to its relation to storage costs, tied-up capital and lost market share when customers change to a faster and more flexible manufacturer.

Nearly 90% of the interviewees agreed that in their particular situation the existence of product variance has a negative impact on order lead time. The amount of complexity-related order lead time is seen as being quite significant. The current average lead time is nearly 11 days (suppliers only), and experts estimate that about 60% of this is due to product complexity. This leaves an average lead time of 4.5 days if complexity is reduced (Fig. 9.6).

9.3.2 Analysis of the Current State

9.3.2.1 Aims and Methods of the New Complexity Cost Model

In contrast to many other existing models that calculate complexity costs in a detailed manner, but which involve a high workload, the goal of this innovative model was to determine complexity costs in a manner based on parameters that can be determined directly from a running production process. The parameters that have an effect on variant-driven complexity are the number of variants produced, the variant drivers (which lead to the particular number of variants) and the number of characteristics each single driver has (Fig. 9.8). Since the same variant driver can have different cost effects depending on the production process, we must be able to evaluate the impact of the same or similar drivers on different production processes.

It is important that the design of the tool should be able to determine changes in complexity costs without the need for calculations for the complete product range, which takes a considerable amount of time and therefore causes significant costs.

Frequently, the decision to introduce or delete a variant has already been made and a new calculation is carried out to confirm the expected results.

The new model should enable OEMs and suppliers to easily check complexity costs for different variant scenarios and make a decision based on the results the model delivers, not vice versa. As there is an integrated correlation between market offer and variety-driven complexity in the companies (Fig. 9.7), the trade-offs between additional sales and option profits and additional (complexity) costs can be figured out and taken as a basis for management decisions.

To achieve the desired goals it was first of all necessary to establish a clear understanding of how the number of variants that leads to the complexity cost is created. Which parameters influence the number of variants that can be created? A combustion engine, for example, can be available in different variants, when it is manufactured with different displacements (e.g. 2.0 l and 2.3 l). In that case the

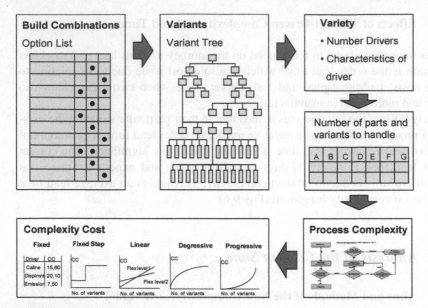

Fig. 9.7 How customer choice affects complexity costs

displacement would be considered a variant driver, which has two different characteristics. The type of available transmission that has to be joined to the engine could be another variant driver. It may also have two different characteristics (manual transmission and automatic transmission). The combination of these different drivers now leads to the number of manufactured variants. It is important to consider that this number of variants does not only depend on the number of characteristics, but also on the way they are combined. Two displacements with two types of transmissions may lead to a maximum of four variants over all, but could also result in only three variants, when, for example, the small displacement engine is only available with a manual transmission. All three important parameters can be found in Fig. 9.8 and are explained and defined below.

After gaining a general understanding of the parameters that led to the creation of variants, it was important to clearly define the relationships among them as a foundation of the model. The definitions for the main parameters on which the model is based (variants, drivers and driver characteristics) are given in the following section.

Variants

In the given context, a product variant is defined as a version of a technical product that differs from another version of a technical product of the same type in at least one area of its technical specification and is created by the combination of different driver characteristics during the production process.

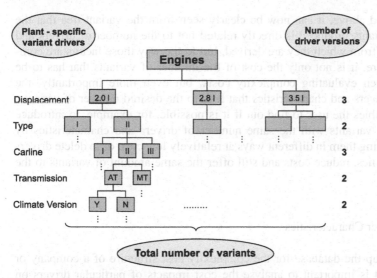

Fig. 9.8 The variant tree consists of drivers and characteristics

The number of variants produced that are considered for the model can be derived from the variant tree of the process analysed (Fig. 9.8).

Variant Driver

A variant driver is a parameter describing a technical specification of a product and leading to the creation of variants as soon as the variant driver has more than one characteristic. The technical properties of an automotive engine, for example, can be described by drivers such as displacement, number of cylinders, fuel type, emission level, car type in which the engine is fitted, etc.

Driver Characteristics

The characteristics that describe a driver are defined as the different technical alternatives that are available for that particular driver. The driver "displacement" for example, can be available in the alternatives 2 l, 2.8 l and 3.5 l and would in that particular case have three characteristics.

Each of these three parameters is represented in the variant tree (Fig. 9.8). The total number of variants manufactured is represented by the number of items on the lowest level of the variant tree.

The number of different drivers can be determined from the number of levels the variant tree shows. On each level (e.g. transmission) the number of characteristics of each driver can be determined.

As described above, it can now be clearly seen from the variant tree that the number of variants produced is directly related not to the number of drivers and characteristics from which they are derived, but to the way those factors are combined. Therefore, it is not only the cost of the number of variants that has to be considered when evaluating complexity costs, but even more importantly, the structure of drivers and characteristics that lead to the desired number of variants. Our model enables the user to find out if it is possible, for example, to introduce and offer new variants with the same number of drivers and characteristics by simply combining them in different ways at relatively low cost, or to delete drivers and characteristics, reduce costs and still offer the same amount of variants to the market.

Impact of Driver Characteristics

When setting up the database for the complexity cost structure of a company or supply chain it is important to analyse the cost impacts of particular drivers on different processes. This allows the effects of variant-driven complexity costs on particular processes, fractions and levels to be calculated for different scenarios in later evaluations. In the process of manufacturing the cylinder head, for example, the driver characteristic "drive train" (rear wheel drive, four-wheel drive) has absolutely no impact. When assembling the wire harness to the engine there might be an impact, because the four-wheel drive needs additional electronic features. The mounting of the oil pan is clearly affected by the "drive train", as the front wheel drive shafts necessitate changes in that area.

The methodology required to come from unknown variant-driven workload to predictable complexity costs requires a set of business cases that need to be divided into different phases. It seems reasonable to define three main phases, which are subdivided into different steps. The main phases are:

- Phase 1: preparation of analysis
- Phase 2: analysis of complexity costs
- Phase 3: composition of complexity cost model

Phase 1: Preparation of Analysis

To prepare the analysis in the first phase, it is necessary to identify variant drivers (i.e. technical features and options causing variants in components and/or products) and their characteristics from the plant-specific variant tree (phase 1a), as described above. Additionally, the organisation has to be structured into reasonable segments, which later represent the basis for semi-structured interviews (phase 1b). Finally, an overview of the head count and costs in these segments is needed (phase 1c), which can be derived from human resource and accounting data.

Phase 2: Analysis of Complexity Cost

In the second phase, the complexity workload and cost in each defined segment relating to the identified drivers, their versions and their impact on the operations within the particular segment have to be evaluated by interviewing the responsible head of the department, shift leader or other experts (phase 2a). Afterwards, the drivers can be ranked by their impact on costs (phase 2b). The results can then be transferred to other parts of the cost chain by extrapolating the data to similar processes or segments (phase 2c). With a sufficient amount of data, it is then finally possible to assign complexity costs to variants, drivers and versions of drivers in any desired part of the value creation chain.

The current-state appraisal includes the analysis of labour costs, investments, and direct and indirect material requirements. For these cost segments, the share of the complexity workload and costs are determined. The input comes from the analysis of work instructions, a head count and ABC analysis of controlling data substantiated by experts' assessment, and an evaluation of component manufacturing and assembly.

Phase 3: Composition of the Complexity Cost Model

The established data set may then be "plugged-in" to the complexity cost model, which then enables the setting of parameters and adaptation of the model to match the specific plant and processes. Based on the existing information, it is possible to separate complexity costs related to the number of variants and those related to driver characteristics (phase 3a). In a final step (phase 3b) the information is put into a matrix that also contains the mathematical correlations between cost and variant-driven complexity. By changing the inputs for variant and variant drivers the matrix displays the changing complexity costs in labour, investments and running materials as a set of output results. It is also possible to view the change in cost per product (e.g. complexity costs per engine) as an output.

9.3.2.2 Results

After carrying out the study and analysing the figures, a comprehensive set of data has been made available showing the levels of complexity and the amount of complexity costs related to:

- The departments included in the study
- The product variety
- The variant drivers and their characteristics
- The quantitative ranking of the drivers' impact

The most informative single figure to be given is usually the percentage of variant-driven complexity costs for the entire plant. Based on the structure of the analysis,

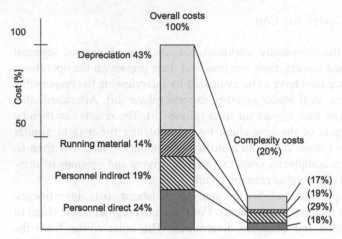

Fig. 9.9 Complexity costs may be one-fifth of the total costs of a plant

it is possible to aggregate the figures to specific cost types, such as personnel costs (direct/indirect), investments and infrastructure (depreciation) and material costs. Some illustrative examples of results from the analysis of a first-tier engine supplier are given in Fig. 9.9.

The variant-driven complexity costs for our sample engine supplier are about 20% of the overall production costs for the total plant (Fig. 9.9). Based on our experience in previous studies, complexity costs in the manufacturing industry, depending on products, processes and organisational structure, typically fall within a range between 15 and 30% of the total production costs. Figure 9.9 also shows where the majority of the complexity costs within the plant are created. While complexity costs for depreciation, running material and direct personnel are slightly below the average percentage of total costs, those for indirect personnel lie close to 30%. In response to this finding, the manufacturer concerned must monitor all variant drivers affecting processes with intense involvement of indirect labour, and can then evaluate the related costs of the resulting variants, and finally of the drivers used to create variants.

In a subsequent process one is then able to analyse which variants would be affected should the most expensive drivers be eliminated. When the potential or actual turnover of these variants is available, the cost and revenue figures can be compared and be the basis for management decisions regarding the profitability of particular variants of the manufactured product. Another opportunity resulting from this information is the potential to test scenarios for the reorganisation of departments where high complexity costs have been identified. It may be possible for the manufacturer to reorganise in such a way as to offer the same variants to the market with decreased production costs.

Another important result of the analysis is the calculation of the complexity costs for the whole cost chain, including all complexity drivers. This allows drivers to be ranked for the whole plant, or for individual production segments (e.g. with high labour costs, as described above) or processes (e.g. assembly) (Fig. 9.10).

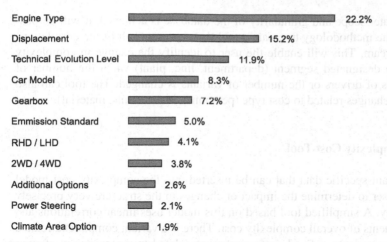

Engine Type	22.2%
Displacement	15.2%
Technical Evolution Level	11.9%
Car Model	8.3%
Gearbox	7.2%
Emmission Standard	5.0%
RHD / LHD	4.1%
2WD / 4WD	3.8%
Additional Options	2.6%
Power Steering	2.1%
Climate Area Option	1.9%

Fig. 9.10 Complexity costs are induced by different drivers

With this information, it is possible to analyse whether the elimination of drivers or reduction of driver characteristics would result in an increased profit. With the support of the tool, it is also possible to check whether (when the number of driver characteristics is reduced), it is more efficient to create the same number of variants with a reduced number of characteristics by combining the remaining drivers in a different way.

Based on the figures, it is possible to predict the effect of changes in the variant and driver structure of the product on the complexity costs. One example from this case has been the decision to house the assembly of the hydraulic pump for the power steering either at the supplier's engine plant or at the automotive OEM's assembly line. From the output of the complexity cost model, it is possible to get clear information regarding how many drivers, versions and variants would be affected and how the complexity costs would change. The assessment can be done for both cases, i.e. increasing and decreasing product variety. These figures can then be used as a basis for taking the decision on the assembly location point. The later a variant is created in a production process, the fewer variants have to be handled upstream from the point of product differentiation, and the less variant-driven complexity costs affect the upstream part of that process. Therefore, in many cases, manufacturers build "late configuration" facilities where a majority of variants can be created shortly before delivering the product. In our pump example we are tripling the number of engine variants, going from 300 to 900, by equipping them with one of three different possible hydraulic power steering pumps.

It can be seen that the wide range of information available from the analysis, which builds the backbone of the complexity matrix, allows the display of even more correlations (variants per production line, driver ranking by departments, driver ranking by process etc.) than shown above. This can be the basis for many different operational and strategic decisions to increase the efficiency of a manufacturing organisation.

When greater scope and granularity of the database is achieved, it will be possible to use the methodology to create a tool that represents all of the correlations in a value stream. This will enable the user to identify the change in complexity costs in each designated segment (department, line, plant) when the number of characteristics of drivers or the number of variants is changed. The tool can also provide cost changes related to cost type (personnel, investments, material).

9.3.2.3 Complexity Cost Tool

The set of plant-specific data that can be inserted into the complexity cost model enables the user to determine the impact of changes in the structure very precisely and efficiently. A simplified tool based on this model uses linear correlations and three components of overall complexity cost. There is a variant component, related to the number variants, a fixed cost component, related to the existence of variants, and a driver component, related to the number of driver characteristics and their impact.

The variant-related component changes linearly according to the number of variants built in the particular production process. The fixed cost component is considered to be a certain percentage of the overall complexity costs, which occurs as soon as the first variant is created (compare Sect. 9.2.3). The driver-related component depends linearly on two factors. It can be controlled individually by the number of characteristics each particular driver has and by the impact of that driver.

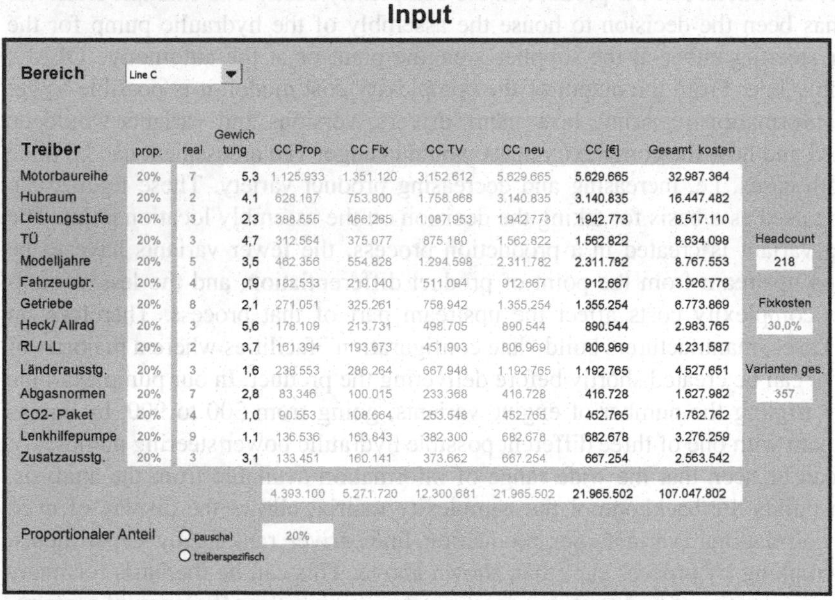

Input

Bereich	Line C ▼									
Treiber	prop.	real	Gewich tung	CC Prop	CC Fix	CC TV	CC neu	CC [€]	Gesamt kosten	
Motorbaureihe	20%	7	5,3	1.125.933	1.351.120	3.152.612	5.629.665	5.629.665	32.987.364	
Hubraum	20%	2	4,1	628.167	753.800	1.758.868	3.140.835	3.140.835	16.447.482	
Leistungsstufe	20%	3	2,3	388.555	466.265	1.087.953	1.942.773	1.942.773	8.517.110	
TÜ	20%	3	4,7	312.564	375.077	875.180	1.562.822	1.562.822	8.634.098	Headcount
Modelljahre	20%	5	3,5	462.356	554.828	1.294.598	2.311.782	2.311.782	8.767.553	218
Fahrzeugbr.	20%	4	0,9	182.533	219.040	511.094	912.667	912.667	3.926.778	
Getriebe	20%	8	2,1	271.051	325.261	758.942	1.355.254	1.355.254	6.773.869	Fixkosten
Heck/ Allrad	20%	3	5,6	178.109	213.731	498.705	890.544	890.544	2.983.765	30,0%
RL/ LL	20%	2	3,7	161.394	193.673	451.903	806.969	806.969	4.231.587	
Länderausstg.	20%	3	1,6	238.553	286.264	667.948	1.192.765	1.192.765	4.527.651	Varianten ges.
Abgasnormen	20%	7	2,8	83.346	100.015	233.368	416.728	416.728	1.627.982	357
CO2- Paket	20%	4	1,0	90.553	108.664	253.548	452.765	452.765	1.782.762	
Lenkhilfepumpe	20%	9	1,1	136.536	163.843	382.300	682.678	682.678	3.276.259	
Zusatzausstg.	20%	3	3,1	133.451	160.141	373.662	667.254	667.254	2.563.542	
				4.393.100	5.271.720	12.300.681	21.965.502	21.965.502	107.047.802	

Proportionaler Anteil   ○ pauschal   20%
                         ○ treiberspezifisch

Fig. 9.11 Straightforward user interface of the complexity cost tool – input module (German version)

Output

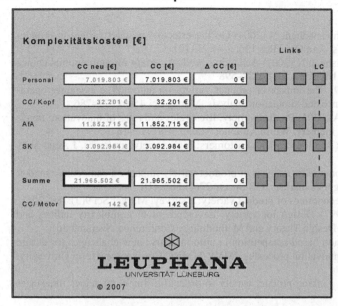

Fig. 9.12 Output module of the complexity cost tool (German version)

It is now possible to evaluate and change the influence of each single variant driver and the overall number of variants manufactured on an easy-to-handle user interface of the tool (Figs. 9.11, 9.12).

9.4 Conclusion

The approach to complexity cost modelling described in this chapter offers a reliable facsimile of a value chain and provides decision support to senior managers. The methodology presented is very effective when applied to processes in the automotive value creation chain. Commercial, technical and organisational decisions can be based on the resulting figures in different processes and environments. Control recalculations with conventional calculation software show that the tool and methodology lead to sound results quickly with little effort. A roll-out to additional processes and industries, including non-automotive applications, is ongoing. The underlying database is being adapted to include additional environmental, economic and infrastructural conditions.

References

Benjaafar S, Kim J, Vishwanadham N (2004) On the effect of product variety in production-inventory systems. Ann Oper Res 126(1–4):71–101

Chakravarty AK, Balakrishnan N (2001) Achieving product variety through optimal choice of module variations. IIE Trans 33(7):587–598

Fisher M, Ittner C (1999) The impact of product variety on automobile assembly operations: empirical evidence and simulation analysis. Manag Sci 45(6):771–786

Holweg M, Greenwood A (2001) Product variety, life cycles and rates of innovation: trends in the UK automotive industry. World Automot Manuf April:12–16

Kahn B (1998) Dynamic relationship with customers: high-variety strategies. J Acad Market Sci 26(1):45–53

Klapper D (2005) An econometric analysis of product variety impact on competitive market conduct in consumer goods markets. OR Spectr 27(4):583–601

Lancaster K (1990) The economics of product variety: a survey. Market Sci 9(3):189–206

Martin MV, Ishii K (1997) Design for variety: development of complexity indices and design charts. ASME Design Theory and Methodology Conference, Sacramento

Midler C (2005) Innovation based competition in auto industry: new challenges for design organization and co-innovation processes. IMVP-MMRC Conference, Hosei University, Tokyo

Pil F, Holweg M (2004) Linking product variety to order-fulfillment strategies. Interfaces 34(5):394–403

Salvador F, Forza C, Rungtusanatham M (2002) Modularity, product variety, production volume, and component sourcing: theorizing beyond generic prescriptions. J Oper Manag 20(5):549–575

Schleich H, Schaffer J (2006) The cost impact of product variety for a 5-day car: the complexity cost model and general rules for the design of BTO networks. ILIPT Project Report, Luneburg

Schleich H et al (2005) State of the art of complexity management. ILIPT Project Report, Luneburg

Thonemann UW, Bradley JR (2002) The effect of product variety on supply-chain performance. Eur J Oper Res 143(3):548–569

Part III
Collaboration

Part III
Collaboration

Chapter 10
Key Principles of Flexible Production and Logistics Networks

Bernd Hellingrath

Fraunhofer Institut Materialfluss und Logistik, Dortmund

Abstract. A key aspect to achieving the target of a 5-day car is the improvement of the flexibility of all physical, informational, planning and control processes in the supply network. This section of the book presents innovative flexible processes for inter-enterprise collaborative planning and execution. These processes support the industry's need for effective and efficient collaboration of planning capacities throughout the supply network. These have been developed to overcome process-related limitations that are prevalent within the automotive industry today. Collaborative planning and execution processes span the whole integrated supply network and are capable of supporting dynamic capacity allocation based on plant-specific constraints. They enhance the capability to adjust capacity to fall within a plant's profitable range. They also take into account the current network context and individual capacity situation in the supply network.

This work was undertaken by contributors to "FlexNet", forming Theme II of the ILIPT project. The overall objective of the work carried out within FlexNet is to define collaborative planning and execution processes and the required supporting ICT systems needed to meet this challenge. The work has the following aims: 100% BTO production of cars, leading to the avoidance of costly stocks and stock-keeping costs at the most expensive point of the supply chain; 100% due date reliability to the final customer; short overall order-to-delivery time (5 days); short answer-time when entering orders and enquiries for a finished car.

The order-to-delivery target time of 5 days required by customer-specific production puts a great demand on the processes of the whole supply and distribution network. At the heart of the approach to the design of processes for planning and executing network operations is the fact that, for the 5-day car, these processes are triggered and operate on the basis of final customer orders. Furthermore, the process design follows the following principles:

- Pure BTO production at the final assembly stage.
- All BTO parts and those BTS parts with a long lead time and high fluctuation of demand are managed by the FlexNet processes; standard parts that can be replenished in a very short time are not considered in detail.
- Special focus is placed on the distributed management of all critical BTO and BTS parts of a car. A BTO part is critical; if there is no capacity to produce this part then the car cannot be produced, i.e. the due date cannot be kept. A BTS part only becomes critical when its stock levels are insufficient to meet forecast order volumes.
- Capacities are the major objects of collaborative planning and execution in the network and are thus defined in explicit terms in agreements and negotiation on the short- to mid-term planning horizon
- Available-to-promise on a network level: when an order for a car is entered, the ILIPT's processes ensure that production capacity is assigned to all BTO parts at the respective plant locations (final assembly, first tiers, second tiers) and that BTS parts at the different plant locations are available.
- Distributed planning enables the tiers to be part of different supply networks
- Distinction between standard orders and fleet orders; only the latter can be produced in advance.

In order to achieve this a distinctive vision has to be maintained across the whole network; the network is divided in two parts: one where all products are manufactured on a build-to-order basis and one where build-to-stock principles are applied. Dependent on the part of the network in question, processes have to act quite distinctively, orientated on flexibility and speed in the BTO part and on efficiency in the BTS part (Fig. 10.1).

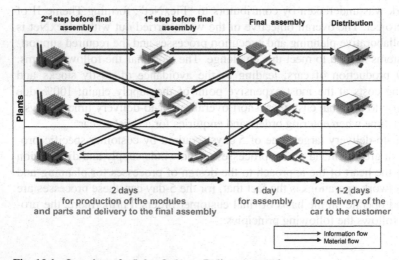

Fig. 10.1 Overview of a 5-day Order-to-Delivery network

The planning processes focus on the improvement of flexibility and the ability of the supply and distribution network to respond quickly. The guiding principle is decentralisation and collaboration in the planning processes. This approach must still leave planning autonomy with the respective company, an indispensable requirement in the highly interwoven automotive industry supply structures. Network partners are only loosely coupled through the exchange of demand and capacity information. Decisions are made through iterative negotiation-based coordination of the different network partners within the boundaries of strategic general agreements.

The major focus of our planning process work is placed on network planning, where the OEM generates a production programme based on the available order and forecast information by distributing production volumes across their different final assembly sites. It is assumed that the OEM knows the agreed capacity bandwidths of suppliers or sub-suppliers, but has no direct access to the detail of their capacity situation. Network planning can thus be seen as the coordination of the OEM's planning of final assembly sites and the local planning activities of suppliers, where more detailed information about the local operations and capacity situation can be incorporated, whilst not being open to the other network partners. This network planning process flows through the network to the sub-suppliers and sub-sub-suppliers until the order decoupling point and a build-to-stock situation is reached. The network planning of each partner is always carried out with an intention not to breach the agreed capacity bandwidths of suppliers. The resulting capacity demands are communicated to the supplier. If the capacity agreements are violated, a defined solution process is initiated, co-ordinating the necessary plan changes amongst the network partners involved, both up- and downstream. This happens through the iterative adjustment of capacity demands and availability, where change requests and proposals are propagated through the whole network. In this way, the compatibility and feasibility of plan changes are checked continuously in the network. The suppliers affected by the plan changes first evaluate if they can react to bottlenecks occurring or under-utilisation within their local production planning. Possible solutions can be the complete or partial acceptance of the requested plan changes. A refusal of the request is also a possible reaction. The analysis of possible response includes the investigation of how sub-suppliers are affected by the new planning alternative in the way described above. Only when all network partners impacted are able to meet the change in demand can the alternative solution be communicated to the company requesting the change. These planning concepts incorporate the costs of using this flexibility offered in the co-ordination process.

The execution processes examined include order management and the sequencing and control of the material flow. The latter processes monitor the material supply through build-to-stock suppliers and handle breakdowns and exceptions. Within order management, the feasibility of delivering the requested car in 5 days is checked at the moment the order is entered into a configurator or ordering system. To allow this, the capacity situation of the whole network is continuously monitored.

Realisation of the described planning and execution processing is only made possible with the development of supporting IT systems. To enable communication among the network partners for planning and execution processes a virtual order bank (VOB) is used. The concept of the VOB is used to support communication between companies, as well as to support the execution of planning tasks, and has been developed especially for this purpose. Although the VOB is a spatially distributed system (due to confidentiality issues), it is an integrated order management and scheduling system. The VOB concept relates customer demand directly to the available capacity of production sites and in a cascading manner throughout the whole production network. It therefore acts as an enabler to collaboration throughout the automotive supply network.

As an integrated order management and scheduling system, the VOB connects customers, dealers, OEMs, suppliers and logistics service providers – across the whole network. The VOB's direct connection of customer demand with the capacity offered in production and logistics of the network companies allows the real-time synchronisation of the network. Therefore, all incoming customer orders are balanced against the available capacities of the final assembly of the OEM, the suppliers and logistics service providers in the build-to-order part of the network. The VOB manages and updates the capacity agreements and restrictions of all relevant companies. The amount of capacity allocated from accepted orders in relation to the agreed capacities is thus transparent for every build-to-order producer. On this basis, new orders can be directly checked against capacity to see if they can be satisfied. If this is the case the order can be booked to the available capacity slot at each plant in a given assembly period.

If the requested delivery date is not feasible, the VOB determines an alternative date, communicating it to the customer, or informing the customer about the selected vehicle options that are the cause of the delay to delivery. In periods of under-utilisation the VOB determines which long-term or fleet orders can be moved or brought forward to achieve the desired level of utilisation in that time period. The special characteristic of the planning support approach provided by the VOB is the fast and uncomplicated connection of the local planning systems used in the network in order to balance demand and capacity across several tiers in the network. The development of interoperability concepts and standards for the seamless information flow between the VOB and among the heterogeneous IT systems is also a working topic.

The following chapters develop the approaches outlined here in more detail and provide an insight into the depth of research undertaken to facilitate the move to build-to-order and a 5-day car.

Chapter 11
Collaborative Planning Processes

Jan-Gregor Fischer[1] and Philipp Gneiting[2]

[1] Intelligent Autonomous Systems, Corporate Technology, Siemens AG
[2] Materials, Manufacturing and Concepts, Group Research & Advanced
Engineering, Daimler AG

Abstract. In this chapter innovative flexible processes for inter-enterprise collaborative planning are presented. The processes support the industry's requirement for both effective and efficient collaboration when planning capacities throughout BTO supply networks. This overcomes current process-related limitations. Collaborative planning processes span the whole integrated supply network. They are developed to support dynamic capacity allocation based on plant-specific constraints and to adjust plants' profitable ranges to the current network context and individual capacity situation. Planning capacities in a collaborative way aims to optimise capacity utilisation corresponding to a current market situation. This chapter further presents computer-supported negotiations as a collaborative approach for decentralised capacity planning. Autonomous software agents serve as mediators between partners and perform negotiations. The concept of agents and their application as negotiators to facilitate the 5-day car vision are detailed in this chapter.

11.1 Approaches to Coordinating "Collaborative Planning"

As vehicles are no longer manufactured by a single enterprise but in complex networks, there is great emphasis on developing inter-enterprise processes that link suppliers and distributors with customers. The companies involved in these networks hold "increasing benefits to the customer" as a common goal. To achieve this goal they need to collaborate with one another. In a build-to-order (BTO) environment this basic but essential need intensifies. For this reason, companies must closely tie supplier production schedules to the assembly schedule of the final product. However, these close ties have to be made at the planning level. This is a basic condition for efficient collaborative processes, which are explained further in subsequent sections of this chapter. The planning processes discussed here include aspects of forecasting along with demand and capacity planning.

Assuming there is willingness for collaboration within a BTO production network, four basic conditions also require consideration during the planning process. These include:

- Process flexibility
- Product mix flexibility
- Volume flexibility
- Efficient exchange of data

Process flexibility is the ability of a company to adapt production processes to meet changing customer needs. *Product mix flexibility* is given, when it is possible to manufacture many different products and variants on the same production line. This makes it possible to smooth the capacity utilisation of the production system despite volatile demands for the individual products. A further condition of a BTO-aligned supply chain is to ensure coverage against unused or inefficiently utilised capacities. This is enabled by *volume flexibility*, which means that the production systems are able to maintain financial viability despite low utilisation levels. This can be achieved, for example, by flexible working time models. Moreover, the basic willingness and capability of partners to exchange sensitive order and process data have to be met by implementing the necessary information and communication technology (ICT).

In addition to the flexibility and pre-requisite ICT capabilities for BTO production a coordination method to shape and steer the inter-enterprise processes needs to be selected. Two basic approaches proposed are hierarchical, centralised coordinated planning and decentralised collaborative planning.

11.1.1 Hierarchical Organised Coordination

One way of coordinating a BTO production network is through a hierarchical collaboration, where a designated company (usually the OEM) defines a common direction; all other companies have to align. This means that the designated company, building on a demand forecast, plans the specification of the technical capacities of the production sites. It also establishes organisational capacity planning options based on that information (negotiation of work/time models with trade unions). As part of a network-wide examination, the OEM's planned production quantities are translated into the required production and warehouse capacities for all suppliers, and are communicated to the partners. This procedure generally functions seamlessly with long planning timelines and stable demand scenarios (e.g. in the ship building industry), as the suppliers can adjust to the conditions as part of their various options and thus ensure an efficient production process.

With high-volume BTO production quantities, this central type of communication and planning results in conflicting objectives. Production planning, which is optimised for the OEM in terms of costs and transport (e.g. with optimal utilisation of the available capacities), places limitations on the efficiency of production of fabricated materials by suppliers. If the OEM wants to agree internal production planning with the BTO supplier, this results in increased complications in setting up agreements because the BTO suppliers may be involved in multiple supply chains. When a supplier is working with several designated companies, a calculation of this type may be impossible due to conflicts of interest. The designated companies may be competing for the production capacities of the individual supplier and are not the only factor affecting supply efficiency. This means that a supply chain cannot be planned by a single company, because the individual parties involved may have conflicting aims. Furthermore, the network-wide process information required for this type of central optimisation is difficult to obtain due to the confidentiality issues of each company. Short-term changes and adjustments to the supply chain (e.g. by integrating a new product) are rendered virtually impossible due to numerous process modifications to be controlled by the central company and to be negotiated individually with each supplier affected. As a consequence, a hierarchically organised supply chain is extremely difficult to implement in the case of a production strategy focussed on BTO across multiple production stages.

11.1.2 Decentralised Coordination

One solution for these organisational barriers with a supply chain (SC) designed for BTO is offered by decentralised coordinated planning. The underlying assumption of this organisational type is that the planning for a single organisation can only be performed by the organisation itself. This planning work then only depends on the planning of the relevant direct customers and direct suppliers of the company in question. The coordination required for this is provided via bilateral negotiations with the relevant neighbours in the SC. This network, consisting of numerous equal and independent partners, can therefore greatly reduce the dilemma of conflicting target functions and the loss of protection of confidential information. The companies plan and negotiate production capacities with their direct suppliers and customers. In this way, the complexity of the SC can be distributed across the participating partners. Due to the close intermeshing of the companies within the supply chain, this also means that it is possible to adjust to any short-term changes and shocks arising, as the changes occurring can be negotiated simultaneously and directly (Fig. 11.1).

Hierarchical Coordination

· High-level planning domain determines
  plans for the entire supply chain

· A focal (dominating) enterprise leads the
  supply chain

→ Centralised planning

Heterarchical Coordination

· Partnership of equals

· Decisions are based on mutual
  agreements

· Negotiation instead of instruction

→ Distributed planning

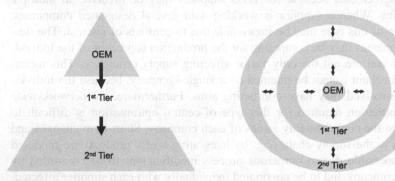

Fig. 11.1 Theoretical approaches for supply network coordination

11.2 Existing Approaches of Collaborative Planning

"Collaborative planning, forecasting and replenishment" (CPFR) was developed to enhance supply chain integration by supporting and assisting joint practices. CPFR seeks cooperative management of inventory through joint visibility and replenishment of products throughout the supply chain. Information shared between suppliers and retailers aids in planning and satisfying customer demands through a supportive system of shared information. This allows for continuous updating of inventory and upcoming requirements, making the end-to-end supply chain process more efficient. Efficiency is created through decreased expenditure on inventory, logistics and transportation across all trading partners. Using CPFR, the "bullwhip" effect can also be avoided, whereby incorrect forecasts from a link within the supply chain relating to the future demand volume mean that its suppliers establish back-up stocks in order to maintain their ability to deliver at all times. If its suppliers utilise the same tactics in turn, an increasing stock of materials is accumulated along the logistics chain. As implied CPFR consists of three phases: planning, forecasting and replenishment. This process is already working in the consumer goods industry, in which the planning is proceeded by a dominating enterprise (e.g. Wal-Mart®).

Another approach dealing with planning issues is Odette's recommendation for supply chain management (SCM) for mid- and long-term "demand and capacity planning" (DCP) (Odette 2005). The goal here is to avoid or detect capacity shortfalls and under-utilisation. The motivation for this recommendation was based on unresolved capacity shortfalls in the automotive industry in the late 1990s.

Within DCP demand forecasts and capacity plans are aligned using a simple comparison between demand and capacity at different stages in the supply network. In the DCP concept each of the participating companies has full access to each other's production plans and to the detailed production programme planning of the OEM. Via bill-of-materials (BOM) explosion and consideration of transportation lead times and initial demand forecasts of the OEM capacity demand has to be calculated at different nodes in the supply chain. These calculations are compared with the participant's capacity planning at each node. In the case of demand and capacity mismatch different "capacity alerts" are given by the DCP system, which is not connected to the local Enterprise Resource Planning (ERP) systems. Capacities in DCP are dynamic and are regarded as a function over time, i.e. different capacity boundaries over time, depending on different flexibility models.

Odette's DCP requires a lot of manual work to define capacity families and the declaration of capacities in the DCP dashboard. While this process is not connected to the local ERP systems, even more challenging are the demand and capacity mapping processes. In the case of conflict, the system will just give a capacity alert to the planner, without providing potential solutions.

In summary, one can say that CPFR does not provide a direct connection to the local planning system of each partner in the network to exchange relevant information between the planning domains. The planning information from the OEM is just pushed into the purchasing network. CPFR is only a single-step, bilateral collaboration process and only supports build-to-stock (BTS) and not BTO supply processes. DCP does not aim to support overall network planning, which is necessary for planning decisions that have an influence at different stages in the BTO supply network. In a BTO production environment, overall feasible production plans, which fulfil the market demand, should be generated. Demand and capacity mismatches should simultaneously be avoided at different stages in the network.

To obtain an overall feasible and mutually agreed demand and capacity plan across the BTO network, it is necessary for the local planning domains to collaborate. This can be done by the decentralised Collaborative Demand and Capacity Planning (CDCP). CDCP affords the above specified requirements and is presented in the next section.

11.3 The Architecture of the BTO Production Network

The stakeholders of the CDCP are listed as follows:

- Marketing/sales department
- External forecast provider
- Final assembly plants (OEM)

- External assembly service provider (EASP)
- BTO first-tier supplier
- BTS first-tier supplier
- BTO second-tier supplier
- BTS second-tier supplier
- Logistics service provider (LSP; inbound and outbound)

This list is limited to suppliers down to the second tier, although CDCP can also be expanded to suppliers from other tiers. The marketing/sales department of the OEM, as well as external forecast providers, are responsible for generating the demand forecasts, whose function is explained in the description of the collaborative planning process. The responsibility of the OEM and support from an external forecast provider are offered here simply as one of a range of common demand forecasts. The inclusion of EASP (External Assembly Service Providers) is a fundamental part of CDCP, as EASP give a considerable degree of volume flexibility. EASP offer additional capacities to several OEMs simultaneously. The OEMs will only set up production and assembly facilities and capacities to cover a defined range (e.g. 75–90%) of their forecasted demand. The remaining required capacity will be provided by EASP. This enables OEMs to run their final assembly plants at a higher overall utilisation level (Fig. 11.2).

Logistics service providers (LSP) are network partners who provide services concerning planning and execution of material flows and logistics. Therefore, they build the interface between suppliers and customers. They may take over multiple functions or process management at each stage of the supply chain. To optimally fulfil their functions, LSP need information about the future demand and future replenishments at an early point in time to plan the required resources and capacities regarding the transportation and logistics services.

All stakeholders involved in the material flow are essentially responsible for designing their own internal company processes (local planning processes). These local processes have interfaces to processes in collaborative planning, but represent

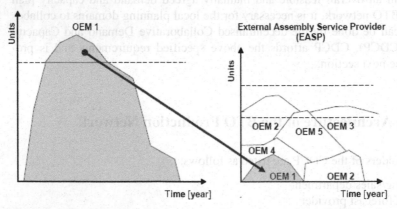

Fig. 11.2 Using capacities of external assembly service providers (EASPs) reduces the necessary maximum capacity of the final assembly plant to be installed

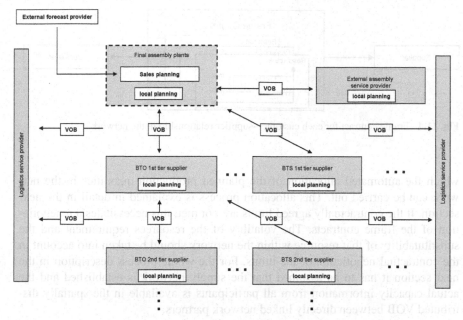

Fig. 11.3 Spatially distributed virtual order banks (VOBs) between directly linked partners

a black box for all other stakeholders in the network. For the processes in network-wide planning, it is necessary to ensure an automated exchange of relevant information between these partners. This link-up is provided by what can be described as a virtual order bank (VOB). The VOB combines, as an integrated order management and scheduling system, customers, dealers, OEMs, suppliers and logistics service providers along the whole supply chain. The basic idea of the VOB is to combine customer demand directly with the capacity available from the production and logistics of a company in the network and therefore enables the synchronisation of the network in real-time. But not every partner within the network is allowed to see all the information. The VOB simultaneously shows relevant information to just two partners with regard to a certain context, e.g. the final assembly plant and the first tier supplier (Fig. 11.3).

The basis for the relevant shared parts of the VOB is provided by a frame contract (Fig. 11.4). This frame contract is made between every customer and its supplier, e.g. OEM final assembly plant and EASP, first-tier supplier and second-tier supplier. The frame contract consists of so-called bucket lists, defining overall minimum and maximum production capacity (for resources producing BTO parts) or overall minimum and maximum stock levels for BTS parts for a certain time period. An agreement can have bucket lists for different plant locations (e.g. the production capacity for all final assembly plants of an OEM) and is stored in the distributed VOB hosted by the partners. Frame contracts are used subsequently for order management to enable 100% due date reliability. The corridors are used during the planning process as maximum areas of scope for flexibility, within

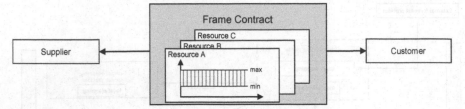

Fig. 11.4 Frame contract for each customer–supplier relationship in the network

which the automated allocation of the planned production quantities in the network can be carried out. This allocation process is explained in detail in the next section. If the contractually agreed limits are not met, this necessitates a renegotiation of the frame contracts. The volatility of the resources requirement and the substitutability of that resource within the network should be taken into account in the contractual negotiation of the limits. For the whole process description in the next section it has to be assumed that the supply network is established and the actual capacity information from all participants is available in the spatially distributed VOB between directly linked network partners.

11.4 The Process of Collaborative Planning

The goal of the collaborative planning process is to determine the capacity demand corridors and the necessary capacity adjustments in the whole supply network (final assembly plants, first- and second-tier suppliers, logistics service providers) for the next planning interval, based on the existing capacity information and on the planned sales volume. In addition to the overall capacity corridor restrictions set in the frame contract, the main trade-off is between ensuring maximum reliability of the production flow (capacity corridors that are as wide as necessary) and economic planning stability for the individual corporation (capacity corridors that are as slim as possible). The planning results for the planning period and each supply net tier are called "capacity buckets".

A single capacity bucket is characterised by the affected BTO resource, the day up to which the capacity is planned and the minimum as well as the maximum production capacity (in production minutes) within the network partners' production facilities. As a result of the collaborative planning process the capacity buckets are defined and stored in those parts of the VOB that are hosted by the corresponding partners. During the execution processes the agreed upper and lower capacity limits of the production systems involved should not be violated (for detailed information see Chap. 12). If it is conceivable that the lower level will be violated, utilisation could be raised by bringing existing long-term fleet orders forward, producing the cars or parts in advance.

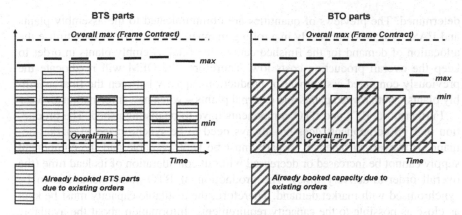

Fig. 11.5 Build-to-order (BTO) and build-to-stock (BTS) buckets

For BTS parts a list of what are called "stock level buckets" has to be planned (Fig. 11.5). A single stock level bucket is characterised by the BTS part, the day to up to which the stock level is planned and the minimum as well as the maximum stock levels. Thus, the main difference between BTS buckets and BTO buckets is that BTS buckets contain quantities of BTS parts whereas BTO buckets contain production capacities to produce BTO parts (e.g. expressed in parts per minute).

Based on these buckets the average capacity offer and the lower as well as the upper boundaries for each supplier in a supplier/customer relationship can be determined. To achieve this, CDCP and as a sub-process, the collaborative forecasting process (CFP) can be used.

The CFP is based on regular (e.g. 18-month) sales planning, which is fed by the volume prognosis from forecasts, volume orders (e.g. fleet orders or other long-term planned volume orders) and historical as well as current sales plans. In connection, forecasts can be provided by external market research agencies, for example. They are derived by means of statistical and mathematical algorithms (such as time series analysis or causal models) based on historical sales data, sales trends and by taking market-specific, demographic, microeconomic and macroeconomic as well as product-specific information into account. Within OEMs there may also be a team of strategic planners, sales planners and production managers who also make forecasts. Adjustments could be conducted biweekly. No forecasts are used for the immediate few days, only real orders. It is possible that the immediate 3 months are forecast on a weekly level and the following timeframe on a monthly level. It has to be pointed out that the planning horizons and the planning frequency can be defined freely by the network partners. The figures mentioned and those coming up are just examples. Based on these data, sales planning can take place, involving dealers and national and corporate sales departments.

As soon as the forecast is generated, the mid-term (timeframe: beyond 5 days and up to 3 months) and long-term (timeframe: beyond 3 months later and up to 12 months) minimum/maximum quantities for the internal plants and EASPs are determined. The previews of quantities are communicated to the assembly plants and EASP concerned biweekly in a rolling manner. The purpose is to optimise the allocation of demand for the finished cars to the final assembly plants in order to keep the overall production costs low. Therefore, the OEM will re-allocate the previously computed assignment of production capacity between the final assembly plants – both owned plants and external plants – when new information arrives.

Furthermore, BTS and BTO components have to be distinguished. The production of BTS components does not always need to be synchronised to market demand. However, sufficient BTS parts must be kept available and the component supply cannot be increased or decreased without consideration of its lead time (the overall order-to-delivery time). The production of BTO components has to be synchronised with market demand. Therefore, the available capacity must be kept as close as possible to the capacity requirements. Information about the available capacity is elementary and must be provided by the network partners. The capacity demand of a partner should always stay between the upper and lower limits, which are defined in the frame contract between two of partners affected.

The production plan has to be communicated in the supply net and divided into the new volume corridors of the internal production sites, the EASP production sites, the BTO first-tier suppliers and the BTS first-tier suppliers. The volume corridor for the BTO first-tier suppliers provides an input for a rough capacity plan at the first-tier BTO supplier's sites. It is held in the VOB along with the volume corridor for the BTS first-tier suppliers. If the internal capacity check of a final assembly plant within the network exposes that there is not enough internal capacity available, including the capacities of the EASP, renegotiations have to take place in which the production volumes will be shifted or the contracted capacity bounds modified.

In addition to the capacities for the critical BTS supplier parts mentioned above, the single partners' VOB nodes contain actual capacity data for the final assembly sites as well as the BTO stakeholders. This includes information about cost structures for car assembly at each site for suppliers over the short term (e.g. the next 5 days) and part of the mid-term horizon (e.g. the next 6 weeks) information. The VOB has to manage customer orders, i.e. order capturing, order storage, BOM explosion for critical parts/components/modules as well as ongoing order status monitoring and exception monitoring and handling. It has to determine delivery dates for each customer order and to manage capacities, complete demand capacity checks as well as capacity allocation to specific customer orders.

During the "local planning processes", the suppliers have to confirm the delivery feasibility based on their available capacities. In this case the demand information is forwarded upstream in the supply chain, cascading to the last critical supplier who starts checking capacities. If the demand can be fulfilled, the supplier confirms the feasibility to its buyer downstream in the supply chain and so forth. If the demand cannot be fulfilled, negotiations are triggered between the partners affected. A more detailed description is given later in this chapter.

11.5 The Local Planning Process

The trigger for the local planning process may be caused by a change in the production preview of a customer, for example. If a modified unit quantity is requested by the customer for the next planning period, a BOM consolidation is carried out within the affected part of the VOB for the relevant supplier (Fig. 11.6). The supplier is then required to determine, as part of their local planning process, whether or not it can manufacture the requested unit quantities efficiently.

This occurs when the supplier optimises their internal capacities with regard to the changed requirements of the customer. The planning process triggered by this can impact the current production structure in place, the shift operations and/or the production processes. As part of the batch size planning to be carried out, customer orders are condensed into individual batches to minimise production costs as well as the costs incurred for warehousing, re-tooling and cleaning the machines. Scheduling and capacity planning is used to determine the earliest and latest deadlines for order dispatch to be able to deliver the requested unit quantities to the customer on schedule. Based on this, detailed planning of the machine utilisation, component sequence, operating material allocations and equipment, as well as tools and the required staff, can be carried out.

This optimisation should normally be observed using the capacity limits agreed with the customer. If this cannot be guaranteed, there is an automated negotiation process within the VOB (Fig. 11.7).

This automated negotiation process within the VOB is carried out by what are known as software agents, and these are covered in more detail below.

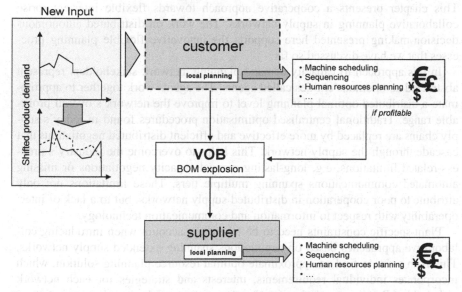

Fig. 11.6 Bill-of-materials (BOM) explosion and local planning

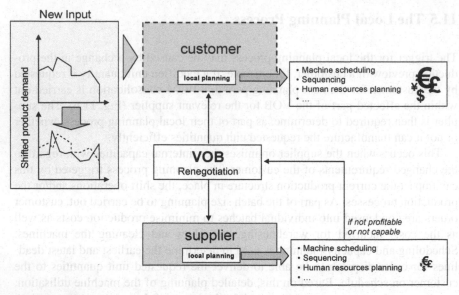

Fig. 11.7 Renegotiation between two network partners

11.6 Autonomous Behaviour and Collaboration in Supply Networks

This chapter presents a cooperative approach towards flexible inter-enterprise collaborative planning in supply networks. The work on distributed autonomous decision-making presented here supports the innovative flexible planning processes that we have discussed so far.

In this approach to capacity demand planning, network stakeholders represent abstracted autonomous entities called agents. The agents work together to approximate a distributed optimal planning level to improve the network's overall profitable range. Traditional centralised optimisation procedures found in today's supply chains are replaced by more effective and efficient distributed negotiations that cascade through the supply network. This helps to overcome the industry's process-related limitations, e.g. long-lasting manual capacity negotiations or missing automated communications spanning multiple tiers. These limitations not only attribute to poor cooperation in distributed supply networks, but to a lack of interoperability with respect to information and communication technology.

Plant-specific constraints need to be taken into account when introducing collaborative approaches based on negotiations spanning extended supply networks. This is in order to find an approximate optimal resource planning solution, which incorporates individual requirements, interests and strategies for each network stakeholder. Resource optimisation, therefore, is not solely aimed at amending an

OEM's or assembly tier's resource planning. In contrast it adjusts the financial and operative range of profitability for the whole network, including its lower-tier suppliers.

11.6.1 Agent-Based Collaborative Planning

Autonomous behaviour is investigated in computer science in the form of what is called "software agents". These entities act independently and make decisions by actively sensing environments and understanding change in situations. An autonomous software agent is defined to be a system "[...] situated within and a part of an environment that senses that environment and acts on it, over time, in pursuit of its own agenda and so as to effect [sic.] what it senses in the future" (Franklin and Graesser 1996).

This definition implies a requirement regarding the kind of behaviour agents display depending on their current and past context. First, an agent must be able to sense its context, i.e. acquire information about its situation within its environment. In addition, an agent is required to act autonomously in order to pursue its client's short- and long-term strategic goals based upon personal constraints and limiting conditions. For instance an agent can represent a computer process integrating with a supply network stakeholder's enterprise resource planning information technology systems to optimise production and logistics resources for certain periods of time. The agent considers the stakeholder's own planning horizon, granularity and frequency to perform local plant capacity optimisation. For instance, one or possibly multiple stakeholder agents can be provided with the information necessary to autonomously plan a 3-month capacity horizon with a daily granularity and a weekly frequency.

In addition, agent behaviour is proactive. Agents orientate themselves by controlling their own flow. Frequently, they act in advance, according to custom expectations or forecasts. For example, a stakeholder's planning agent that expects a capacity bottleneck for a certain timeframe would try to modify the configuration of a plant's internal resources. The agent could incorporate unused capacity or investigate changes to resource utilisation according to constraints defined by the company's strategic department.

Up to now, agents may be seen to be self-centred and self-serving. Indeed, under normal conditions they would not be well-suited to being employed in inter-enterprise planning scenarios as, so far, they cannot cooperate. Hence, an important collaborative aspect of an agent's attributes is "reactivity". Without the ability to react to other entities' signals by deriving and executing context-sensitive decisions, a software agent is unable to integrate short-term situational changes into its decision-making procedures. In reality, agents are required to quickly react to other agents' decisions. For instance, an unforeseeable capacity shortage of a critical supplier cannot usually be neglected and must be solved in a collaborative manner by a reaction to the new situation. In distributed supply networks this can be achieved by carrying out negotiations in order to redefine available

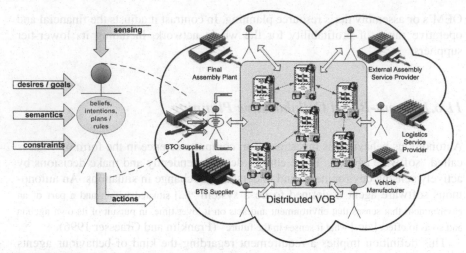

Fig. 11.8 Agent-based collaboration in supply networks

supply and demand, leading to a partial or even complete reconfiguration of network capacities.

The term collaboration also implies that agents demonstrate adaptive behaviour. Adaptability suggests an ability to take action whilst respecting the interests of the agent's environment, e.g. by forming agreements on production capacity utilisation and signing corresponding contracts. By handling such long-term situational changes, agents are called to adapt to their environment.

Software agents that understand and adapt to their actual context and have the ability to reason and react to situational, environmental and logical changes in accordance with their individual decision-making processes are said to be "intelligent" (Wooldridge and Jennings 1995). They encourage a novel approach towards collaboration in distributed supply networks. Figure 11.8 shows intelligent agents following pre-specified "desires" that they try to achieve by executing goal-orientated tasks such as negotiating capacity agreements and production capacity–demand agreements with suppliers and consumers. These may be located within the specific supply chain or externally.

The process of defining the desires agents keep to when negotiating with other agents is a complex task. It requires formal (e.g. logic-based) declarations of individual goals to be pursued that are dependent on changing situations. So-called desires represent an approximated or exhaustive search for an in extremis arbitrarily complex trade-off relation between effort and cost. For instance, a supplier's agent might have the "desire" to maximise delivery reliability by improving plant utilisation and to increase resource capacity whilst minimising associated production and logistics cost at the same time. Another "desire" could be the increase in a plant's long-term profitability.

The autonomous, proactive, reactive and adaptive characteristics of software agents are founded upon the agent's ability to understand its task. It must consider constraints, possible alternative procedures and environment, and it is this information cognition (understanding) that is critical to the entire agent methodology.

The collaborative functionality of intelligent agents depends on the agents' ability to commonly understand individually processed as well as communicated information in an application domain. In the case of supply network logistics and collaborative planning processes, information comprises terminology such as network, product and planning concepts as well as related dependencies, relations and properties accessed from distributed semantic data repositories. For instance, agents negotiating production capacities must understand the meaning of related capacity bucket concepts, corresponding relations to the terminological concepts of planning period, horizon and granularity, and additionally production concepts such as products and resources and the concepts of logistics service providers. Based on this common understanding of required terminology and definitions, agents act within a particular scope. This scope is defined by the agent's client – in general a supply network partner – on an individual basis. The client restricts the type and course of possible actions the agent can take for different situations. This enables agents to act in awareness of changing situations within their domain. They generate representative models of the logistics world, which they observe, in order to create individual "beliefs". On the one hand, these beliefs can incorporate information about the agent's own client, e.g. capacity agreements with other partners, a plant's line capacities, shift models, resource restrictions and other data fetched from local data repositories. On the other hand, beliefs integrate information about other agents' clients that is not deemed to be confidential. These external data can be queried by agents over the network, and may include a partner's product portfolio or the cost of modifying production capacities, to give two examples.

Agents pursue their own desires using both internal plant-specific and externally requested information together with defined terminology. This is accomplished by executing pre-defined or dynamically generated plans and by applying rule-based approaches to declare agent intentions. These are related to the required steps an agent must execute in order to reach its goals. For example, an agent may aim to find a solution to overcome a capacity bottleneck that cannot be solved by local resource optimisation. Its intentions could include a plan or rules for executing individual steps during capacity negotiations with partners' agents, both upstream and downstream in the supply chain. During agent negotiations carried out to solve the capacity bottleneck situation the initiating agent might suggest a modification to the current production capacity assignment configuration, e.g. by shifting capacities to another partner that can provide equal production capabilities.

In summary, intelligent software agents incorporate terminological domain-specific information, as well as continuously changing data sourced from the supply network, in order to pursue the desires of their clients. Clients include stakeholders on different abstraction levels, such as the OEM, its final assembly plants, the external assembly service providers, BTO and BTS suppliers, as well as logistics service providers. Instead of collecting and analysing all data arising from the network at a central place (e.g. a database hosted by the OEM) in order to apply central optimisation algorithms, a collaborative agent-based approach employs distributed negotiations that take confidential and classified plant data into account.

11.6.1.1 Role and Design of Agent-Based Negotiations

Agent-based negotiation can be applied to various cases in supply networks. In particular negotiations can be used to accomplish coordination and optimisation tasks that span a minimum of two partners. To reiterate, the main benefits of using negotiations instead of central optimisation are the ability to first reduce data complexity and second to hide classified partner information. Data complexity is reduced, since for most scenarios heterogeneous information sourced from distributed network partners does not need to be aggregated in a central data repository for subsequent analysis, interpretation and optimisation. Data heterogeneity represents a major drawback of information integration. In addition, partners' classified information, such as plant capacity, profitability and strategic decisions, are not disclosed to other network partners, since information is processed in a decentralised way and thus is hidden from other entities operating in the network.

The negotiation paradigm may be applied to different cases found in the context of supplier–consumer relationships. In summary, agents serve as mediators between partners and perform negotiations on an autonomous level (Fig. 11.9), finally delivering to legal representatives an updated negotiation outcome – an agreement between the according participants. This negotiation agreement is then validated and signed by the contractors' representatives.

Negotiation agents represent any of the following parties: manufacturer (OEM), one or multiple assembly plants, external assembly service providers, BTO and BTS suppliers, as well as logistics service providers. During negotiations, the respective agent must have access to information about other participants, the scope and subject of negotiation and an individual, private strategy to be followed. For instance, a manufacturer's agent might be provided with information about the network structure comprising relationships between plants and resources, product BOMs, the manufacturer's production programme for each location, and production and transport cost differences between sites. In addition, data must be available

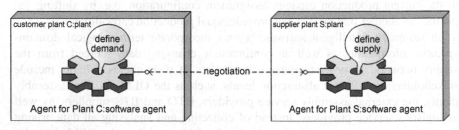

Fig. 11.9 Negotiating agents (UML deployment view)

on defining lead times for BTS parts, adjustment measures including quantifications of costs of capacity adaptation incurred, detailed customer order information and some strategy, such as to decrease costs, while increasing planned capacity.

Before a negotiation can be performed the initiator must first detail the subject of negotiation by specifying its identifier, negotiable properties, as well as any additional descriptions. A final assembly plant agent may trigger a negotiation about its suppliers' production capacity. The respective subject (a certain product with a certain name as its identifier along with some descriptive information) is defined by a list of fixed features such as product dimensions, material, colour and processing quality, as well as by a collection of negotiable properties. These properties may comprise the amount of product or capacity required, the production timeframe or flexibility requirements with regard to call-off policy, and most importantly, the price per unit.

The negotiation process also depends on a software-based negotiation "venue" and on a negotiation interaction protocol that must be followed by all participating agents.

The "venue" represents a distributed, virtual market place with which agents of supply network partners may register to participate in hosted negotiations. In order to enable fair negotiations the "venue" controls the initiation, sequenced execution and termination of negotiations carried out over the distributed network infrastructure. Dependent on the type of negotiation, the "venue" also declares a protocol that specifies rules for the initiation and termination of negotiations, participation and for the submission and withdrawal of proposals. These may include, inter alia, known automated negotiation protocols comprising one-to-one (bargaining), one-to-many (e.g. open-outcry or English auction) and many-to-many relationship (e.g. continuous double auction) (Bartolini et al. 2006).

The following section sets out an example of a negotiation for supplier production capacity initialised by an agent of a final assembly plant. The negotiation comprises a custom protocol that permits a number of agents to participate. The agents can represent first-tier BTO suppliers producing the same modules.

The protocol may state that supplier proposals must not be issued before the assembly plant agent has submitted its one and only proposal specifying a concrete request for production capacity. Furthermore, the protocol specification could specify that proposals may not be withdrawn from negotiation once they are submitted and that supplier proposals must exceed each other in each submission step. The termination rule could be defined to close negotiation after a defined period of inactivity, at a pre-determined time, or if a supplier proposal exactly matches the capacity requested by the initiator. The latter case reflects one option of forming a negotiation agreement. If no supplier proposal is submitted that satisfies the request, the protocol could ask that the final proposals of each supplier are evaluated for best fit conditions. On the outcome of this evaluation multiple agreements would be formed between the final assembly plant agent and the corresponding suppliers. The exact strategy for choosing supplier proposals when forming this agreement depends on rules that are specified by the initiating agent when starting the negotiation.

In the course of negotiation (i.e. initialisation, opening, participant admission, proposal submission and negotiation closing), participants are notified about negotiation events. During negotiation failures can also occur. For example, a participant might not follow the protocol defined by the negotiation or an agreement cannot be formed as the result of the negotiation. In any such case, the corresponding participants must be informed about these failures.

An example of the negotiation protocol mentioned above is the contract-net protocol. It is defined by the Foundation for Intelligent Physical Agents (FIPA) to be used for agent interaction during negotiations (FIPA 2002). In summary, it allows a participant to issue a call for proposals and provides partners' responding proposals, which can be evaluated for selection by the initiator. The initiator then informs all bidders of the new contract status, i.e. acceptance or rejection of each bidder proposal. The protocol ends if either no bidder proposal is accepted or when all partners with accepted proposals confirm the agreement to be achieved.

In addition to the specification of a protocol controlling the course of events during a negotiation each participant follows its own strategy. Hence, automated negotiating systems provide agents that are able to act and react autonomously, understand their application domain as well as the subjects of negotiation, react to other agents' proposals, submit proposals of their own and adapt to the outcome of a negotiation. Agents derive their negotiation strategy from information on the individual stakeholders they represent. To reiterate, required negotiation data comprise capacity information, market and competition strategies as well as pricing policies, among others. In general, this information is classified and must not be made available to other network partners, let alone other competitors.

Figure 11.10 describes the negotiations during collaborative capacity planning. Negotiations can be used to decouple local planning process steps and thus hide classified data from other agents. This is achieved by "black-boxing" local plant processes from the collaborative process using negotiations. In this context, a "black-box" refers to the abstraction of local plant planning logic, i.e. processes residing within the box of each network partner. Hence, only the behaviour of the "black-box" to the outside world is important for negotiations between agents during collaborative planning. This behaviour is represented by one or multiple agents of the corresponding supply network partner. Consequently, the collaborative planning process consists of a series of negotiations (e.g. on distributing capacity demands), which comprise negotiation subjects providing properties that would not represent internal plant information. For instance, cost data can be provided to other negotiation parties without publishing the plant's internal process details.

Figure 11.11 illustrates the use of cost indicators as outputs of a stakeholder's internal negotiation strategy. In this example agents negotiate capacity agreements that define contracted minimum and maximum limits of reserved capacity per period.

Here, the curves represent a dependency between an adjustment of previously contracted upper capacity bounds and the corresponding additional cost. In general agents will not be provided with static cost functions. Instead, they compute complex multi-dimensional cost relations involving a variety of influencing factors.

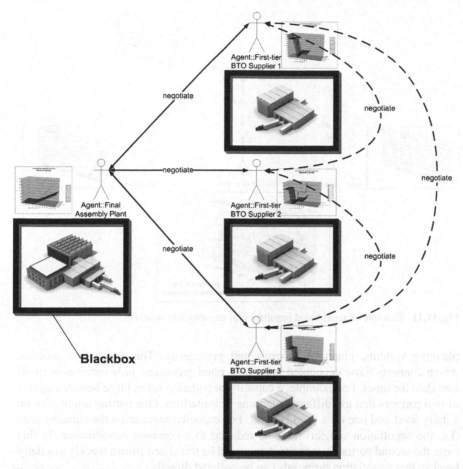

Fig. 11.10 Agent-based negotiation example

These relations are generated dynamically dependent on the agent's private strategy, its plant's individual capabilities and constraints as well as the agent's own last proposal and current competing proposals. Cost relationships can be used as a global criterion for distributed capacity optimisation or for increasing plants' profitable ranges or for both aims simultaneously. Furthermore, agents can recognise internal plant knowledge and hold information about consumer–supplier relationships across multiple manufacturers' supply networks. This is especially useful if the corresponding stakeholder produces for different manufacturers at the same time.

Another criterion corresponds to the abstraction of process details of traditional capacity and capacity demand optimisation at the local plant level. Collaborative planning does not necessarily require synchronised, sequenced process execution throughout a supply network. In contrast, each stakeholder can be left to independently perform local plant planning runs on an individual schedule using individual

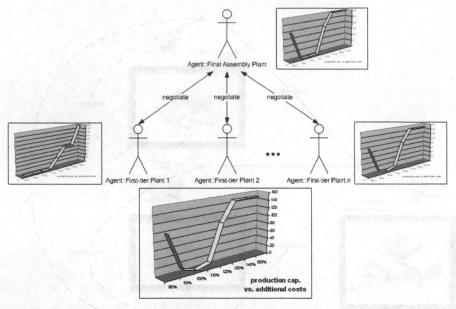

Fig. 11.11 Example of agent-based negotiation of capacity adjustments

planning systems, planning horizons and granularity. The negotiation process, which connects those decoupled individual plant processes, only receives normalised data for input. For example, a capacity negotiation takes place between agents of two partners that use different planning granularities. One partner might plan on a daily level and one on a weekly level. For capacity negotiation the capacity data (i.e. the negotiation subject) must be reduced to a common denominator. In this case, the second partner's capacity data would be translated from a weekly to a daily level so that negotiation proposals can be collated directly.

To date, supply negotiations have generally been considered as independent of each other. Hence, a negotiation between a set of participants did not influence the contractual situation of other network partners. However, in modern supply chains negotiations may be dependent on each other. For example, negotiations that change capacity configurations may require subsequent negotiations reflecting changed conditions by involving other stakeholders in the network that may have been affected. This type of process might span multiple cascades and involves recursive models of contracting. The following section explains this property by means of two scenarios that occur quite frequently in supply networks today.

11.6.1.2 The Recursive Property of Negotiations in Supply Networks

This section presents details on negotiations comprising more than two tiers of a supply network, and explains recursion as a necessary property for collaborative processes.

Recursion is the definition of an operation in terms of itself. In the capacity planning context a negotiation "N" between two partners "A" and "B" might affect an existing capacity agreement between one of "A" and "B" and one or multiple partners of the set "P", where "P" contains all partners that have a direct supplier–customer relationship with "A" or "B". In this case "N" is dependent on at least one additional negotiation, which is recursively executed. "N" is concluded successfully only if all recursively dependent negotiations are executed successfully. If at least one of the recursively invoked negotiations fails then "N" fails.

Our first scenario reflects negotiations for assigning demand for production capacity in the network, initially based upon the current market demand for OEM products. During the periodically executed overall planning process, all of the stakeholder agents concerned are required to cooperate in order to find a matched distribution of production capacity among production resources. In general, collaborative negotiations can replace the traditional OEM planning processes that are characterised by centralised optimisation. A negotiation procedure starts when market demand data, representing the expected number of goods produced by product models for single periods of a normalised planning horizon, is available to the OEM. After analysing all of the partners directly affected, negotiation is performed in a recursive manner, triggered by the initiating stakeholder – usually the manufacturer in this scenario. Thus, the affected negotiation partner (e.g. a final assembly) first negotiates with all of its suppliers (e.g. first-tier plants) and again those recursively negotiate with all of their respective suppliers (e.g. second-tier plants) before the initiator is provided with the negotiation result. Generally, multiple stakeholders are involved during this process, including logistics service providers, in order to contribute to the overall negotiation result. The collaborative process is concluded if partners on all levels of recursion agree on the subject of negotiation.

In the capacity demand distribution scenario, the process is initiated by an agent of the OEM that negotiates with partners' agents to find an overall matching assignment of production capacity demand to available network capacities. On each network tier (final assembly tier, BTO and BTS tiers) required capacity is computed according to a BOM explosion for the final products, BTO modules and components as well as BTS parts. Computed required capacities are then related to stakeholders' production capacity demand. Figure 11.12 provides an illustrated example.

First, the manufacturer and the final assembly tier negotiate a distribution of capacity demand according to the OEM's forecast output demand. The forecast output demand is calculated from historic data for each product model for each period within the planning horizon. During recursive negotiations the output demand is related to the capacities of final assembly plants as well as any available external assembly service providers. In addition to this assignment, these plants negotiate required capacities with BTO suppliers in the first tier, who in turn initiate negotiations on the next tier below. An assignment of capacity demand to the final assembly tier is only successful if dependent capacity demand assignments are negotiated on all lower tiers. Essentially, this reflects the recursive property of the negotiation

process, which continues until all partners contributing to the production of the
respective product are assigned a valid demand for production capacity.

If any negotiation step impairs a contract with a third party, the recursion must
be extended to include an additional corresponding negotiation step. For instance,
a new capacity contract between plants 4 and 8 depicted in Fig. 11.12 might imply
a strong increase in assigned production capacity for plant 8, which could lead to
a violation of an existing agreement between plants 3 and 8. In this case, a sup-
plementary recursion step must be included that renegotiates the agreement affected. The result of this negotiation will in turn influence the overall process of
finding an approximate optimal solution for all partners. Additionally, contracts
arranged between the partners of this manufacturer's supply network and the
wider network must also be taken into account. Hence, multiple networks may be
involved during recursive negotiations.

The overall goal of such negotiations is to smooth the planned utilisation of
production capacities throughout the whole network, in turn reducing production
costs. The collaborative negotiation process ends by determining agreements on
assigned capacity volumes for all negotiation steps.

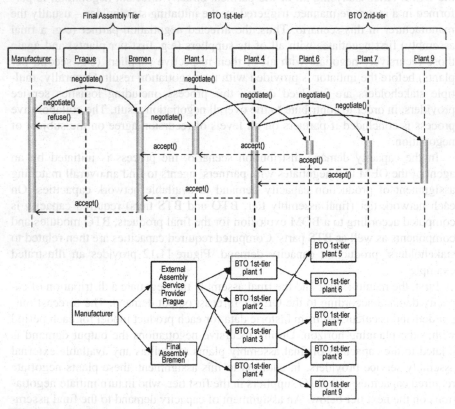

Fig. 11.12 Example of a recursive negotiation process (*top*: UML Sequence diagram, *bottom*:
supply network structure)

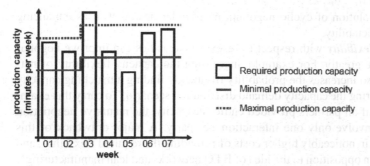

Fig. 11.13   Capacity agreement violation example for weeks three and five

In the second scenario, directly connected network partners negotiate agreements for reserved production capacities in the form of minimum and maximum capacity boundaries for each plant. The scenario reflects negotiations initiated where either required capacity agreement for supply and demand of reserved production capacity does not yet exist between two partners in a supplier–customer relationship, or where at least one such current agreement cannot be kept between two stakeholders in the network.

Both situations imply that a plant is no longer profitable according to its present capacity agreements. This means the plant may violate a previously negotiated agreement, which would require the initiation of a process for negotiating capacity agreements among all the supply network partners affected.

A capacity agreement violation arises if a partner is assigned capacity demand exceeding its contracted capacity boundaries. A concrete scenario for this process is an invalid assignment of capacity demand during the first negotiation scenario. In general, an exception is caused when capacity bounds are breached by at least one partner in the network (see Fig. 11.13). In order to resolve such a situation, negotiations for raising or lowering the respective plant's capacity boundaries are triggered for all related periods.

Again, negotiations are performed in a recursive manner. However, the initiator is not necessarily the manufacturer in all cases. Instead, each stakeholder that faces a possible capacity agreement violation can trigger negotiations with its partners.

11.6.1.3 Increasing Performance of Negotiations

In a worst case scenario, performing recursive negotiations in supply chains implies negotiations involving all partners in multiple supply networks. Due to performance considerations this might not be acceptable in cases that involve many partners participating in tightly connected supply chains. Therefore, the runtime of negotiations, and most importantly the number of negotiation steps executed, should be reduced in order to deliver faster response times. Possible solutions comprise an increase in plant capacity flexibility, the introduction of parallel negotiations, resolution of cyclic negotiation redundancies, as well as enhancing negotiation predictability.

Expanding flexibility with respect to reserved capacities can improve negotiation performance greatly. For example, increasing the capacity flexibility of network partners also increases the probability of quickly finding a matching capacity configuration during the capacity demand distribution scenario. To carry this example to extremes, if all partners provided endless capacity, the recursive negotiation process would involve only one interaction per party. A major drawback of this approach results in noticeably higher costs of providing the required flexibility, and in fact is placed in opposition to the idea of BTO networks and lean manufacturing.

The second approach does not necessarily require an additional increase in capacity flexibility. By analysing negotiation dependencies, those negotiations found to be independent can be run in parallel. In order to achieve this *parallel negotiation processing*, a directed data dependency graph is created as negotiation recursion proceeds through the network. Two negotiations, one of subject "1" and one of subject "2", are called independent if and only if the negotiation of "1" does not depend on the outcome of the negotiation of "2" and vice versa (see Fig. 11.14a). If a negotiation of subject "1" depends on the negotiation of subject "2", then "1" must not be negotiated before "2" (see Fig. 11.14b).

Detecting independencies between negotiations enables parallel processing, which noticeably improves performance of the collaborative process. In many cases negotiation steps taking place in different sub-trees spanning branches of negotiations can be executed independently.

Another similar approach aims for *identifying redundancies* in the course of negotiations. Redundancies most likely arise when negotiations are mutually dependent, so that negotiations can become invalid because information resulting from one negotiation was not taken into account when executing another. This would require a repeat of all processes affected, adding effort that could have been saved by analysing corresponding dependencies. In general, a mutual dependency between two negotiations "1" and "2" exists only if a dependency cycle interconnects the corresponding negotiations (see left side of Fig. 11.15).

Figure 11.15 shows that the cycle is resolved by introducing a super node that contains all interdependent negotiation nodes. The semantics of such a super node defines a combined negotiation of all corresponding subjects. In the case of additional interdependencies by multiple negotiation cycles, this process is repeated recursively so that all nodes and existing super nodes forming a cycle are resolved to constitute a new super node. An example involving two process steps is depicted in Fig. 11.16.

The result displayed in the lower right corner of this figure denotes a combined negotiation of all subjects of single nodes. The recursive procedure of identifying and resolving interdependencies between negotiations is repeated until no more cycles (between nodes and/or super nodes) exist. In general, the fewer super nodes are analysed in a negotiation dependency graph the less synchronisation of negotiations is required between partners of the supply chain, which in turn results in higher process performance.

a) negotiations of 1 and 2 are
independent from each other,
parallel negotiations are possible

b) negotiation of 1 depends on
the outcome of negotiation of 2,
negotiations must be performed sequentially

Fig. 11.14   Data dependencies between negotiations

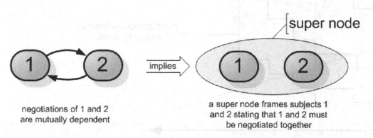

negotiations of 1 and 2
are mutually dependent

a super node frames subjects 1
and 2 stating that 1 and 2 must
be negotiated together

Fig. 11.15   Generation of super nodes for combining negotiations

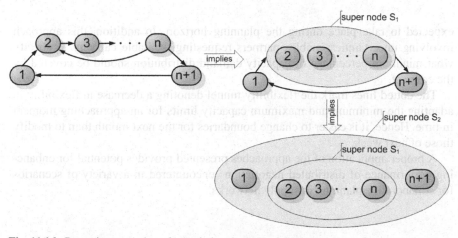

Fig. 11.16   Recursive resolution of negotiation dependency cycles

The last approach presented represents ***enhancing predictability*** of negotiations and thus refers to the agents' internal negotiation strategies. Figure 11.17 shows the progression of a plant's lower and upper capacity borders defined for a consumer plant, as well as capacities assigned to an arbitrary product over time.

As indicated in Fig. 11.17, probability distributions (displayed rotated right) are assigned to values of the capacity function at different points in time, e.g. representing the execution of collaborative planning processes. The aim of introducing probabilities for the planning of capacities is to estimate a plant's required capacity in the future. The more precisely the area below the probability curves is distributed, the more accurate the planning prediction is, and the less re-negotiation is

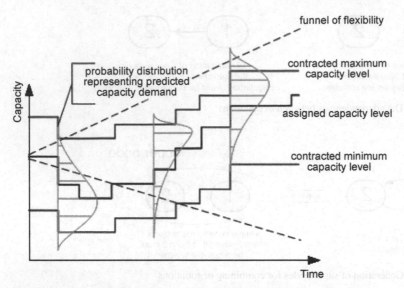

Fig. 11.17 Using probability distributions for estimating capacity progressions

expected to take place during the planning horizon. In addition, this approach involving uncertainties enables partners requesting available capacity to specify what minimum percentage of capacity demand distribution should be covered by the contract.

The dotted lines mark the flexibility funnel denoting a decrease in flexibility in adapting the minimum and maximum capacity limits for an approaching moment in time. Hence, it is easier to change boundaries for the next month than to modify those of next week.

A proper application of the approaches presented provides potential for enhancing performance of distributed negotiations encountered in a variety of scenarios for collaborative planning in supply networks.

11.7 Conclusion

The consideration of both autonomous behaviour and collaboration as presented in the form of negotiations are critical to this approach. As well as the introduction of intelligent agents communicating and negotiating with each other whilst encapsulating sensitive information and the complexity of stakeholder's internal processes, this chapter presented two scenarios to replace today's central capacity optimisation process by distributed negotiations. The first scenario explained the course of steps to be taken to negotiate capacity demand that is initially sourced directly from a manufacturer's demand forecasting system. By analysing a product models' bill-of-materials the corresponding output demand is translated into a capacity demand for individual modules, components and parts and related to the tiers within the supply chain. Indeed, this is the starting point for negotiations taking place over the network and possibly involving external stakeholders belonging to other OEMs. The distribution of negotiations across the network structure implies a recursive negotiation property, which was presented in this chapter. The second scenario covered negotiations regarding capacity agreements. It shows the procedure that is required to define contracts within the profitable range of plants, which refers to the frame between the minimum and maximum capacity limits for a product for a specified time. The final section presented approaches suitable for increasing the performance of negotiations in supply networks.

The following chapter focuses on collaborative execution processes. These comprise some alternative approaches to order tracking, order feasibility checks directly invoked by final customers, as well as the coordinated order booking and sequencing. Questions on the underlying information infrastructure, the most important aspects of the interoperability of IT systems, are still left open. Interoperability of information technology is a major enabler for the introduction of negotiating agents with collaborative planning processes, since it provides the foundation for a semantics-based common understanding of application domain and negotiation-specific terminology.

References

Bartolini C, Preist C, Jennings NR (2006) A software framework for automated negotiation. HP Laboratories, Palo Alto

FIPA (Foundation for Intelligent Physical Agents) (2002) FIPA contract net interaction protocol specification. Available http://www.fipa.org

Franklin S, Graesser A (1996) Is it an agent, or just a program? A taxonomy for autonomous agents. Third International Workshop on Agent Theories, Architectures, and Languages. Available at http://www.msci.memphis.edu/~franklin/AgentProg.html

Wooldridge M, Jennings NR (1995) Intelligent agents: theory and practice. Knowl Eng Rev

Chapter 12
Collaborative Execution Processes

Joerg Mandel

Fraunhofer-Institut für Produktionstechnik und Automatisierung [IPA],
Stuttgart, Germany

Abstract. This chapter will explain new processes designed to reduce the lost time between order entry and the scheduling of vehicles to the appropriate final assembly lines. Very often, this takes days and weeks. New order management and related systems must be able to directly process vehicle orders entered by the dealer or, in the future, by the customer themselves and assign them to suitable plants. This system, which substantially reduces the information flow times, is the "virtual order bank", abbreviated to VOB, representing the central unit of the new order management system. Due to plant capacity constraints the VOB must not accept unlimited numbers of orders. It is necessary to first define the capacity of the final assembly plants and all BTO suppliers and have these data stored in the VOB. These capacity data do not refer to fixed values, but to capacity margins indicating minima and maxima, which are adapted for each product every 6 weeks. Based on these margins in the supply chain, the capacity buckets of the final assembly plants are defined on a daily basis before being assigned to orders arriving in the VOB. Apart from directly booking the capacity buckets, the network also buffers fluctuations in demand, making it more flexible. It is only possible to calculate a reliable delivery date after capacity buckets have been defined and an integrated available-to-promise (ATP) process put in place. The chapter will also show processes for capacity management to determine requirements and match them with the capacity available at the relevant companies across multiple levels of the supply chain. In selecting the plants, relevant factors such as costs, capacity and distance to final customer must be considered. Only then can the appropriate assembly plant for each individual customer order be processed and cars built and delivered within 5 days.

12.1 The Current Situation in the Automotive Industry

One of the major requirements of a build-to-order (BTO) environment is flexible processes, i.e. the speed at which a company can make decisions and alter schedules to meet customer needs (Holweg and Pil 2001). However, vehicles are not manufactured in a single enterprise, but in complex networks. Thus, the companies must focus upon both linking customer requirements directly to production and inter-enterprise processes that link suppliers and distributors to the company. Furthermore, for complex industries, component buffers are not an option. Instead, companies must closely tie their suppliers' production schedule to the assembly schedule of the custom products. The traditional forecast-based mass production systems of the automotive industry do not fit these requirements and cannot meet sophisticated customer needs.

12.2 The Role of the BTO/BTS Border in a Pure BTO Manufacturing System

In order to reach the goal of producing all cars based on final customer orders within a short OTD lead time, new technological innovations will be needed to support real time communication and computation. Systems must be able to interact without downloading or uploading activities and without overnight batch processing, resulting in seamless communication flows between the partners in the value chain. Not all partners of the value chain will have to be included in the real-time communication flow. Who needs what information at what time will be

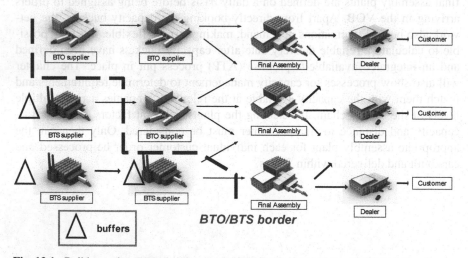

Fig. 12.1   Build-to-order (*BTO*)/build-to-stock (*BTS*) border in an customer driven network

determined by the customer order and the build-to-order/build-to-stock (BTO/BTS) separation point. The BTO/BTS border will be one of the essential strategic variables within the future automotive industry, representing the interface between those stakeholders producing to stock and those producing/assembling to customer order (Fig. 12.1). Physically, this border represents the last buffer in the supply chain where parts are stored that do not belong to a specific order. The position of the BTO/BTS border varies among different types of supply chains and is strongly influenced by factors such as production technology, production lead time for the components, degree of modularisation of the product, product type and structure, the distance between buyer and supplier, as well as the OEM's postponement strategy and the delivery lead time demanded by the customer.

12.3 Order Management for a Pure BTO Manufacturing Environment

In a pure BTO manufacturing environment the order management process is only triggered by a customer order and not by forecasts. When designing flexible processes it is essential to characterise these customer orders. Three different order types are subsequently depicted. In a pure BTO system with a short order to delivery time, the main differentiators are the available time between the customer order and the required delivery date and the number of cars.

12.3.1 Order Types

Short-term orders make up the largest share of the overall demand volume. Such an order is issued by a customer who orders a single car. This customer either orders very late – for instance 5 days before delivery – or makes late changes to the configuration of an ordered car. Thus, the production of the car and its BTO components can only start shortly before the due date; the second tiers can manufacture the BTO components 4 days and the first tiers only 2 days before delivery. This leaves 1 day for production and 1–2 days for final delivery. Therefore, the earliest and latest production dates are the same for this type of standard customer order.

Long-term orders have a lower share of the overall demand volume. They differ from the standard orders by having an earlier fixation point for the ordered configuration. This fixation period is at least 2 weeks before delivery. Thus, fixed long-term orders can be produced in advance in order to smooth capacity demand peaks and gaps and load level production. Consequently, long-term orders have an earliest production date and a latest production date. This applies to all production tasks: the final assembly as well as the production of related components by the first and second tiers.

Fleet orders also have a lower share of the overall demand volume. They differ from standard orders as, like long-term orders, they have an earlier fixation of the ordered configuration. This fixation period is usually several weeks up to several months before delivery. Furthermore, fleet orders differ in quantity, from a single car to orders of up to several hundred cars for car rental companies, for example. Hence, fleet orders can usually be produced well in advance; furthermore, the production of the cars of a single fleet order can be split amongst several final assembly plants.

12.3.2 Order Management

Operating without forecasts and with only three types of orders requires a seamless ICT infrastructure. This substitutes the current isolated systems at different stages of the order management process and integrates not only order management, but processing and scheduling into one run. Such an ICT infrastructure will eliminate the current long information flow lead times. To realise this concept, a semi-centralised system, the virtual order bank (VOB), is required. This system, which can capture, process and schedule orders, takes the relevant capacity of supplier and product information into account (Fig. 12.2).

Only after the capacity buckets have been defined (see the Chap.), and with an integrated available-to-promise (ATP) component, is it possible to determine a reliable due date for the delivery of the vehicle. The task of the capacity management function is to determine the capacity requirements and match them with the capacity available at the relevant companies across multiple levels of the supply chain.

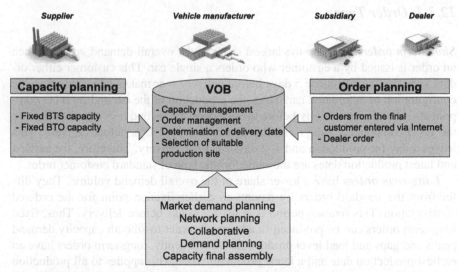

Fig. 12.2 Build-to-order (*BTO*)/build-to-stock (*BTS*) border in a customer-driven network. *VOB* virtual order bank

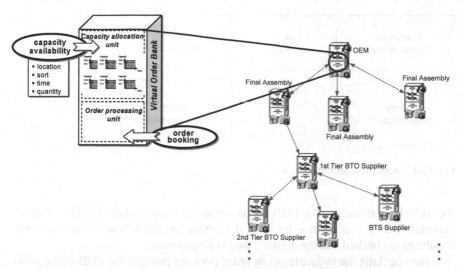

Fig. 12.3 Virtual order bank (VOB) synchronisation in a BTO network

In selecting the plants, the VOB considers relevant factors such as costs, capacity and distance to the final customer before choosing the appropriate assembly plant for each individual customer order. Finally, the creation of assembly schedules includes the creation of daily assembly schedules for each plant as well as the creation of delivery schedules for the BTO suppliers.

The stakeholders concerned have to be able to monitor the order situation in real-time. Thus, they have to have direct access to the VOB, if required. This gives them more transparency for future orders and increases their flexibility, giving them more time to optimise their processes. However, security mechanisms mean that each partner can only get access to the information it is allowed to monitor, e.g. agreements on capacity, agreements on stock levels etc. To fulfil the tasks mentioned above, the VOB has to be linked to all relevant ICT systems in the value chain, such as a vehicle manufacturer's ERP system and car configurator and suppliers' ERP systems. Furthermore, the relevant data, such as the capacity of a BTO supplier, have to be updated continuously through all VOBs in the network Fig. 12.3.

12.3.3 Order Entry

The starting point (Fig. 12.4) for the order management process is an order entry. In this process, the capacity information that emerges from collaborative planning forms the outline structure for customer orders. These outlines will fill the predefined capacity buckets and then be populated with real orders. The customer can either configure his/her desired car at the vehicle manufacturer's web-based car configurator or at a dealership. The configured specification is then forwarded to

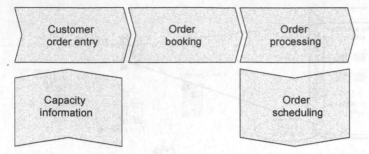

Fig. 12.4 Order management process

the vehicle manufacturer's VOB as an enquiry, where, after a simple bill-of-material (BOM) explosion – for critical parts/materials/modules – the assembly feasibility is checked and the delivery date is determined.

After checking the capacities in an order booking process the VOB of the vehicle manufacturer begins with the selection of a suitable production site. To choose the best production site, the process needs information on collaborative capacity planning and the actual situation at all production sites. In addition to the actual capacity of the production site, the capacity information for BTO and BTS suppliers is important when making this selection. To ensure that the delivery date, i.e. 5 days, can be met, a feasibility check is done immediately after the enquiry for the critical capacities and critical parts/materials/modules required (Fig. 12.5).

At the end of a shift a set of short-term orders and some fleet orders are stored in the VOB. The VOB starts checking for free capacity. If there is any capacity left throughout the final assemblies or external assembly service providers, long-term

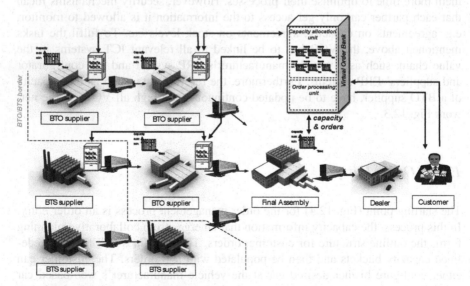

Fig. 12.5 Feasibility check in a BTO network

orders or fleet orders will be added to better utilise this spare capacity. The VOB interrogates all orders that have an earliest and latest production time and can pull forward orders that are between these two time stamps.

12.3.4 Assembly Scheduling

The captured orders are assigned to a suitable assembly plant by examining plant capacities, other orders and the distance to the customer. In this case, there is a trade-off between the optimisation of shop floor operations and outbound logistics costs. At the end of each shift the assembly schedule for the same shift 4 days in advance is forwarded to all internal plants and the external assembly service provider. After receiving the daily/shift schedule each plant then determines its own assembly sequence, based on its restrictions (Fig. 12.6).

The sequencing at final assembly and at the external assembly service provider starts when an unsequenced set of orders is received from the vehicle manufacturer's VOB. The difference between a vehicle manufacturer's (VM) own final assembly and an external assembly service provider (EASP) is that an EASP produces different car models from different VMs, which means that they may receive orders from numerous VOBs.

Each individual order in the set of sequenced orders is divided into BTO and BTS parts. The decision whether a part is BTO or BTS and whether it is sequenced or not, is based on relevant factors such as transportation time, lead time, variants, volume and value. This classification lies behind the production BOM. The call-offs from the final assemblies are sent to the BTO and BTS suppliers of the first-tier level. After receiving the sequenced orders from their customer the BTO first tier start its sequencing. The order of the final customer cascades down the supply chain until the last BTO supplier is reached.

Furthermore, the logistics service provider (LSP) who is responsible for outbound logistics from final assembly receives a pre-shipment notice containing the vehicle type and destination 3 days in advance. This notice gives the LSP enough time to prepare the deliveries.

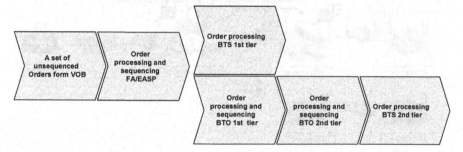

Fig. 12.6 Assembly scheduling in a BTO network

12.3.5 Call-Off to the BTO Suppliers

As soon as each assembly plant has received its 3 working days advance, a BOM explosion is conducted by its ERP system and the call-offs for the first-tier BTO suppliers are generated. In this case the supplier receives the module/component specification and a delivery time window. If the parts/modules/components have to be delivered in sequence for assembly, the sequencing information will also be forwarded to the supplier. This daily/shift call-off will automatically be the pro-duction/assembly schedule for the first-tier suppliers. As soon as the first-tier BTO suppliers have received the VM's call-offs, they conduct a BOM explosion and determine the call-offs to their BTO sub-suppliers. These call-offs are transmitted via electronic data interchange (EDI) immediately after receipt and BOM explo-sion (Fig. 12.7).

The short-term demand allows both VMs and suppliers to adjust their personnel capacities 4 days in advance by varying the daily working hours and shifting pro-duction volumes of both vehicles and parts/components across their plants. Also, if a short-term bottleneck occurs, the increased flexibility allows an alternative supply solution to be quickly found. The seamless call-off information flow to the suppliers is of crucial importance in this process. The whole process, starting with forwarding schedules to the plants, BOM explosion, call-offs to first-tier suppliers and second-tier suppliers, has to happen within minutes, in order to give the sup-pliers the time they need for their internal scheduling. In this case, the deployment of EDI is a prerequisite for inter-enterprise communication.

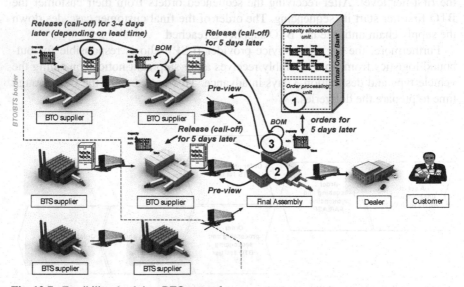

Fig. 12.7 Feasibility check in a BTO network

12.3.6 Replenishment of BTS Parts and Materials

Suppliers who are not able to replenish their parts/components/modules within 3 days will produce to stock. In this case, a vendor-managed inventory concept is applied between the customer and supplier in order to make warehousing as lean as possible and to give the supplier some flexibility in optimising its production. The stocks are located either on the customer's site or in a supplier park and are fed to the line as required. There are no stocks of finished goods at the supplier's plant (Fig. 12.8).

The BTS suppliers do not get call-offs from their customers in the same way as BTO suppliers. By having real-time access to the customer's inventories and usage rates and based on the forecast demand, the suppliers can plan the replenishments with respect to the agreed minimum/maximum stock levels. ICT systems have to provide the supplier with real-time transparency over stock levels, which might be located some distance from the warehouse.

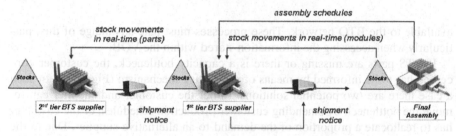

Fig. 12.8 Replenishment process between BTS suppliers and their customers

12.4 Exception Handling

In many cases, disruptions within a BTO network have an influence on the overall supply chain, and potentially on the final customer. Many of today's concepts for emergency planning for sequenced parts can be adopted, as interruptions have a major influence on transportation, suppliers and the final assembly facility. However, current concepts mainly depend on lead-time buffers that allow enough reaction time to reproduce and transport goods. In a pure BTO network a car is produced and delivered within a short 5-day delivery time. The underlying cost and efficiency aims are undermined if a day's worth of parts stock are stored at the final assembly location, to ensure redelivery in the case of an emergency. Therefore, processes need to be defined that assure continuous production in the case of an emergency and do not influence the lead time of the overall product. But how can a continuous production flow be ensured without buffers to the customer? The answer is it cannot. The aim is to reduce these buffers to a minimum, with the help of new technology. Fast, seamless information flow increases the reaction time

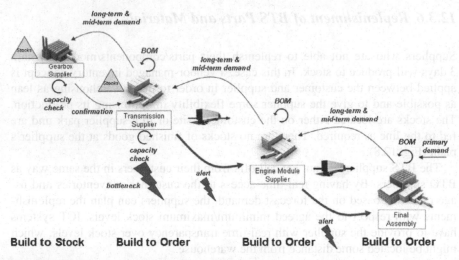

Fig. 12.9 An alert to the partners concerned is generated

available to the BTO network. These processes must take advantage of this, particularly when accessing the information stored within the VOB.

If BTS parts are missing or there is a capacity bottleneck, the customer concerned has to be informed by means of an alerting mechanism (Fig. 12.9). In such a case there are two potential solutions. Either the customer and supplier can remove the bottleneck by amending customer production schedules, or the customer has to reallocate a proportion of the demand to an alternative supplier. Due to the increased standardisation of parts/components and multi-sourcing strategies of buyers, an alternative supplier can be found quickly.

A number of emergency situations can arise during the production of the car. A summary of the emergency processes is shown in Fig. 12.10.

If a production error occurs at the final assembly site, the possible reactions are very different to a production failure at a supplier and they are quite limited. As with other processes, the first step is to identify the nature of the error in order to evaluate possible reactions. The production failure information is forwarded to the

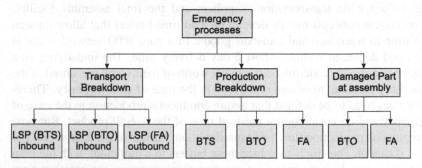

Fig. 12.10 Emergency processes tree

VOB in order to be processed and forwarded to all stakeholders within the supply chain. Within the VOB, the system first identifies the number of vehicles that have failed to be produced on time and second the number of vehicles that will be affected until the point at which the problem is resolved.

In the case of a short-term error, the emergency process results in a production delay, which might be resolved with an increase in tact times, or longer shifts to make up for the lost time. In the case of a long-term issue the likely solution is to reallocate demand to different production sites. If production is not possible at a different production site, the VOB forwards the information to the dealer and the final assembly site. There is a requirement for vehicle production to take place within 3 days. This reallocation option should only be chosen if the production breakdown is likely to last, as the customer may have to be informed of a delay.

12.5 Execution Prototype

The aim of the execution prototype is to demonstrate the processes and functions described in a software environment. Order management, call-off and exception handling are integrated into use case scenarios to support the work of implementing this prototype. The execution prototype demonstrator will only deal with sample data defined in these use cases and thus will not be connected to the productive ICT systems of companies.

The sub-goals outlined are:

- To give an overview on the overall systems concept
- To show the described data formats and protocols
- To provide interface concepts and interface requirements

12.6 Module Deployment and the Software Required

The execution prototype demonstrator is software-based. A model of the value creation network was stored in an SQL database. The database was managed using an MS-SQL Server or a Microsoft SQL Server Desktop Engine. The implementation of the execution prototype core component uses Microsoft .Net framework, versions 1 and 2. In addition to the framework, the execution prototype uses the order-to-delivery simulation component (Fig. 12.11). This component includes the VOB functionality of the execution prototype and calculates the dynamic behaviour of the production network. An http server is also integrated into the core component and is used to communicate with related graphical user interfaces (GUIs) as well as with the planning prototype.

The execution prototype and VOB simulator have been constructed and are operational. The application of the concepts described has been shown to work in

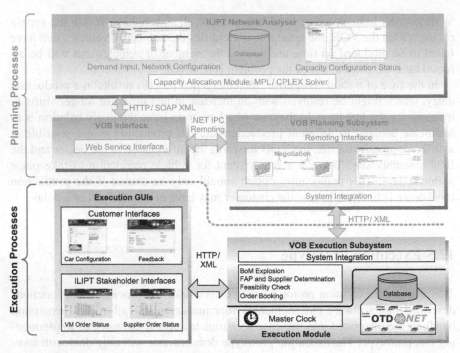

Fig. 12.11 Architecture of the execution prototype

this simulated environment. The prototype enables the demonstration of use case scenarios depicted in Chap. 14.

Evolving from these case studies, the relevant aspects of such a system and its interfaces within this prototype have been used to demonstrate the concepts to industry professionals.

12.7 Conclusion

Today's isolated legacy systems at vehicle manufacturers are not capable of meeting the requirements of a pure BTO manufacturing environment. They were originally developed to support the requirements of specific departments in the function-orientated companies of the past.

In a pure BTO manufacturing system with short OTD lead times the synchronisation of the supply chain is of crucial importance. Customer orders have to be linked directly to assembly planning and supplier scheduling without any time delay and distortion by means of an inter-enterprise order management system, which is referred to as a VOB. In this case, the vehicle manufacturer, as the focal company, has to synchronise its assembly operations with suppliers and distribution logistics. In order to meet all the requirements of a pure BTO manufacturing environment, a VOB has to provide a number of functionalities, such as management of customer orders, delivery date calculation, capacity management including demand-capacity checking, and the selection of suitable assembly sites, by taking all relevant constraints and costs into account.

The main functionalities are integrated into the execution prototype that demonstrates the behaviour of the VOB and order processes under different system states. The prototype has been used for demonstration activities to show the concepts to industry professionals and will facilitate the adoption and transition to BTO.

References

Holweg M, Pil FK (2001) Successful build-to-order strategies: start with the customer. MIT Sloan Manag Rev 43(1):74–83

Chapter 13
Functionalities of Supporting IT Systems: Current Situation, Future Requirements and Innovative Approaches

Jan-Gregor Fischer[1], Markus Witthaut[2], Michael Berger[1]

[1] Intelligent Autonomous Systems, Corporate Technology, Siemens AG
[2] Fraunhofer-Institut für Materialfluss und Logistik, Germany

Abstract. Information technology (IT) issues are a major enabler of the new logistics concepts necessary to manufacture and deliver a built-to-order car in within 5 days. Unfortunately, the current IT infrastructure and systems applied within the automotive industry do not fulfil the requirements posed by new collaborative processes. The existing legacy systems were originally built for a "different world" of IT development, specific tasks (not integrated) and where technology was associated with central control. Thus, supply chain partners currently depend on software applications exchanging information mostly on the basis of proprietary data schemes and interfaces using non-standard transportation and application protocols hampering customised system integration. New collaborative interaction approaches, however, require the flexible integration of IT systems from different organisations. The resulting IT infrastructure must be interoperable and has to follow a distributed architecture. On the basis of a common understanding of content and the meaning of information transferred by computer systems, intelligent IT technologies such as ontology-based data interoperability and service-orientated design principles support stakeholders in coping with the increased run-time complexity of supply chain management amongst integrated network partners.

13.1 Process Requirements of Information Technology

The companies in an automotive production network, covering vehicle manufacturers and their supplier tiers, seek to improve their supply chain competitiveness by offering better service at lower cost. From a logistics point of view, the service improvement is driven by two main factors, radically reduced lead times of a few days and 100% delivery reliability, ensuring that promised due dates are kept. With today's production processes, this can only be achieved by stocking cars through the vehicle manufacturer. However, such an approach ties up capital at the costliest stage of an automotive supply network. Another substantial drawback of a make-to-stock approach is the risk of producing the wrong type of finished car. This risk is of high relevance to the European car manufacturers because they have a very high number of car variants, which makes the forecast of the exact customer demand an extremely difficult task. Production to order avoids these drawbacks of tied-up capital and the risks of manufacturing car configurations that are not requested by the market. By shifting the order penetration point even further up-stream the supply chain will avoid these disadvantages. Currently such a build-to-order (BTO) approach, spanning the suppliers, can only be realised through long order-to-delivery times. After order intake the final stage of the supply chain (the final assembly plants) performs its production and material supply planning. The result of this planning is the input for the first-tier suppliers, which themselves plan their production to determine their material demand, which is the input for the second tiers. Thus, a vehicle order can only reliably be confirmed when all involved parties have completed their production planning.

Only new production planning and order management approaches can achieve the two major aims of improving customer service while keeping within cost boundaries. From an IT perspective the two elementary requirements of these new processes are reducing the lead time for information processing, together with guaranteeing order delivery by managing the production capacity of all BTO suppliers. From an order management perspective, this means that the IT system dealing with customer requests gives, within only a few seconds, a reliable answer regarding the production and delivery feasibility of a certain car configuration to the respective point of delivery. This means that the IT system of the automotive production network must manage the capacity of all resources – not only the production capacity of the final assembly lines, but also the capacity of all facilities producing the BTO components at the respective first and second tiers. Only through this network-wide capacity management of BTO resources is it possible to compute feasible due dates for vehicle orders. Another key requirement in this context is a short response time with respect to vehicle orders. The response time on order enquiries – can a car configuration be delivered to a particular location on a specified date? – and the time taken by the subsequent order booking may not exceed a few seconds.

Therefore, it is necessary that the local IT systems of the supply chain stakeholders are able to communicate and interoperate in an efficient manner. Interoperability is thus a key IT enabler for the new production planning and order management approaches. In the following sections we will provide a short examination of how the new requirements with respect to production planning, order management and interoperability are fulfilled in the automotive industry using information technology that is currently available.

13.2 Capabilities of Current IT Systems

The existing IT infrastructure applied within the automotive industry is diverse. Each OEM has its own IT architecture for order intake, production planning and communication with its suppliers. The suppliers themselves apply individual solutions for the same tasks. Interfaces between these different systems are only partially automated. Figure 13.1 outlines the main IT systems involved in such an automotive supply chain from dealer systems through to supplier systems. The arrows indicate where communication links are required. It is the proliferation of these communication links (using different protocols and exchange formats), as well as the number of different systems, that often cause delay in information processing.

Legacy systems were originally built for a "different world" of IT development, specific tasks (not integrated) and where technology was associated with "control". IT systems follow company or even plant internal individual functional requirements and are therefore not driven by true customer order fulfilment, thus inhibiting smooth information flow. The extent of IT legacy means that the ability of automotive production networks to move to a build-to-order environment is severely limited. Material and plant optimisation drive the vehicle manufacturers' business process. IT systems are currently geared towards the purchasing and production aspects of material supply and inbound logistics (pull to production) rather than towards flexibility to respond to individual markets (pull to customer de-

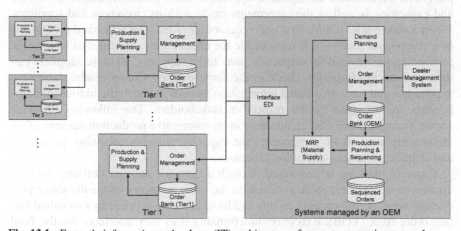

Fig. 13.1 Example information technology (IT) architecture of current automotive networks

mand). There is some evidence of a slowly growing emphasis being placed on removing internal stovepipes and increasing system visibility.

Batch processing also represents the barrier of major IT systems to a 5-day car. The current configuration of a vehicle manufacturer's systems results in individual mainframe systems updating once a day, processing batches or "buckets" of orders in a time-intensive cycle. The respective part requirements are then forwarded daily to the first tiers. However, for the most part they perform their production planning on a weekly basis. Subsequently, the production planning results are transmitted to the second tiers, which also perform their planning infrequently (weekly or even only monthly). This means that there cannot be a real-time order interface with the customer. Dealers have to make the choice to either sell cars that are available (stocked cars) or make a vague promise to the customer regarding the delivery date of a specially configured car that is not in stock and for which it is unsure if the tiers can supply the required BTO parts in time.

13.3 Distributed Information Management

The requirements of future IT mentioned above include functionality for a network-wide capacity and order management. Given the further constraint that the vehicle order booking steps call for a response of a few seconds, centralised management of all information would be the obvious solution. However, such a solution would neglect a principal aspect. Almost all tiers supply to different OEMs; hence, a centralised IT solution would need to manage the capacities and orders related to several vehicle manufacturers. Eventually, such a system would lead to a world model of the automotive industries. This alone calls for a distributed but integrated information management approach. Each partner manages its orders and capacities as well as the agreements made with its customers and suppliers. When a stakeholder (OEM, first tier) receives an order – for a finished car (OEM) or a component (tiers) – it is responsible for checking its capacities as well as for forwarding the demand for respective parts to its suppliers; all these suppliers can then perform the feasibility check in the same manner. Consequently, these new approaches call for a distributed information system in which the individual components are managed by the respective stakeholders. The following Fig. 13.2 shows a sample distributed architecture in an automotive production network. This architecture consists of components for capacity planning (planning processes) and order management (execution processes).

Capacity planning is performed through a network analyser defining the production capacity that is made available to the other partners of the automotive production network. Information on this available capacity is stored in a so-called virtual order bank (VOB). Execution functionality uses user interfaces for the final customer to configure and order cars as well for monitoring components for OEMs and tiers. A company communicates with its suppliers and customers over its VOB interface.

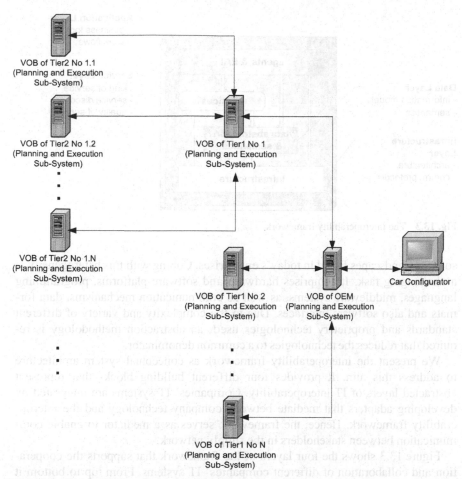

Fig. 13.2 High-level IT architecture for automotive production networks

13.4 Interoperability Framework

To reiterate, interoperability is of substantial relevance to successful cooperation in supply networks. In order to enable the industry to improve today's performance and keep up with market's rapid changes computer systems of different companies must be able to interact and cooperate seamlessly. Hence, interoperability represents the foundation for highly efficient cooperation in distributed supply chain management processes.

In general, it is unlikely that all partners of a supply network will agree on common information and communication technology (IT) for agent negotiation, service discovery and interaction, data formats and low-level communication protocols. This is due to the previously discussed heterogeneous computing environments and

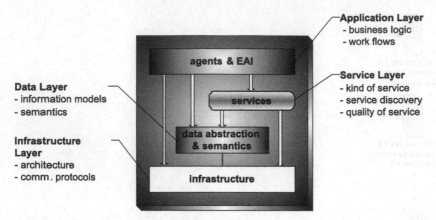

Fig. 13.3 The interoperability framework

software landscapes found in today's enterprises. Coping with this heterogeneity is a challenging task. It comprises hardware and software platforms, programming languages, middleware systems, as well as communication mechanisms, data formats and also software interfaces. Due to the complexity and variety of different standards and proprietary technologies used, an abstraction methodology is required that reduces the technologies to a common denominator.

We present the interoperability framework as conceptual system architecture to address this aim. It provides four different building blocks that represent abstracted layers of IT interoperability. Companies' IT systems are integrated by developing adapters that mediate between company technology and the interoperability framework. Hence, the framework serves as a mediator to enable communication between stakeholders in the supply network.

Figure 13.3 shows the four layers of the framework that supports the cooperation and collaboration of different companies' IT systems. From top to bottom it comprises interoperability tiers for applications, services, data and infrastructure.

13.4.1 Application Interoperability

The top layer of the interoperability framework refers to the processing of exchanged information by collaborating applications in order to support business processes. Hence, in this case interoperability focuses on seamlessly executing work flows across enterprise borders within a supply network. These work flows are not restricted to IT systems running within the information infrastructure of a single company. In contrast, they may logically be distributed across computing systems of multiple partners. For each company, manually triggered applications as well as autonomous software agents, may cooperate to share resources, negotiate, decide, plan, optimise and schedule, amongst others. They are highly specialised to support the business logic required for collaborative processes.

Autonomous software agents are software processes that collaborate in an application domain. They initiate work flows, react to changes in the application domain and computing environment. Multiple agents can form an agent-based software system that provides agent cooperation by supporting their collaboration, by coordinating their actions and by supporting agent life-cycle management. As presented in a previous chapter, software agents can be employed with collaborative planning processes to perform capacity negotiations between network partners.

Frequently, new software must be integrated into the existing computing infrastructure and established applications. This is much more complicated than it seems at first sight. Often these applications do not even provide interfaces for direct integration. Instead, they can only be addressed by developing additional interfaces. Integrating these legacy applications is the aim of enterprise application integration (EAI). EAI not only concentrates on integrating applications, but also on incorporating existing company processes and data. It focuses on enabling the collaboration of applications that are not necessarily designed to cooperate and interact with each other. Hence, EAI supports the integration of applications and local back-end systems of companies (e.g. customer relationship management, dealer communication, enterprise resource planning, production planning and control, warehouse management systems) into the distributed architecture. A task of the application layer reflects EAI to support IT interaction between applications of multiple companies.

13.4.2 Service Interoperability

Services decouple functionality defined on the application layer by the access mechanism to the functions, in order to support work flow management of business processes. Interoperable services are not only required for accessing external systems (e.g. detailed schedulers or car configuration software), providing basic functionality for optimisation or scheduling, but also for storing and retrieving data (e.g. from distributed VOB nodes or a partner's legacy database management system). Other important interoperability services enforce security guidelines, transform information, monitor IT systems and synchronise distributed data stores. Services connect to each other in order to dynamically form an information processing system for a specific task. For this, they offer functionality to other services and/or applications via common interfaces. Services are generally classified according to their extent of specialisation and their re-usability by different applications.

Application support services process one or more specific tasks in the supply chain management application domain (e.g. order booking or analysing order feasibility). Since application services generally rely upon functionality they do not provide on their own, they connect and interact with management services as well as generic services. Management services have the ability to execute tasks that are not directly dependent on concrete business processes. In spite of this, they provide domain-specific functionality to application services. Sample tasks for management services are searching in databases that are distributed over networks of collaborating companies, orchestration of services forming complex ad hoc systems, negotiating between business partners and network partner trust management. Generic services are designed to be completely reusable independent of a particular application domain. They encapsulate certain aspects of IT systems such as storage, basic security mechanisms, data format transformations, information aggregation, data replication, logging and auditing.

In order to enable services to find and interact with each other dependent on the functionality with which they are provided, they are identified using service descriptions. Semantic descriptions about service functionality, the call parameters required, as well as type and format of the information returned, support dynamic service discovery and automated service composition, which enable service interaction and collaboration.

13.4.3 Data Interoperability

Interoperability deals with interpreting information that has been exchanged. In general, information managed in companies' distributed data repositories is not founded on a standardised terminology. There is an obvious need to describe the semantics of the data exchanged in a formal way to establish a common understanding not only between two partners, but within the whole network. Ontologies can be used to define these semantics that provide a means of relating data from different companies to each other.

In computer science ontology refers to a conglomeration of structured and organised concepts that describe knowledge in selected observable domains of interest. Concepts are represented according to their mutual dependencies and relationships, as well as by their particular properties.

Ontologies represent a powerful mechanism to synthesise knowledge of application domains in terms of classes of individuals and their relationships together with their properties. It is possible to distinguish between terminological knowledge regarding common concepts and assertional knowledge referring to information about individuals in the world.

Ontologies enable semantics-based data interoperability in supply networks. Terminologies of different companies can be related to each other by defining relationships between different concepts and relations (e.g. equivalence relationships). This process is called ontology merging and enables data instances to be integrated and correctly compared with each other on the basis of a common understanding of shared terminologies.

Besides the advantages presented with respect to information integration, ontologies benefit from their reasoning capabilities. Reasoning is the process of making implicit knowledge explicit. In general, ontology-based reasoning is similar to logic-based inference (i.e. the process of deriving new information from existing information), especially when considering description logics.

Description logics cover a family of languages that are employed to declare knowledge about concepts and their hierarchies with respect to an application domain. Their basic structure is set up by concept classes, binary relations between individuals of classes and individuals. Concept classes encapsulate the common properties of a set of individuals and specify general relationships between them. Furthermore, language constructs are used to define the structure of particular description logics. These are intersection, union, transitivity, complement, enumeration, disjunction, equality, inequality, etc.

Comparing description logics with the constructs of ontologies it can be seen that the two structures are quite similar.

Concept classes can be mapped onto ontology classes, binary relations onto ontological properties and individuals onto ontological entities. This enables description logic-based reasoning over ontologies. Hence, description logics can be used to evaluate the satisfiability and consistency of the ontology concepts modelled, class subsumption and the validation of instances, i.e. individuals of a concept class. For a comprehensive insight into the theory of description logics the reader is referred to Baader et al. (2003) and Calvanese et al. (1998).

13.4.4 Infrastructure Interoperability

The bottom level of the interoperability framework refers to the IT infrastructure that performs basic communication tasks. It considers heterogeneous centralised and decentralised software architectures and various kinds of protocols according to the ISO/OSI (International Organization for Standardization/Open Systems Interconnection) reference model (ISO/IEC 7498-1:1994).

13.5 Modelling Data and Semantics

A common understanding of information exchanged between the partners of an automotive network is a mandatory constituent of the new planning and execution processes. Current practice for reaching this understanding is the application of business standards such as Odette or OAGIS (Open Applications Group Integration Specification 2007), which standardise the messages – e.g. delivery forecasts or purchase orders – that are exchanged between the network partners. This approach has a major drawback: the semantics of the messages is described in an informal manner and leaves room for interpretation. The consequence is that the meaning of a message or a message element is only informally described and not attached to the message itself. As a result, the establishment of an automated information exchange between a customer and supplier is both time-consuming and error-prone.

As mentioned above, new ontology-based approaches provide a better solution for the efficient and reliable exchange of information between network partners. Through an ontology-based approach to information exchange a higher degree of flexibility can be introduced. Therefore, the companies involved in the automotive industry have to apply and extend a common *interoperability model*. This model consists of semantic definitions of domain-specific concepts, their properties and the relationships between the concepts. The interoperability model must define the semantics in the following areas for supporting new capacity planning and order management processes in a BTO production network:

- The planning ontology consists of the concepts needed for the implementation of planning processes namely, network-wide and plant-specific capacity planning.
- The execution ontology contains concepts needed for the order-to-delivery process management.
- The structure ontology contains the common elements needed for both the planning and the execution ontology.

13.5.1 Structure Ontology

The concepts of this type of ontology (see also Fig. 13.4) are needed both for capacity planning and subsequent order management. The following briefly explains the concepts required:

- An automotive production network consists of several different *Organisations* such as the OEM, first-tier suppliers of parts delivering directly to this OEM, the second-tier suppliers and so on.
- Each of these organisations contains at least one *Plant*, but may have several plants. The "supplies_to" relationship is used to describe the customer–supplier relationships between the different plants.
- A plant consists of at least one *Resource* needed for the production and/or warehousing of the final vehicle (in this case it is a final assembly plant) or parts for the assembly of the vehicle (supplier plant).
- It is important to note that different *Products* (parts or different car models) can be produced on the same *Production_Resource*. However, these products may have different production times in the various production resources. Consequently, it is necessary to maintain the *Resource_Consumption* needed to produce a product with a production resource.
- *Inventory_Resources* are needed for the management of parts that are stocked.

These core concepts are applied in the planning and execution ontology described in the following.

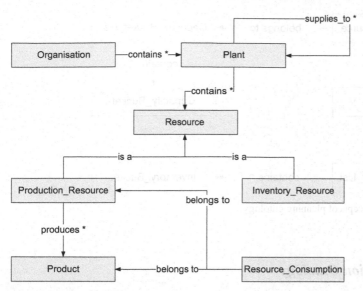

Fig. 13.4   Main concepts of structure ontology

13.5.2 *Planning Ontology*

The planning ontology concepts are needed for the network-wide planning of production capacity – for the finished car and the parts that are BTO – and the inventory level of parts that are still kept in stock. The following briefly presents the main concepts (see also Fig, 13.5):

- The planned available inventory of a certain product at an inventory resource for a specified date can be described by a so-called *Inventory_Bucket*. This concept describes the minimum and maximum number of products that will be made available in stock on the particular date. An *Inventory_Bucket_List* is then simply an ordered list of these buckets. These concepts can be used to match inventory levels for a product with future demand.
- The offer and demand for production capacity is related to a production resource. Consequently, a so-called *Capacity_Bucket* specifies the minimum and maximum available and/or required production capacity for this resource at a certain date. The *Capacity_Bucket_List* is thus accordingly an ordered list of capacity buckets.

In the capacity and inventory planning of an automotive production network, these lists are the topic of negotiations between the organisations involved. The results of these planning processes are bucket lists that are used for feasibility and order booking tasks.

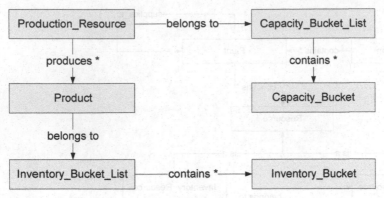

Fig. 13.5 Main concepts of planning ontology

13.5.3 Execution Ontology

The execution ontology concepts are needed for network-wide order management. This includes the feasibility check for all production and inventory resources needed to manufacture a car configuration requested by a customer. The following briefly presents the main concepts of the execution ontology (see also Fig. 13.6):

- There are two types of orders: a **Vehicle_Order** is the ordering of the car by a customer. The configuration of this car is described by a list of features. A **Production_Order** is the order for the final assembly of the car or the production of a part needed for a car.

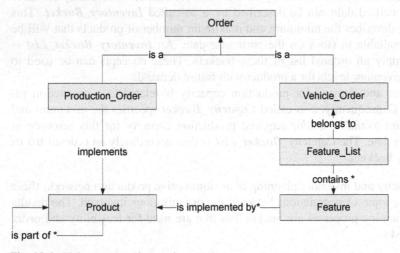

Fig. 13.6 Main concepts of execution ontology

- Each production order builds exactly one **Product**. For reliable order booking, it is necessary to check that there is sufficient capacity in the affected production resource and sufficient stock available at the related inventory resource (see also Sect. 13.5.1 above). The BOM information of a product is modelled with the "is-part-of-relationship" between products.
- There is a many-to-many relationship between **Features** and products. Some features affect many products. For instance, the feature "interior colour" determines some of the other products: seats (through the colour of the seat cover), interior panels, cover of the steering wheel, and so on. Then again, a product can implement several features. For example, a seat could implement the features "leather interior", "seat heating" and "colour red".

13.6 Systems Architecture

Through the application of the previously described interoperability framework the high-level systems architecture presented can be further detailed (see Fig. 13.7). Using this architecture, the local stakeholders of an automotive supply chain can extend their current infrastructure to realise the advanced requirements of new collaborative planning and execution processes. This happens using a three layered approach:

- On the bottom layer we find the current and likely remaining local legacy systems needed for planning (e.g. sequencing of production orders) and execution (e.g. production order monitoring) that must communicate with the systems from other network partners. The necessary functions – especially with respect to order intake and order status reports – must be wrapped through an interface to these legacy systems. Due to the heterogeneity of the local IT systems applied, each organisation of an automotive production network has to build their respective interfaces.
- The middle layer is built by ontology-based software supporting the new collaborative planning and execution processes. Here, we find semantic information on the products (finished cars as well as their parts), their features, the organisations involved and their plants and resources, agreements on capacities and inventory levels, as well as vehicle and production orders managed by the production network. However, each organisation sees only its intersection of this network: OEMs maintain the production capacities of their final assembly plants and their agreements on capacities and/or stock levels with their specific first-tier supplier. Each first-tier supplier in turn keeps its related information in this middle layer. This information is on the resources of the first tier, which are bound to the respective agreements made with all customers of the tier, as well as the agreements made with all suppliers (= second tiers).
- The communication functionality between the different organisations of an automotive production network is finally grouped in the collaborative process

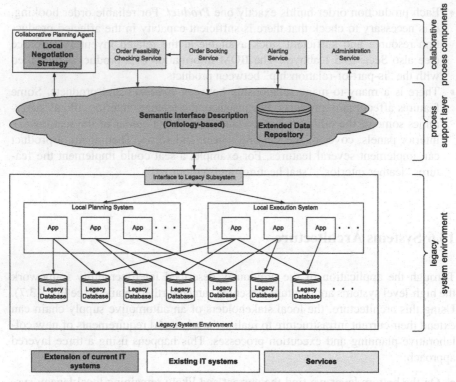

Fig. 13.7 Proposed system architecture for interoperable automotive production

components layer. In this layer we find the services and their interfaces for collaborative planning and collaborative execution. The latter consists of services for feasibility checks, order booking and alerting services. Finally, services for administrative tasks are needed as well.

13.7 Concluding Remarks on IT Functionality

We have presented an IT approach that will support new collaborative planning and execution processes, covering the core elements of the IT solution required from a functional perspective. Non-functional requirements such as answer times, systems administration and transaction security pose further important obligations for IT solutions, especially considering the distributed systems architecture proposed. Consequently, there is a demand for testing this approach in further pilot cases, covering many different network organisations and with a realistic systems load of vehicle orders and associated production orders. Once this has been achieved, a functioning IT system may be implemented to support the vision of the 5-day car.

References

Baader F, Calvanese D, McGuiness DL, Nardi D, Patel-Schneider PF (2003) The description logic hand-book: theory, implementation, applications. Cambridge University Press, Cambridge

Calvanese D, Lenzerini M, Nardi D (1998) Description logics for conceptual data modeling. In: Chomicki J, Saake G (eds) Logics for databases and information systems. Kluwer, Amsterdam

ISO/IEC 7498-1 (1994) Information technology – open systems interconnection – basic reference model: the basic model

OAGIS® 9.0 (2007) Open applications group integration specification. Available from http://www.openapplications.org

Chapter 14
Modelled Scenario Examples for Planning and Execution Processes

Stefanie Ost and Joerg Mandel

Fraunhofer-Institut für Produktionstechnik und Automatisierung [IPA],
Stuttgart, Germany

Abstract. In this chapter scenarios demonstrating the planning and execution processes within a pure build-to-order production network will be explained. The model uses demand data provided by industry and taken from a real car's life-cycle, projected forwards to the years 2015 to 2022. All scenarios modelled are based on a specific product structure for a modular car. The product structure is broken down into defined modules such as the engine, the front, the door and the modular cockpit. The modules come in different variants and consist of different BTO parts sourced through the supply chain. Within the given time period it is assumed that all variants of the product structure remain available to the final customer after they are launched. The model scenarios presented show how the network tools respond to demand and capacity requests within the supply chain. Automated reconfiguration and capacity allocation are used to ensure vehicles can be delivered to customers within 5 days.

14.1 Basic Structure for the BTO Demonstrator Model

An explanation of the basic structure of the model used for demonstration of the feasibility of the build-to-order production network is necessary to set the context for the scenarios described in this chapter. The demonstration scenario approach consists of simulation-based evaluations of the collaborative planning and execution processes. The model uses real demand data for a new product launch sourced from industry and is set in the period 2015 to 2022 (Fig. 14.1). To realise a pure BTO production network, several system states, aligning market demand and available capacities at the plants and suppliers, are necessary. The year 2015

Fig. 14.1 Forecast monthly sales for the year 2015 and beyond

serves as the initialisation phase for the production network and contains only demand data for a single launch vehicle, the limousine (this style is also variously referred to as a "saloon", "notchback" or "sedan"). From 2016 to 2022 all four variants of the modular car (station wagon, hatchback, limousine and convertible) are made available to the customer.

The product structure of the modular car used in the demonstration model is shown in Fig. 14.2. In addition to the body variants mentioned, other modules such as the engine, the front end, the doors and the modular cockpit make up the product structure. The modules themselves come in different variants and consist

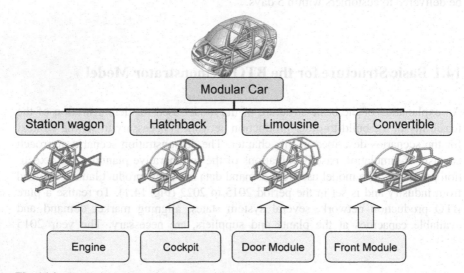

Fig. 14.2 Product structure of the modular car

of BTO parts sourced through to the third-tier level of the supply chain in the demonstration cases.

A pure BTO production network consists of BTS and BTO suppliers from different tier levels. The OEM's final assembly plants (FAPs) and external assembly service providers (EASPs) are utilised. The proposed final assembly network applied in the basic model structure here consists of three OEM plant locations and one external assembly service provider. Geographically, the plants in the demonstration model are located in Bremen, Turin and Toulouse and the external assembly service provider is located in Prague. Each plant is the priority one assembly location for a subset of different markets. The plant locations and the related markets are shown in Fig. 14.3.

The allocation of order to the first-tier suppliers for the different modules depends on the location of the related final assembly plant. The engine module, for example, is produced both in Magdeburg, Germany and in Marseille, France. The plant in Magdeburg supplies the final assembly plants in Bremen and Prague and the production in Marseille supplies Toulouse and Turin.

Based on this model structure, the demonstration examples and scenarios will be explained in the following paragraph.

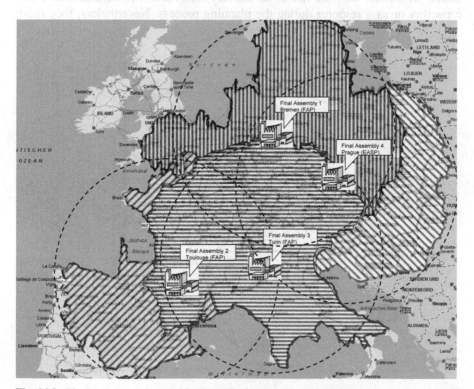

Fig. 14.3 Final and external assembly plant locations and market relations

14.2 Scenario Examples for the Collaborative Planning Process

Collaborative planning comprises joint planning and local planning operations. Local planning is executed by each plant (FAP, EASP, first- and second-tier plant) and pursues the optimisation of the plant's internal processes, such as planning which component to produce on which line. It represents all internal planning activities at the plant level. This planning utilises precise information on local operations and more detailed insights into the available and used capacity of each plant, which is not disclosed to network partners. Hence, local planning is initiated by a single plant and is executed more often than collaborative planning.

Local planning is required to utilise plant capacity. It is updated in order to fulfil customer demand or when demand cannot be fulfilled due to orders exceeding the plant's capacities, to highlight that another production plant needs to be identified that can produce the production volumes required.

Collaborative planning is initiated by the issue of forecast data of the vehicle manufacturer and plans for the optimal allocation of demand to production capacity across the supply network. Figure 14.4 shows the partners involved in the collaborative planning process. The model assumes that logistics service providers are very responsive to fluctuations in transport capacity demand, even at short notice. Therefore, they are not to be actively integrated into negotiations about capacities or as a resource during the planning process. Nevertheless, they must

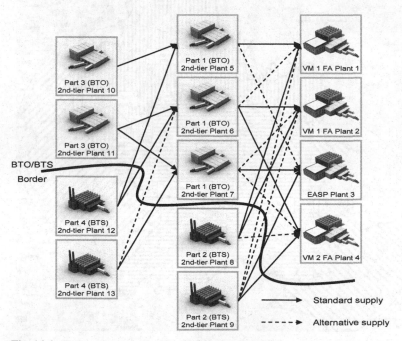

Fig. 14.4 Partners involved in the collaborative planning process

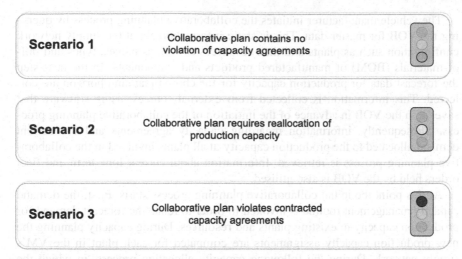

Fig. 14.5 Scenarios of the collaborative planning process

receive information on prospective transport volumes at an early stage to be able to enhance their own local planning activities.

Some examples for initialising this planning process are:

- A new product needs to be introduced into production.
- The overall capacity in the network needs to be increased.
- The overall capacity in the network needs to be reduced.
- Local negotiations do not lead to a result.
- The network structure needs to be significantly changed.

In a nutshell, collaborative planning creates a plan that contains mutually agreed capacity requirements derived from production volumes allocated to final assembly and production plants. The intent is not to violate contracted capacity agreements. The resulting capacity buckets are directly stored in those parts of the VOB that are hosted by the corresponding partners. Furthermore, planning is considered to be a rolling task.

Figure 14.5 demonstrates three common scenarios for collaborative planning. Note that these scenarios do not explicitly define data management operations since they refer to internal system functionality, which is independent of the implementation of the collaborative planning processes.

14.2.1 Collaborative Planning – Scenario 1

The first collaborative planning scenario reflects the desired normal case. Collaborative plans that are executed do not require reallocation of production capacity demand or renegotiation of capacity agreements.

The vehicle manufacturer initiates the collaborative planning process by querying the VOB for master data. This is static information about the supply network configuration such as plants and their resources as well as names, types and bill-of-materials (BOM) of manufactured products and components. In the next step the forecast data for production capacity for the chosen planning horizon are collected. This information is collected from external sources, using software that saves it in the VOB in advance of the initiation of the collaborative planning process. Subsequently, information regarding capacity agreements and the current demand allocated to the production capacity at all plants involved in the collaborative planning process is retrieved. Information about current long-term and fleet orders held by the VOB is also utilised.

At this point the initial collaborative planning process starts. First, the demand capacity management routine is invoked in order to relate the forecast demand for production capacity to existing plants and resources. During capacity planning the new production capacity assignments are computed for each plant in the VM's supply network. During the following capacity allocation process, in which the demand for production capacity is logically allocated to individual locations, the collaborative plan is developed.

The resulting information regarding a time-distributed allocation of production capacity demand to the individual plants in the supply network is now forwarded to the VM in order for them to accept or reject it. If it is rejected the process would have to start once more. If it is accepted the VOB ensures that:

- Production capacity agreements are not violated
- No lead time constraint violation appears

In this first scenario the VOB finds the new collaborative plan to be consistent and stores the work allocation to the network. Each stakeholder in the network is subsequently informed by the appropriate VOB about the demand allocated to its production capacity. Figure 14.6 shows the scenario described.

14.2.2 Collaborative Planning – Scenario 2

The second scenario for the collaborative planning process describes how a VM allocates production capacity among multiple final assembly plants in order to better utilise its own resources.

If, for example, the capacity of a VM's final assembly plant is increased, the capacity allocated to an external assembly service provider located near to it may be decreased in order for them to fully utilise their resources. Capacity increases can be made by:

- The employment of additional personnel
- Investment in or technical modification of additional production equipment

An initiator for this scenario is frequently the introduction of a new car or facelift model into the network. This usually leads to an investment. In some cases the selected plant may need to shift production volumes from other products to other plants in order to make capacities available.

In the scenario proposed, shown in Fig. 14.7, plant n of VM 1 exceeds its maximum value of capacity buckets. Fortunately, plant 1 of the same VM has unused capacity available and both plants are supplied by the same BTO and BTS suppliers.

In terms of the planning process, this scenario closely resembles scenario 1 except for a single functional call. Instead of advancing capacity demand data among a single plant's capacity buckets, the demand is reallocated to plant 1, which results in a modified volatile capacity demand and a consistent collaborative plan (see Fig. 14.8). Compared with the inconsistent plan, both plants are now producing their vehicles within the capacity frame that has been negotiated.

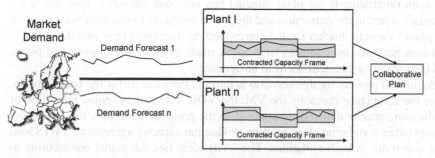

Fig. 14.6 General procedure of the first collaborative planning scenario

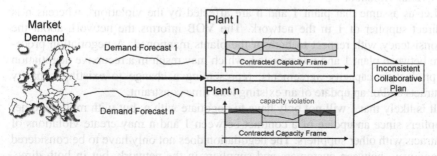

Fig. 14.7 Inconsistent collaborative plan due to capacity violations

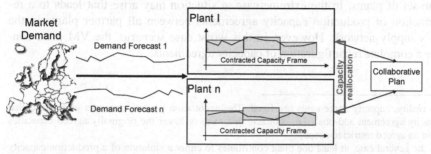

Fig. 14.8 Reallocation of violated capacity

14.2.3 Collaborative Planning – Scenario 3

For all functionalities in collaborative, local and execution planning it is crucial that the VOB contains valid and up-to-date capacity agreements. The VOB checks for violations of valid capacity agreements between plants and lead time constraints of critical BTS suppliers before storing the production capacity allocation. For each violation case the VOB raises an exception, which in turn triggers the collaborative, recursive procedure for renegotiating production capacity agreements, described previously in this book.

Violations imply that the new required capacity breaks either the upper or the lower threshold values that have been previously negotiated[1] with a plant over a certain timeframe. If the plant affected has only one customer, then the upper boundary refers to the maximum and the lower boundary to the minimum value of the plant's capacity bucket for the corresponding timeframe (e.g. one day). If this violation happens, negotiations have to be made between the appropriate plants and the capacity agreement needs to be updated.

Again, the process for this scenario is initially the same as for the first scenario. After the acceptance check by the VM, the VOB stores the accepted allocation of production capacity demand. In contrast to the previous scenarios, the VOB now detects either a violation of an existing production capacity agreement, a BTS lead time constraint or both violations. The VOB identifies the plants contributing to the violation and analyses the situation of these individual plants in the context of the overall violation.

Let us assume that plant 1 and n are affected by the violation[2], whereas n is a direct supplier of 1 in the network. The VOB informs the network about the inconsistency with respect to both of the plants mentioned. A negotiation procedure between plant 1 and n is initiated, which may result in a recursive negotiation of production capacity agreements, representing a change to existing capacity contracts and/or an update of an existing lead time constraint.

It is likely that 1 will not only have to negotiate with n, but with multiple other suppliers since an update of a contract between 1 and n may create violations of contracts with other suppliers. The negotiation does not only have to be considered as one-way, between customers and suppliers in the network, but in both directions. This bi-directional negotiation behaviour is not necessarily restricted to a sub-set of plants. In the extreme case, a situation may arise that leads to a renegotiation of production capacity agreements between all partner plants in the VM's supply network. However, in this worst case scenario, the VM must consider a complete reconfiguration of capacity agreements.

---

[1] In reality, capacity agreements are fixed. The negotiation process concerns what are called "capacity agreement add-ons", i.e. contracts that raise or lower the originally agreed capacities within an agreed restricted scope.

[2] In the general case, at least one plant contributes to either a violation of a production capacity agreement or a lead time constraint of a BTS product.

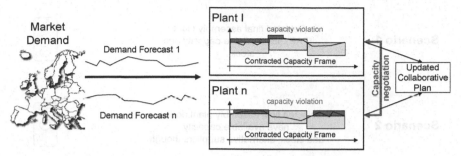

Fig. 14.9 Renegotiation between several plants due to violated capacity agreements

In general, the negotiation process results in an updated consistent collaborative plan, which is made persistent in the VOB (see Fig. 14.9).

14.3 Scenario Examples for the Collaborative Execution Process

The execution scenario examples described in this chapter aim to give a general overview of the functionality of the collaborative execution processes. In order to guarantee delivery reliability of 100%, a new order entry process for customer orders has been developed. A substantial and key factor within this process is the spatially distributed virtual order bank (VOB), which hosts network-wide capacity buckets and stock level information. For every incoming standard customer order the VOB performs a feasibility check in order to guarantee 100% delivery reliability.

The feasibility check incorporates a subset of different functionalities like capacity buckets and stock level checks. Therefore, the VOB must perform a BOM explosion for the critical parts of the specific car and must take into account the supply strategies of the production network.

To demonstrate the functionality of the VOB and the feasibility check at different states of the underlying production system three scenarios were defined as follows (Fig. 14.10).

These scenarios are realised in the production network structure as shown in Fig. 14.11. The pictured network structure consists of different customer regions, FAPs including external assembly service providers and first- and second-tier BTO and BTS suppliers. Figure 14.11 shows the different standard supplier relationships between production units and alternative supplier relationships in the case of a production unit having no additional capacity or stock. In order to achieve short delivery times, the VOB will try to dispatch a standard order to the final assembly plant located nearest to the customer. This behaviour is achieved through the integration of different customer regions into the abstract network structure.

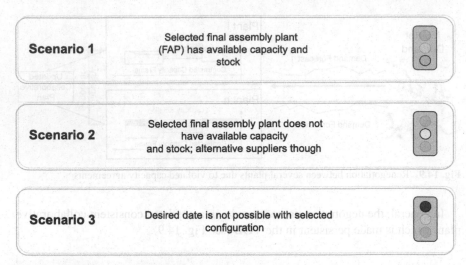

Fig. 14.10 Scenarios of the collaborative execution process

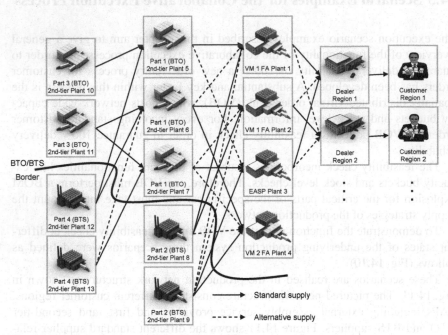

Fig. 14.11 Abstract network structure for execution case study examples

In order to show the relationships between the different actors within the execution case studies, Fig. 14.12 shows them using a unified modelling language (UML). Within the network the VOB plays a key role, with requirements placed upon it, defined as follows:

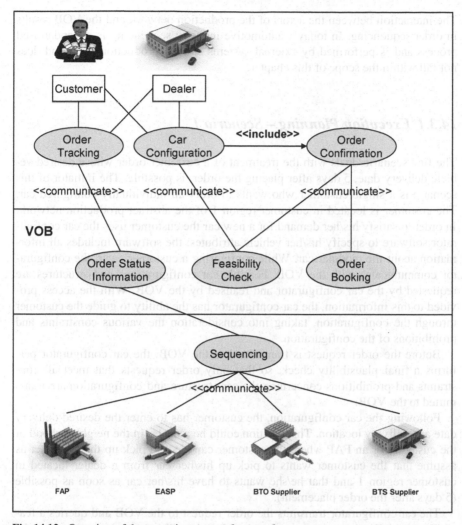

Fig. 14.12 Overview of the execution case study example

- Feasibility check
- Order booking
- Order status information

The external actors for the process scenario demonstration are customers and dealers, who generate the order load within the system. Actors in the production network are FAPs, EASPs, BTO suppliers and BTS suppliers. For dealers and customers the following requirements have been identified:

- Car configuration
- Order confirmation
- Order tracking

The interaction between the actors of the production network and the VOB results in order sequencing. In today's automotive industry sequencing is a standardised process and is performed by external systems at each production plant and does not fall within the scope of this chapter.

14.3.1 Execution Planning – Scenario 1

The first scenario deals with the treatment of a customer order whose desired vehicle delivery date, 5 days after placing the order, is possible. The initiator of this scenario is a single customer who wants to order an individually configured car. The customer is located in customer region 1 of the abstract production network. In order to satisfy his/her demand for a new car the customer uses the car configurator software to specify his/her vehicle attributes; the software includes all information about the modular car. While configuring a customer's car, the configurator communicates with the VOB. Feasible car configurations and structures are requested by the car configurator and realised by the VOB. With the access provided to this information, the car configurator has the ability to guide the customer through the configuration, taking into consideration the various constraints and prohibitions of the configuration.

Before the order request is transmitted to the VOB, the car configurator performs a final plausibility check, so that only order requests that meet all constraints and prohibitions concerning the car structure and configuration are transmitted to the VOB.

Following the car configuration, the customer has to enter the desired delivery date and delivery location. This location could be a dealer in the neighbourhood of the customer or an FAP where the customer can elect to pick up their car. Let us assume that the customer wants to pick up his/her car from a dealer located in customer region 1 and that he/she wants to have his/her car as soon as possible (5 days after the order placement).

The car configurator transmits the order request to the VOB and queries a feasibility check for the customer order. At this point, all information concerning the product configuration, desired delivery date and desired delivery location will be transmitted to the VOB.

In order to perform a feasibility check for the customer order, the VOB has to perform a set of functions. The first step of the feasibility check is the determination of a possible FAP or EASP. Since the available production and distribution lead time is very short for a standard order, the VOB will determine the nearest FAP or EASP to the delivery location desired by the customer. In this scenario final assembly plant 1 is determined to be the nearest one. This functionality also includes a check for available capacity buckets at the assembly resource identified. Therefore, FAP 1 has enough capacity for the assembly of the desired car.

The next step of the feasibility check is the execution of a BOM explosion for the desired car, which is needed to determine all BTO and critical BTS components.

The results of the FAP determination and the BOM explosion are the prerequisites for supplier determination. Each FAP or EASP has prioritised suppliers for each component.

Based on the abstract network structure, supplier determination will identify first-tier plant 5 as the supplier for the BTO component 1 and first tier plant 8 as the supplier for the critical BTS component 2 within this scenario. For the critical second-tier plants, the VOB will determine second-tier plant 10 as the supplier for BTO component 3, which is required for the assembly of BTO component 1 and second-tier plant 12 as the supplier for BTS component 4, which is also required for the assembly of BTO component 1.

The final step of the feasibility check is the capacity and stock level check, wherein the VOB validates the available capacity buckets for each determined BTO supplier plant and the stock levels of the critical BTS suppliers. In this scenario, the nearest FAP and each priority 1 supplier has enough free capacity buckets or stock levels to accept the customer order, so that the car can be delivered on the desired delivery date and at the correct delivery location.

The VOB transmits the results of the feasibility check back to the car configurator. This makes it possible for the customer to validate and book the order. Since all customer requirements are met in this scenario the customer will confirm the order, which is then transmitted to the VOB[3]. In the scenario described, the feasibility check will still be successful, so that the customer order can be booked in the VOB. Therefore, the capacity buckets and stock level information in the VOB will be updated and the information will be distributed in the network.

14.3.2 Execution Planning – Scenario 2

The aim of the second scenario is to demonstrate VOB functionality during the feasibility check when a standard supplier does not have enough capacity to produce a desired BTO component.

A customer located in region 2 wants a new car and configures it using the car configurator, described previously. Once the configuration of the individual car is finished, the car configurator performs a final feasibility check. Then, the customer enters the desired delivery location and delivery date. As mentioned in the first scenario, the customer places a standard order 5 days ahead of the desired delivery date. In this scenario, the VOB determines final assembly plant 2 to be the nearest assembly plant that has enough free capacity for the assembly of the desired car.

---

[3] In reality, no capacity buckets will be reserved based on external order requests. This avoids service attacks in the system and makes another internal feasibility check necessary after the order booking is received.

After the execution of the BOM explosion for all BTO components of the car and the critical BTS components, the VOB has the ability to determine the prioritised suppliers for each relevant component. The determined suppliers are:

- First-tier plant 6 for BTO component 1
- First-tier plant 8 for BTS component 2
- Second-tier plant 11 for BTO component 3
- Second-tier plant 13 for BTS component 4

This time the capacity and stock level check reveals a capacity bottleneck at one of the BTO suppliers identified, in this scenario first-tier plant 6. All other suppliers that are not related to first-tier plant 6 return their available capacity buckets.

The capacity bottleneck at the first-tier supplier necessitates the determination of an alternative supplier for BTO component 1 by the VOB. In the scenario considered the alternative supplier for this component is first-tier plant 7. The VOB now determines all suppliers affected upstream in the value creation process, which are:

- First-tier plant 7 for BTO component 1
- Second-tier plant 10 for BTO component 3
- Second-tier plant 12 for BTS component 4

The change to an alternative BTO supplier without changing the desired delivery date is only possible when the distribution lead times between the alternative suppliers and their customers (n-1 tiers, FAPs, EASPs) are similar. Furthermore, double sourcing has to be a part of the frame contract.

The following feasibility check will be successful and the results are transmitted to the car configurator by the VOB.

Like the first scenario, the order booking functionality of the VOB will subsequently be used for order confirmation by the customer. Once more, the feasibility check will be performed by the VOB and will still be successful, so the customer order can be booked in the VOB. Therefore the capacity buckets and stock level information in the VOB will be updated and the information will be communicated throughout the network.

14.3.3 Execution Planning – Scenario 3

The third scenario aims to demonstrate a network state in which a desired delivery date of a customer is not feasible due to capacity bottlenecks. In this case, the VOB functionality will determine the earliest possible delivery date for the customer's car concerning all allocation alternatives within the network. In addition to the suggestion of a new delivery date, the VOB will determine the critical components that are the reason for the bottleneck. Therefore, the customer is also offered the possibility of changing the configuration of their car to meet a desired date.

The first steps of the order entry process correspond to the previous scenarios. The customer in this scenario is located in customer region 1, which means that final assembly plant 1 is the nearest to them. The capacity check for this plant will show that no capacity buckets for production are available. Therefore, the VOB identifies final assembly plants 2 and 3 as alternative assembly plants. The capacity check for these plants returns available capacity buckets only in final assembly plant 2.

After the execution of the BOM explosion, the suppliers affected will be determined by the VOB and are as follows:

- First-tier plant 6 for BTO component 1
- First-tier plant 8 for BTS component 2
- Second-tier plant 11 for BTO component 3
- Second-tier plant 13 for BTS component 4

Since they are similar to the suppliers affected in the second scenario, the capacity and stock level check will determine a capacity bottleneck at the first-tier supplier plant 6, which has no available capacity buckets for the production of BTO component 1. All other suppliers that are not related to the first tier plant 6 return available capacity buckets.

As an alternative supplier of BTO component 1, the VOB verifies first-tier plant 7 and, as affected suppliers upstream the value creation process, second-tier plant 11 for BTO component 3 and second-tier plant 13 for BTS component 4.

The capacity and stock level check will show that the first-tier plant 7 has no available capacity buckets for the customer order. This means the desired delivery date is not feasible.

In order to make a suggestion for an alternative delivery date and to provide the maximum transparency to the customer, the VOB will determine the critical part or component that causes the bottleneck. In this example, the critical component is BTO component 1.

In the next step, the VOB determines the earliest possible delivery date for the customer order. This is a complex function of the VOB, because all valid combinations of alternative supply chains within the network have to be proven. The VOB has, for each day, predefined capacity buckets. The first possible day for the delivery is 5 days ahead of the order booking. If this day is not possible due to capacity constraints the 6th day is checked. The result of this functionality is the earliest possible delivery date of the car specified by the customer order, taking all network possibilities and restrictions into account.

The VOB will transmit the information collected regarding the critical component and the earliest possible delivery date to the car configurator. The customer may now accept the suggested delivery date or change a critical component in the specified car (e.g. a special variant of the engine) in order to keep the desired delivery date. In this scenario, the customer accepts the suggestion for the change of the delivery date. After the order confirmation is performed, the VOB finally executes the booking functions corresponding to the other scenarios.

14.4 Conclusion

For a 5-day car to become reality, processes coordinating the allocation, adaptation and balancing of capacities need to be radically redesigned. Currently, market demand and production capacity utilisation do not match. Sometimes demand can be much higher than the production capacity utilised. This results in longer delivery times. If delivery times are to be shortened to, for example, 5 days, this problem can be solved by the creation of a network. It is here that collaborative planning and execution processes become necessary. This helps to balance regional fluctuations in demand and solves problems caused by changes in demand due to new modules or face-lifts. In order to reduce the complexity of the interaction between the network participants, an agent-based planning approach has been proposed for the semi-automation of negotiation processes. These agents are given the autonomy to trigger negotiation and to make decisions within a defined option space. Some decisions, however, e.g. investment into additional capacity, needs to be decided by the management.

Scenarios describing what kind of collaborative planning and execution events can happen and how the system reacts have been demonstrated using the prototype software and processes described within this chapter. Application of these processes will facilitate the realisation of the European 5-day car.

Part IV
Validation

Chapter 15
A BTO Reference Model for High-Level Supply Chain Design

Katja Klingebiel

ebp-consulting GmbH, Stuttgart, Germany

Abstract. The automotive industry today faces massive challenges in vehicle discounts and unsold finished vehicle stock due to its outdated build-to-stock (BTS) production system. A build-to-order (BTO) approach can offer a solution to this challenge. Consequently, many innovative concepts have been developed, discussed and piloted in the context of process, structure and product redesign. Nevertheless, all players in an automotive supply network are continuously challenged by the question of which of these concepts is relevant and promising in their individual setting. The main driver of the work presented here is to provide a comprehensive and simultaneously efficient supply chain design process that provides decision support for the transition to a more flexible and stable BTO strategy for all partners in an automotive supply chain. This is accomplished by the provision of a BTO reference model for high-level network and process design. With its classification framework and proposed design procedure, this model acts as a guideline for a supply chain design process for automotive BTO networks that yields cost, time and accuracy benefits.

15.1 Introduction: Motivation and Objectives for a Build-to-Order Reference Model

The automotive industry reacted to stagnating markets and increasing competition during the 1970s and 1980s with a shift from production orientation to market and product orientation. Yet, the first chapters of this book show that with manufacturers providing a proliferation of models to differentiate themselves from competitors, whilst operating production plants with financial objectives and work load targets, the production concept of "lean management" has never reached the market: long planning cycles and consequently long response times to market demand changes are traded in for local efficiency. Thus, the European automotive industry is currently typified by "stock push" vehicle supply, whereby the majority of vehicles are sourced from existing finished goods inventory in the marketplace. This requires holding stock at the most expensive point in the supply chain, finished goods.

Hence, the European industry faces massive challenges in vehicle discounts and unsold finished vehicle stock due to its outdated production system. A build-to-order (BTO) approach may offer a solution to this challenge. Consequently, many innovative BTO concepts in the context of process, structure and product redesign have been developed and have already been discussed in the preceding chapters. Furthermore, many of the accompanying actions and recommendations, as well as requirements, risks and barriers, have been identified.

Nevertheless, for the industry the transition to a BTO system is not as simple as it may seem: conventional build-to-forecast systems are capable of smoothing the wide swings in seasonal demand patterns (Economist 2004). The higher the BTO share in production and the lower the customer-requested order-to-delivery lead times, the more reactivity is required in the automotive chain. Processes, resources and structures need to be completely flexible to react to short-term fluctuations in demand. Thus, the strategic switch will require re-engineering of the complete automotive value stream from the material producers to the end-consumers through a cost-optimised system delivering what the customer really wants without delay.

The implementation of these new concepts will require the integration of many actors and departments. With the automotive industry being one of the most important industry branches in Europe, many jobs are at stake. Consequently, all players in an automotive supply network are continuously challenged by the questions of which of the new concepts are relevant and promising in their individual setting. Furthermore, as adaptations and their effects on given networks and processes are unknown, it is unlikely that the European automotive industry will implement these new concepts ad hoc. Existing structures act to limit radical change, instead forcing changes to be implemented step-by-step.

Consequently, it is only plausible that with any new and promising concept the evaluation of applicability constitutes the first essential step before implementation. In every case, it has to be verified that a new concept transforms the original production or logistics network system into a more efficient one – and that this transition yields economic advantages (Kuhn and Hellingrath 2002). Thus, the challenge of integrating BTO concepts into traditional automotive production and logistics networks can be seen in the assessment of implications on the performance of the overall network system. A comprehensive and simultaneously efficient supply chain design process is needed.

The driver of the work presented here was to develop this process and an underlying methodology that provide this kind of decision support for the transition to a more flexible and stable BTO strategy for all partners in an automotive supply chain. It aims to answer three leading questions:

1. Which BTO concepts are relevant for application in a given industry case, i.e. problem area?
2. How may the transition to a BTO system by application of BTO concepts take place, i.e. what does the BTO to-be system in a given problem area look like?
3. How may customer-orientated performance and overall profitability for a BTO to-be system be assessed?

It is obvious that automotive networks are not suited to testing fundamental and strategically initiated redesigns within the already complex processes and structures. Typically, models mirroring the relevant aspects of processes and structures are being constructed to gain insight into the future design of a supply network system and to ensure that process and structural design scenarios may be backed up by experiments and analysis.

Models that support and facilitate this evaluation step by providing a framework for a selected concept, e.g. BTO, may be characterised as setting points of reference for the future desired industry transition. Thus, the cornerstone of our work in answering the above questions is the provision of a *BTO reference model* for high-level BTO network and process design.

15.2 The Idea Behind a Build-to-Order Reference Model

Whereas the preceding chapters may have given the reader a detailed insight into the innovative concepts behind BTO, the concept of a BTO reference model may be new. The basic purpose of a reference model is that it is a model on the one hand and that it serves as a reference for other models on the other hand.

This definition seems to be a fairly trivial one. It appears not to be accurate enough to define the concept of a reference model. Looking at practical applications as well as academic work, we notice several totally different concepts of reference models resulting from this colloquial use of the term "reference":

- Models that are used to compare different applications or software systems are called reference models.
- Models resulting from successful projects and thus giving a reference for future projects are called reference models.
- Models defining frameworks and sets of building blocks for different applications are called reference models.

In any case the term "reference model" is used more widely and several authors have either developed reference models or worked on their systematic analysis and categorisation. In particular, the third class of reference model is gaining importance in industrial practice, especially in production and logistics. Below are two important reference models of this type:

1. The SCOR model defines a framework for supply chain management as it links business process, metrics, best practices and technology features (Supply Chain Council 2006).
2. The ARIS reference models facilitate the specification and customisation of SAP R/3 systems by provision of standard SAP R/3 business process models and building blocks (Scheer 2002).

Hence, having identified the underlying specifications and definitions of a reference model given in practice and in the literature, we chose to define a reference model accordingly:

"A reference model depicts structures, attributes, relationships and behaviours of objects for a given domain. It is represented in a general, reusable and applicable form, so that specific application models can be created by adaptation and modification. It serves as a recommendation and framework for future modelling and design tasks."

What does this imply for a BTO reference model that attempts to answer the three questions posed in the previous section?

Answering the first question of identification of relevant BTO concepts in a given problem area, a reference model needs to characterise identified relevant aspects of BTO concepts. This characterisation of concepts has to provide transparent insight into targets and fields of action, as well as into answering BTO concepts for the automotive industry. Thus, it also provides indications of promising practice for the specific application in a given industry use case. The characteristics depicted may be called classifications; this part of the reference model may be called a "classification model for BTO concepts in automotive networks".

According to the definition of a reference model the BTO reference model will do more than support the identification of fields of action and promising concepts. BTO concepts shall be presented in such a way that to-be models for specific application in an industry case can be derived to answer the second leading question: what does the BTO to-be system in a given problem area look like? Hence, the reference model needs to represent universally valid and applicable, i.e. recommended, aspects of the BTO concept. In this regard, the reference model gives holistic and accessible insight into innovative BTO concepts to support and facilitate their application.

Consequently, besides the classification of concepts, the integration of concepts represents a substantial part of the BTO reference model. These integrated BTO reference concepts serve as a recommendation and template for the adaptation to the specifics of an industry-driven use case. As new requirements and new concepts will emerge continuously, the ability to expand the integrated and classified concepts is one major requirement for the construction of the model itself.

Answering the third leading question, the "developed" BTO to-be concepts shall be assessable with regard to logistics costs and performance, thus allowing the user to identify risks, barriers and the potential for implementing given BTO concepts in an application or problem area. Hence, suitable evaluation methodologies need to be chosen accordingly. The reference model has to provide the basis for the recommendation of an evaluation method, but it must not be tailored to a specific methodology. Decision support shall be manifested in a procedure model and an application guideline completing the model.

Summarising this argument, the BTO reference model contains three partial models:

1. A classification framework for BTO reference concepts to support the identification of relevant concepts for application in a given industry case, i.e. problem area
2. An integration framework for BTO concepts and the integration of selected BTO concepts into this framework to support the development of a BTO to-be system for a given problem area
3. A procedure model and application guideline for the migration of theoretical BTO concepts into real-life implementation

These three partial models shall be discussed and presented in the following.

15.3 A Classification Framework for Automotive Build-to-Order Concepts

In order to characterise the BTO concepts, it is necessary to take a closer look at the motivation for the strategic shift in the European automotive industry.

The previous discussions in this book point out that the automotive market has shifted from a supply-driven market to an aggressive demand-driven market. Under these conditions, the European automotive industry has to strengthen its focus on the customer: customer orientation is being seen as one of the key competitive factors today and the main driver for BTO (Corsten and Gabriel 2004). Thus, the BTO target system for suppliers and OEMs is customer-orientated.

Consequently, based on industry awareness of a stronger orientation towards BTO, the point of departure in an industry case can be characterised by the non-achieved customer-orientated targets. An in-depth analysis of as-is processes and structures may identify the actual associated problem areas. Hence, any industry case striving to implement BTO is characterised by non-achieved customer-orientated targets and associated problem areas.

Each innovative BTO concept is aiming at a specific customer-orientated target. Furthermore, it touches specific fields of action that are dependent on certain problem areas. Thus, it may be concluded that relevant and promising concepts for an industry case are those that face the non-achieved targets and problem areas identified. By matching these BTO concept characteristics and application area characteristics, relevant and applicable concepts can be narrowed down. A BTO classification framework should contain criteria based on customer-orientated targets and problem areas, i.e. associated fields of action.

15.4 A BTO Target System

The customer-orientated target system for the automotive industry has been identified by several authors (Stautner 2001; Verein Deutscher Ingenieure 2000; Supply Chain Council 2006). To summarise those targets that will indicate the future success of the European automotive industry with regard to customer orientation, the target system comprises the following:

• Delivery lead time: providing the customer-desired lead time from order to delivery to the customer.
• Delivery capability: offering the desired product in the desired configuration on the desired delivery date to the customer.
• Delivery flexibility: providing the customer with the possibility of changing the ordered product configuration for as long as possible.
• Delivery reliability: meeting the promised delivery date.
• Delivery quality: meeting the promised quality.
• Information capability: providing order status transparency during order processing.

The preceding chapters in this book discussed the point of departure in the automotive industry and were able to show that these targets are not met sufficiently today.

Several studies identified desired vehicle order-to-delivery lead times of not more than 3 weeks. The 3DayCar study (3DayCar Research Team 2003) showed that young customers especially, i.e. less than 25 years old, desire a short lead time: a higher-than-average number, 84%, claim an adequate delivery time of not

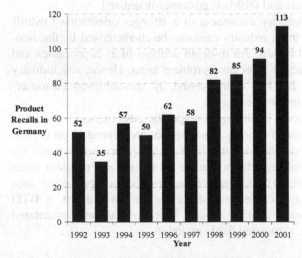

Fig. 15.1 Product recalls in Germany (Automobilproduktion 2006)

more than 2 weeks. Yet, the typical average delivery time in Europe is more like 40–60 days (Miemczyk and Stone 2005).

And more than half of all vehicles sold are from stock. Due to high product variance these stocked cars seldom meet the customer-desired configuration (Baumgärtel et al. 2006), often forcing the customer to accept a different product. Furthermore, quality problems and product recalls are still prevalent. Figure 15.1 illustrates the situation in the German market.

Furthermore, the automotive industry is usually not able to give an exact delivery date to the customer, allow changes to ordered vehicles or just to keep to the promised delivery date. Consequently, a Japanese, not a European, OEM, Toyota, led the industry as the most satisfying brand in Germany in 2005 (see Table 15.1).

Table 15.1 Customer satisfaction index (CSI) study 2005 – winner per segment (JD Power and Associates 2005)

Segment	Model
Small car	Toyota Yaris Verso
Lower medium	Toyota Corolla
Upper medium	Toyota Avensis
Executive/luxury	BMW 5er
Sports car	Mercedes-Benz CLK
MPV	Toyota Corolla Verso
SUV	Toyota RAV4

15.5 BTO-Related Problem Areas and Associated Fields of Action

The automotive industry is torn between increasing dynamics on one side and increasing complexity on the other. Automotive manufacturers as well as suppliers are orientated towards financial targets; this drives forecast-driven and not demand-driven processes within traditional function-orientated structures.

Build-to-order requires a continuous and holistic adaptation of complex production and logistics networks to constantly changing environments and markets. This demands flexible and fast-adapting business processes (Beckmann 2004). The current processes cannot cope sufficiently with the challenges faced, as they are based upon the optimisation of single sections. Hence, customer-specific BTO products are integrated into these kinds of systems with difficulty. And the necessary adaptability is mostly achieved by building up inventories.

Accordingly, seven key problem areas and associated fields of action can be identified for a transition to BTO.

15.5.1 Process Flexibility

Manufacturing and production of automotive products constitutes one of the most capital-intensive industry branches. This is mainly caused by production technology for the vehicle itself. A reduction of the capital employed is difficult due to the level of technology and automation in vehicle production. Nevertheless, the given capacities are not utilised optimally; prevalent amongst the symptoms is high over-capacity.

Hence, the primary objective requires the optimal and flexible utilisation of the given capacities, i.e. the achievement of process flexibility that meets the requirements of volatile and competitive markets. Today's dominating production constraints, which hinder the flexible processing of customer orders, need to be minimised. Also, product structures have to be developed that support flexible and customer-orientated production processes.

Product complexity can be identified as another of the major structural causes of a lack of process flexibility. On the one hand, product complexity is caused by external variance, i.e. the high level of customer-offered choice in vehicle configuration. As European OEMs identify individualisation of vehicles as a key success factor, variance necessarily has to cover those vehicle attributes that are relevant for competition within the targeted market segments. Every non-competition-relevant and non-distinguishable variant must be reduced to a minimum. This also implies that product complexity in the non-perceptible part of the car needs to be avoided.

The resulting simplification and standardisation of automotive products provide the basis for achieving the process flexibility required by BTO. By reduction of internal product complexity, a higher level of flexibility can be guaranteed, e.g. as vehicles may be configured to the specifics of an order in a late production step. Also, platform strategies and other modular product structures support product standardisation and thus help to reduce internal complexity. Hence, future BTO product structures are characterised by a high level of modularity.

However, our focus of supply chain design focuses on process and network design, not product design, so we will regard product concepts as basic mechanisms to support process-wise and structural BTO concepts.

To gain process flexibility, the simplification of network complexity is primarily relevant. The various and numerous relationships of production locations to their global and local suppliers need to be reduced as far as possible to make inter-business relationships manageable. Product standardisation and product complexity reduction constitute a driver for this. By product modularisation and shifting of value-creating steps down the supply chain, a producer can limit the number of direct relationships to a few key suppliers. This allows much better control, and it is anticipated that remaining suppliers will be able to react to changing ambient requirements more flexibly.

Nevertheless, it is obvious that the main advantage lies in the process-related implications of structural simplification, not the structural change itself. Thus, structural simplification acts as a basic mechanism to support the process challenges.

15.5.2 *Process Stabilisation*

Market pvolatility destabilises processes, but may be controlled and managed by introducing process and production flexibility. Ways of limiting this destabilising influence on processes is being addressed by BTO concepts.

Process stability is improved by decoupling processes from customer orders and hence from market-driven volatility. Modularisation of automotive components and of the vehicle itself is one major driver of these decoupling mechanisms. It supports the order-anonymous preproduction of standardised modules with low variance (example: painted bodies). Furthermore, modularisation often supports the late differentiation or configuration of products, thus implicating the production of standardised products in preceding processes. This implies higher process stability for these processes.

But a lack of demand and subsequent planning stability will destabilise all subsequent processes. It is obvious that reduced planning revisions and associated information would mean that processes are less exception-sensitive and thus delivery reliability is improved. Stabilisation of processes by robust planning is a key field of action. An associated aspect is a "frozen" planning period as it guarantees suppliers stability within a certain timeframe. An important example of this is the BTO "pearl chain" concept, which aims to maintain a "once planned" order or production sequence (pearl chain) throughout the complete production and distribution process (Weyer 2002).

Nevertheless, it is necessary to design these stable time windows according to the delivery times desired by the customer. This applies to suppliers as well as OEMs.

15.5.3 *Process Standardisation*

Process complexity today often prohibits the stabilising of processes. Hence, BTO concepts strive to standardise processes by implementation of uniform mechanisms and support of standardised ICT systems. This allows not only for stabilisation, but also for simplification of supporting processes, e.g. the associated material supply processes of standardised production processes. Both performance and cost advantages are anticipated.

15.5.4 *Process Acceleration*

Stable and standardised processes are not the complete answer: long planning cycles, batch processing and the early freezing of programmes prevent short-term customer orders from being integrated into the planned or frozen programme.

Consequently, these orders have to queue up at the end of the order pool – often behind dealer-specified stock orders. The direct result is long delivery lead times for the customer. Furthermore, other products are being built without triggering customer orders (i.e. demand) (Stautner 2001). Recent studies were able to identify the current average delays in order processing (see Fig. 15.2 below).

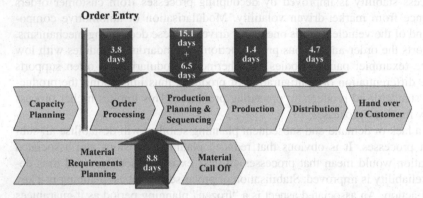

Fig. 15.2   Average delay in processes related to order processing (Holweg and Pil 2004)

In order to provide short order-to-delivery lead times to the customer, process lead times – especially information processing lead times – need to be reduced dramatically. As Fig. 15.2 reveals, the most important processes that need to be accelerated are the production planning and sequencing processes.

15.5.5 Process Transparency

As a result of technological developments the European automotive industry will continue to remain complex and not all processes can be standardised or stabilised – especially not by decoupling from market demand. Thus, suitable mechanisms are needed to ensure that all relevant market and demand information is communicated to all the parties and processes involved without delay. In particular, the direct communication of planning results will guarantee that all partners gain transparency concerning market and customer demand. This increased quality of demand information also increases the forecast accuracy, thus providing inventory level benefits.

The OEMs need to refrain from forecast-orientated mechanisms like volume commitments and quota arrangements in order to maximise the utilisation of real market information. In addition, on suppliers' demand planning of BTO parts and modules has to be mainly based on order information.

15.5.6 Process Synchronisation

Yet, these fields of action are not sufficient to avoid negative effects like the bull-whip effect or the Forrester effect in the supply chain. Process transparency and process acceleration between supply chain partners need to be synchronised in order to allow for the optimal and prompt processing of information and provision of an adequate supply of material, parts and modules. Supported by innovative IT systems new interoperable communication technologies need to be realised. These may offer the non-interrupted information flow that provides the basis for real-time queries and fast order processing.

Process synchronisation is often realised by collaboration. Previous chapters presented planning and control processes that realise a synchronisation of demand and capacity planning by real-time collaboration based on a so-called virtual order bank.

15.5.7 Failure Prevention and Reaction

The named fields of action may standardise, synchronise, accelerate and stabilise processes and increase process transparency. Yet, failures endangering reliability and lead times may still occur. Process transparency alone does not allow the prompt reaction to unforeseen events that endanger delivery reliability and process stability. To protect the value chain against breakdowns and exceptions, continuous monitoring of processes is indispensable. But furthermore, innovative BTO concepts contain mechanisms to immediately alert partners of potential breakdowns. Failure prevention remains a relevant action to realise a profitable and feasible BTO network system.

15.5.8 Summary and Conclusion

Automotive BTO networks need to manage the order processing and product delivery of customer-specific vehicles in a few days. Innovative and highly integrated structures and processes characterise this new strategic orientation of build-to-order. The required flexibility of automotive production implies robust, fast and transparent processes across all logistics centres. The point of departure for all kinds of innovations is the network-wide, integrative and target-orientated design of the supply network and its processes. The implementation of BTO concepts is necessary to guarantee the achievement of the six relevant customer-orientated targets. Key factors are the underlying processes, product and network structures within seven identified fields of actions:

1. Process flexibility
2. Process standardisation
3. Process stabilisation
4. Process acceleration
5. Process transparency
6. Process synchronisation
7. Failure prevention and reaction

As stated before, the classification framework of a BTO reference model should allow the identification of relevant BTO concepts for an industry application by matching the concept-specific characteristics with those of the application area. Together with the target system identified, these fields of action provide the basis for the targeted classification. Whereas BTO concepts may be classified by targets and focussed fields of action, the application area in an industry case is characterised by non-achieved customer-orientated targets and problem areas directly associated with the seven fields of action.

It is necessary to analyse whether this classification includes all characteristic aspects to deduce the potential applicability of a specific BTO concept. One classifying factor is clearly still missing. Assume that two BTO concepts have the same targets and fields of action, e.g. the improvement of delivery reliability by standardisation of processes. Whereas the first concept is aimed at standardisation of production processes, the second concept is aimed at standardisation of demand planning processes. An industry case – characterised by poor delivery reliability due to complex and thus not properly managed production processes should consider the application of the first concept; yet, not necessarily the second one. Hence, the classification of all process related fields of action completed by an additional specification of the process context.

This scope is naturally given by the scope of supply chain management, as logistics comprises the design, planning and execution of all material and information flow in a supply network. Based on this definition, we defined the building blocks of process classification accordingly.

Material flow processes can be divided into production, storage, transport and test processes (Kuhn and Hellingrath 2002); planning processes into demand planning, supply network planning, supply planning, production planning, distribution planning, inventory planning and transport planning (see Kuhn and Hellingrath 2002; Fleischmann and Meyr 2002). Execution processes mainly comprise control and management processes like the BTO-relevant order management process. A further information flow process is communications (Fig. 15.3).

The question remains whether the basic fields of action of product standardisation and structural simplification need to be detailed by classification of product and structure.

Successful product-related BTO concepts have been developed and piloted within enterprises, but have been highly specific to individual products and technologies and also to their position in the supply network. In these cases, we know that the product concepts may only be transferred with great difficulty into other

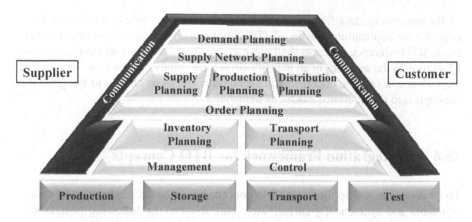

Fig. 15.3 Build-to-order-relevant process types

product environments. However, the associated process fields of action are more readily re-applied. Hence, it is highly relevant to classify the process type affected by the concept, as this constitutes a relevant point of focus for the BTO reference model.

Structural aspects are commonly divided into topology and organisational and communicational structure. Topology focuses on the dimensional arrangement of resources like machines, assembly lines, equipment and warehousing. Organisational structures represent the responsibilities for processes and decisions that are manifest in the roles of organisational units. The technical communicational structure comprises the interfaces of information systems. BTO concepts may be based on one or more of these areas. It is noted that the individual organisational concepts may differ greatly from each other, e.g. topological and communicational. In an analogy with the product structure concepts, the structural concepts only provide the basis for process-related BTO concepts. Thus, a further refinement of the classification is not reasonable as we solely regard the field of action of structural simplification as a basic mechanism to support the process-related concepts to be classified (Fig. 15.4).

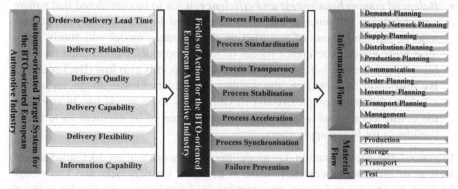

Fig. 15.4 Classification framework for BTO concepts

By transparent classification of reference concepts, a source of common knowledge for the application domain of BTO supply chain design is provided. Furthermore, BTO reference concepts shall serve as a recommendation and template for the adaptation to the specifics of an industry-driven case study or "use case" in a second step. Hence, a suitable documentation method needs to be chosen to integrate the concepts into the reference model, as demanded.

15.6 An Integration Framework for BTO Concepts

To be re-usable and applicable a BTO concept has to be documented in an understandable and syntactically correct way. This leads to the following postulations on the documentation methodology (Klingebiel and Miemczyk 2006):

- *Common terminology*: in order to be understandable and applicable, the documentation form should reflect the common terminology of the application domain of supply chain design. To support the dissemination and re-use of a reference model, it is helpful if the form of documentation is commonly known and widespread in the application domain.
- *Easy to use*: the form of documentation should allow for the easy construction of the reference model and its subsequent application models. Graphical toolsets supporting this procedure are of special importance if complex models are constructed. Hierarchical language concepts support systematic access to reference models and facilitate their step-wise adaptation.
- *Modularity*: to allow for adaptation and reuse in as many cases as possible, it is useful to represent the reference model with a modular structure. This serves more than just the purpose of easily generating application models. Reference models always represent subjective know-how of the application domain. Therefore, if a reference model is used by an individual outside the group of creators, it may be necessary to expand or modify the reference model. Modularity may support this issue: parts of the model can be removed, replaced, expanded or modified without influencing the rest of the model.
- *Balanced comprehensiveness and efficiency*: it has to be considered that a more comprehensive reference model is more efficient for constructing application models, but the process of constructing the reference model will surely introduce more effort. A reference model of low complexity will broaden its flexibility of application, but its possible decision support is also low (Becker 2004). By restriction of scope, i.e. the number of problem classes to be supported, a balance between complexity and flexibility has to be aimed for. In any case, the construction process of a reference model has to be efficient in terms of usefulness for future application.

In general, the documentation, i.e. modelling, of network and product structures as well as of business processes, is used to transport explicit facts and knowledge between businesses and users, thus making its knowledge available for external application. But the processes of a logistical network, as well as their interrelation, structural embedding and performance are complex. It is extremely complicated to present every aspect in a single chart, diagram or view. Thus, many modelling languages do not cover all aspects, i.e. emphasise only the structural or process aspects. As a result many holistic approaches combine different kinds of modelling languages or at least different views on the modelling constructs. Also, the level of abstraction varies considerably.

A detailed analysis shows that for our purpose the Unified Modelling Language (UML – a standardised specification language for object modelling) activity diagrams are suited best. The UML model allows us to interpret UML actions as logistics tasks or processes. The modelling constructs of UML activity diagrams were customised for application in the defined domain of BTO supply chain design.

Before the customised UML could be applied to document promising BTO concepts a supporting modelling environment needed to be chosen. We selected the GME (graphical modelling environment) (Klingebiel and Miemczyk 2006). The GME is database-based and allows the integration of many modelling languages including the graphical representation of the associated modelling elements.

Today, the BTO reference model incorporates concepts under development across the ILIPT project as well as commonly known concepts, e.g. the "late configuration concept", which we will use as an example of the documentation of BTO concepts.

This concept is based on the decoupling of BTS and BTO processes through late configuration of products by, for example, a logistics service provider. The production of standardised components that are configured in a later state in the supply chain can help to keep production and inventory costs low. The high-level view of the concept's documentation as integrated into the reference concept is depicted in Fig. 15.5. Each modelling element is detailed at lower levels of this representation.

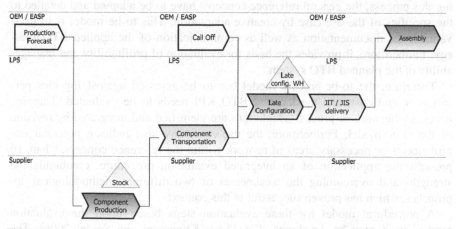

Fig. 15.5 High-level view of the integrated reference concept "late configuration"

Consequently, this modular and easy-to-use documentation is applied not only to the reference concepts, but also to develop a to-be model. The underlying supply chain design procedure is discussed next.

15.7 Supply Chain Design Procedure Model

When deciding on a BTO strategy, clarification is needed of which reference concept and aspects best accomplish the transition to a more flexible and stable logistics system. Thus, the first step is to specify the scope of the targeted built-to-order strategy. This is obtained by identifying the essential business segments, including suitable product and customer groups.

Furthermore, each of these business segments can be characterised by its BTO motivation, which naturally translates into BTO objectives. It is advisable to methodically map these strategic BTO objectives into measurable key performance indicators by application of a strategic KPI (key performance indicator) system designed for this task.

The next step is the analysis of the current system, especially its processes. By analysis of those KPIs chosen as BTO objectives, the performance gap between the current and the envisioned system becomes comprehensible. Thus, the scope and aim of the supply chain design process are determined.

Subsequently, concurrent documentation of processes and structures is conducted to identify weak points and problem areas. Here, the BTO strategy and objectives serve as means of restraining the scope of this task. Based upon those results, fields of action and relevant reference concepts can be identified. By taking into account the existing restrictions and opportunities, a limited number of applicable reference concepts are selected for further analysis.

These identified and applicable reference concepts are integrated into the "as-is system" and the "to-be model" for each industrial use case that is developed. During this process, the general reference concepts have to be adapted and detailed to the specifics of the use case by creative adaptation. This to-be model is in itself valuable for documentation as well as communication of the applied BTO strategy. Furthermore, it provides the basis for evaluation of profitability and manageability of the planned BTO system.

Therefore, the to-be-process model has to be assessed against logistics performance and cost, and the targeted BTO KPI needs to be evaluated. Thereby, necessary network or process adaptations are identified and integrated by revision of the to-be-model. Furthermore, the use case may also indicate potential improvements or necessary areas of re-work within the reference concept. Chap. 16 presents the application of an integrated evaluation procedure, combining the strengths and overcoming the weaknesses of two different methodological approaches, which has proven successful in this context.

A procedural model for these evaluation steps based upon the evaluation method used may be developed (Fig. 15.6) (Klingebiel and Seidel 2006). The

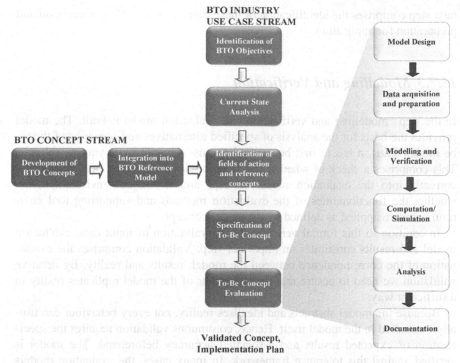

Fig. 15.6 Supply chain design procedure model (Klingebiel and Seidel 2006)

general method for proceeding within an evaluation step is separated into the phases model design, data preparation, modelling and verification, scenario computation/simulation, analysis and documentation. Parallel to these steps, the models, scenarios and results are continuously validated, which may initiate a loop back to previous steps to refine the model and even the scope, objectives or reference model concept.

15.7.1 Model Design

During this phase, a general overview of the evaluation model is developed, which is in the most part related to the to-be model, but may vary in the details, depending on the specifics of the evaluation tool applied.

15.7.2 Data Acquisition and Preparation

Usually evaluation of concept feasibility requires more complex model data; hence, the to-be model must be completed with additional data. Therefore, the next step comprises the identification of these data, as well as their acquisition and preparation for application.

15.7.3 Modelling and Verification

In the step "modelling and verification" an evaluation model is built. The model provides the basis for the analysis of specified alternatives and scenarios of the to-be model. Thus, it has to first be verified against semantic and syntactic failures. This comprises a check of whether the to-be model concept has been transferred correctly into the evaluation model concept and the model environment, i.e. whether the functionalities of the evaluation methods and supporting tool environment were applied as defined by the model concept.

In addition to this formal verification, the validation of input data, evaluation model and results constitutes an important step. Validation comprises the examination of the correspondence between the model, results and reality. By iterative validation we need to ensure that the behaviour of the model replicates reality in a sufficient way.

Because the model abstracts and idealises reality, not every behaviour can usually be studied in the model itself. Hence, continuous validation requires the specification of expected results and associated accuracy beforehand. The model is verified against this tolerance framework. In many cases, the evaluation models replicate stochastic effects parameterised by distributions. The characteristics of a distribution are replicated best when many random samples are executed. Consequently, it is often necessary to execute a high number of test runs within the validation phase to gain reliable insight into the model's behaviour. The outcomes of these test runs have to be discussed with industry experts.

15.7.4 Computation and Simulation

In the next phase, alternative scenarios are defined. The evaluation model is altered as specified and outcomes computed or simulated. Often, the outcomes of one run cause new questions and thus scenarios. Thus, evaluation is often executed in systematic trials where previous results inspire new test runs.

15.7.5 Analysis

The results of the test runs are measured by the previously defined KPIs. Thus, the resultant data, which often comprises basic measures, is aggregated to KPIs. Most evaluation methods and tool environments provide interfaces for external analysis tools that allow specific and individual processing of result data.

15.7.6 Documentation

In the documentation phase the evaluation results are summarised and prepared for interpretation and discussion. As this documentation constitutes the basic information source for the later decision regarding which BTO concepts to apply, in reality, a clean and accurate analysis of results is necessary.

This basic procedure gives a guideline for the supply chain design process within the context of BTO concepts. Within several industry-driven projects the application of this basic procedure has proven its efficiency. Some of those are outlined in the following chapters.

15.8 Summary and Outlook

Valuable starting points for supply chain design tasks are provided by BTO reference concepts. These concepts cover process architectures from the strategic level to a short-term operational level based on product-structural and network-structural changes. They recommend product standardisation and associated collaborations between network partners. Network structures are affected concurrently, i.e. the partner relations and locations of plants, hubs, etc. The aspects involved range widely and the three aspects of network structure, product structure and process are strongly interconnected.

By the classification of objectives, targets and fields of action within the BTO-orientated European automotive industry, the BTO reference model presented resolves the question of which concepts to apply in a given problem area.

The provision of integrated concepts allows the model-based development of a specific BTO to-be system for a given problem area. Thus, the reference model presented facilitates and supports the transition to a BTO system by the simulated application of BTO concepts.

With its proposed design procedure, the model also acts as a guideline for a supply chain design process for automotive BTO networks that yields cost, time and accuracy benefits. The evaluation experiments will provide data on anticipated performance, costs and risks of adapted and applied BTO concepts, thus demonstrating the customer-orientated performance and overall profitability of proposed BTO systems.

References

3DayCar Research Team (2003) Towards a Customer Driven System, 3DayCar Summary Report. Available via http://www.3daycar.com

Automobilproduktion (2006) Qualität: Rückrufaktionen in Deutschland. Available via http://www.automobil-produktion.de/themen/00149/index.php

Baumgärtel H, Hellingrath B, Holweg M, Bischoff J, Nayabi K (2006) Automotive SCM in einem vollständigen Build-to-Order-System. Supply Chain Manag 1:7–13

Beckmann H (2004) Supply-chain-management. Springer, Berlin

Corsten D, Gabriel C (2004) Supply-chain-management erfolgreich umsetzen, 2nd edn. Springer, Berlin

Economist (2004) Fighting back. Economist 2 September, pp 8–10

Fleischmann B, Meyr H (2005) Advanced planning, In: Stadtler H (2005) Supply chain management and advanced planning, 3rd edn. Springer, Berlin

Holweg M, Pil FK (2004) The second century: reconnecting customer and value chain through build-to-order. MIT Press, Cambridge

JD Power and associates (2005) Toyota ranks highest in overall customer satisfaction in Germany for fourth consecutive year, Munich, 29th June 2005. Available via http://www.jdpa.com/studies_jdpower/pressrelease3.asp?ID=2005088

Klingebiel K, Miemczyk J (2006) Deliverable 7/8.3.2 – BTO reference model. Information Societies Technology (IST) and (NNP) Joint Programme, Project: Intelligent Logistics for Innovative Product Technologies (ILIPT), Project number: IST/NMP-2004-507597

Klingebiel K, Seidel T (2006) Deliverable 8.4.2 – Methodology for the dynamic inter-enterprise network process design and evaluation. Information Societies Technology (IST) And (NNP) Joint Programme, Project: Intelligent Logistics for Innovative Product Technologies (ILIPT), Project number: IST/NMP-2004-507597

Kuhn A, Hellingrath B (2002) Supply-chain-management. Springer, Berlin

Miemczyk J, Stone G (2005) Deliverable 7./8.2.1 – White Paper: holistic BTO network design vision for the future of BTO-networks in 2015. Information Societies Technology (IST) and (NNP) Joint Programme, Project: Intelligent Logistics for Innovative Product Technologies (ILIPT), Project number: IST/NMP-2004-50759720357701

Scheer A-W (2002) ARIS – vom Geschäftsprozess zum Anwendungssystem. Springer, Berlin

Stautner U (2001) Kundenorientierte Lagerfertigung im Automobilvertrieb. Deutscher Universitätsverlag, Wiesbaden

Supply Chain Council (2006) Supply-chain operations reference-model. SCOR version 6.1 overview, Pittsburgh. Available via http://www.supply-chain.org. 19.07.2006

Verein deutscher Ingenieure (2000) VDI-Berichte 1571: Innovationen in Logistikstrukturen der Automobilindustrie. VDI-Verlag. Düsseldorf

Weyer M (2002) Das Produktionssteuerungskonzept Perlenkette und dessen Kennzahlensystem. Helmesverlag, Karlsruhe

Chapter 16
Rapid Supply Chain Design by Integrating Modelling Methods

Thomas Seidel

4flow AG, Berlin, Germany

Abstract. Supply chain design is a task that has to both be economically efficient and provide viable design results. Therefore, the demand placed on supply chain design is for modelling methods that rapidly and efficiently provide key performance indicators for the networks assessed to a desired level of granularity and to a sufficient degree of realism. Several such modelling methods exist that each divergently fulfils these requirements. Of special interest in automotive supply chain design are static scenario comparison and dynamic simulation. The former, because it is easy to comprehend and straightforward to apply; the latter, because it is so powerful and its results are so desirable. Integrating both methods leads to significant advantages when modelling during supply chain design. The combined approach taken provides quick and efficient evaluation of a large number of scenarios at a suitable level of granularity and provides a higher degree of realism than just applying a static scenario comparison. Static modelling provides an adequate initial assessment of supply chain design and reduces the number of scenarios to a set of viable networks. These are automatically transferred to a dynamic simulation where assessment at a detailed level of granularity and lower level of abstraction takes place. Thus, all possible network alternatives are assessed quickly. The user gains efficiency and speed, whilst having the best data and key performance indicators possible for decision-making.

16.1 Introduction

16.1.1 The Challenge at Hand

Today's companies demand flexibility and reactivity in all business areas. Competition stems from all over the planet. A customer's loyalty is challenged by unprecedented transparency of alternative offers and value for money. Global logistics systems mean products of choice are only hours or days away. This nurtures the customer's wish for ever-more individualised products of the highest quality – from the cheapest provider. We consequently see a change from a sellers' to a buyers' market.

Technological progress in all business areas and the success of information and communication technologies (ICT) accelerate business processes in an unprecedented way. An immediate consequence is the curtailment of product life cycles, the time span available for the amortisation of investments on product and production facilities, as well as for generating profits. Companies have to place their products on the market faster in order to achieve a sufficient return on investment.

Companies actively addressing and mastering these challenges gain great potential from these abilities. To face global competition, local wage advantages are utilised. Tasks not part of the company's essential value creation processes are outsourced. Core competencies are focussed on. High product individuality is offered to the customer and, in combination with the cost reductions realised, this leads to improved overall performance.

Enabling business processes by means of IT is perceived as a chance to implement closer contact and information exchange with suppliers, service providers and customers for accelerating order fulfilment.

These challenges and opportunities directly impact upon the prerequisites corporate planning faces: building and managing global value creation networks with a large number of participants. The goal is the rapid, high-quality and efficient satisfaction of customer wishes under ever-decreasing timescales.

The requirement placed on supply chain management is apparent: it needs to rapidly address the demand for flexible, reactive and near optimal design of value creation networks.

16.1.2 Goal

This challenge is faced in supply chain design, the long-term design of value creation networks. Supply chain alternatives need to be goal-orientated, i.e. achieve performance goals at minimal cost. Simultaneously, nothing less than the realistic depiction of the value creation network is demanded in order to minimise uncertainties in decision-making and to ensure the feasibility for implementation of the proposed alternative. At the same time the effort for this design task should be minimised and thus the design process made efficient, leading to results in the shortest time possible.

The supply chain design process is supported by a number of methods and corresponding software. Prerequisites equally comprise the rapid and efficient modelling and evaluation of design alternatives. Some modelling methods have proven their applicability in everyday practice. Among them are mathematic optimisation, static scenario comparison, dynamic simulation and system dynamics. All of these fulfil the prerequisites posed to different degrees according to their ability to consider system characteristics.

The goal of this chapter is to present the work performed on creating a method of supply chain design that better fulfils the prerequisites posed. This is achieved by a combination of the methods mentioned in an integrated method that overcomes the limitations of individual methods.

16.2 Supply Chain Design

This section presents a short overview of supply chain design basics.

16.2.1 Framework Conditions of Supply Chain Design

Supply chain design is under the influence of ever-increasing globalisation and increasing customer wishes, as well as technological progress, especially in the field of IT (Christopher 1998; Bovet and Martha 2000; Corsten and Gabriel 2002; Konrad 2005; Kuhn and Hellingrath 2002). These framework conditions have ultimately lead to the development of supply chain management, as previous management concepts could not sufficiently cope with these challenges (Fig. 16.1).

Globalisation has been caused by a liberalisation of world trade. This has led to increasing customer demand. An immediate consequence of globally extending value creation structures and the increasing number of value creation partners to

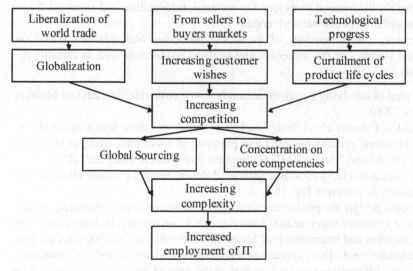

Fig. 16.1 Framework conditions of supply chain design

be coordinated is an increase in complexity. The number of locations, as well as the number of interrelating processes, grows significantly. Goal-oriented design of these networks is the original task of supply chain design and this complexity can only be managed efficiently and effectively by the utilisation of IT.

16.2.2 Defining Supply Chain Design

Supply chain design comprises the long-term-orientated, strategic area of supply chain management (Chopra and Meindl 2004), while supply chain planning covers the functions and processes of planning the supply chain and supply chain execution covers the functions of the execution level (Holthöfer and Lessing 2004). The objective is the design of value creation networks under consideration of logistics strategy, supplier selection, supply relationships and location problems (SCM Competence and Transfer Center 2003). Supply chain design as an independent task determines framework conditions for the consecutive phases of supply chain planning and supply chain execution (Beckmann 2004; Hellingrath et al. 2004; Persson and Olhager 2002). Its goals are provided by a superordinated supply chain management strategy and the associated goal of efficient and rapid satisfaction of the customer wishes (Beckmann 2004; Chopra and Meindl 2004; Kuhn and Hellingrath 2002; Seidel 2006).

Supply chain design can be defined as the company-spanning design of value creation and logistics networks for a strategic time horizon starting from the products and their associated markets.

Supply chain design:

- Determines the general structure of the network
- Determines the general processes for material, information and financial flows within and in-between network nodes
- Determines the dimensions of locations, processes and relations based on planned volumes of raw materials and finished products as well as capacity resources

with the goal of satisfying custumer demands at low costs (Hellingrath and Mehicic Eberhardt 2006).

The tasks of supply chain design can be grouped into three levels of detail, i.e. network planning, process planning and planning of functional areas, as shown in Fig. 16.2 (Wolff and Nieters 2002; Hellingrath and Mehicic Eberhardt 2006).

Within these levels, questions regarding different fields of supply chain design are addressed, as shown in Fig. 16.3.

In process design the production strategy of customer order-orientated (make-to-order) or customer order-neutral (make-to-stock) production is determined. The number, location and interrelation of locations for production and logistics are part of the structure field. Their capacities are dimensioned in the field of resources. Harmonising different system loads is part of the field of interfaces.

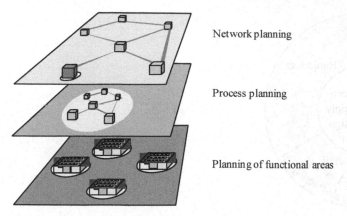

Fig. 16.2 Levels of detail in supply chain design (Wolff and Nieters 2002)

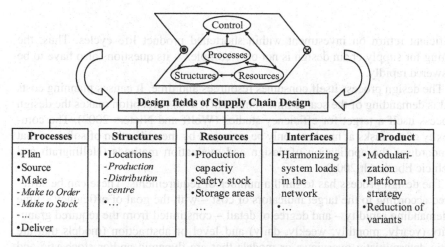

Fig. 16.3 Design fields of supply chain design (Hellingrath and Mehicic Eberhardt 2006)

Product design is also very important (SCM Competence and Transfer Center 2003; Levy 1997; Handfield et al. 1999). A product, through its structure and thus interrelated production processes, as well as through its physical properties, contributes greatly to the determination of all other fields of design (Corsten and Gabriel 2002; Stock and Lambert 2001).

16.2.3 Requirements for the Supply Chain Design Process

It has become clear that supply chain design is immensely complex. Additionally, its framework conditions put time pressure on all partners involved. Designing and evaluating network alternatives has to be performed rapidly to ensure

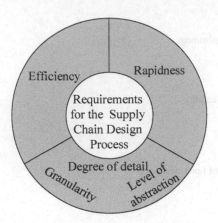

Fig. 16.4 Requirements for the supply chain design process

sufficient return on investment within shortened product life cycles. Thus, the setting for supply chain design is not only complex – its questions also have to be answered rapidly.

The design process itself consumes resources and time. It causes planning costs and is demanding of the scarce time commodity. This perception makes the design process itself a target for efficiency studies (Wolff and Nieters 2002). The complexity of the tasks at hand can only be managed by the utilisation of software that standardises the application of design and evaluation methods (Hellingrath and Mehicic Eberhardt 2006).

The design process has to fulfil a number of requirements. These can be structured according to the target indicators of cost – with the goal of efficiency – time – demanding rapidity – and degree of detail – constituted from the required granularity (yearly, monthly, weekly, daily) and level of abstraction (models that are static deterministic overviews or models that are dynamic and/or stochastic and reflect reality more closely). These requirements are presented in Fig. 16.4.

The degree of detail required for key performance indicators (KPI) is variable over the course of time (Seidel 2006). In the early phases of the design process, KPIs of low granularity and/or a high level of abstraction may be sufficient for decision-making. In later stages, the KPIs need to be in greater detail. Thus, the granularity of KPIs needs to increase and their level of abstraction decrease.

Simultaneously, the number of feasible design alternatives is reduced over the course of time. While at the beginning of the design process a large number of scenarios seems feasible, on the basis of successive evaluations their number decreases steadily until only a few remain that are assessed with detailed KPIs.

Thus, the requirements of the supply chain design process are clearly defined. In the target triangle of costs, time and degree of detail, the necessary KPIs for evaluating network alternatives have to be generated as efficiently and rapidly as possible and with the required detail.

16.3 Modelling and Evaluating Value Creation Networks

This section presents the basics of modelling and several modelling methods. These methods are then evaluated for their suitability for the supply chain design process.

16.3.1 Modelling

In order to enable system analysis a system has to be described or mapped at a chosen level of detail. The reasons for this are that the complexity of systems prohibits their comprehensive penetration and analysis, it is difficult to gain insights by directly investigating the system, and the efforts regarding time and resources or the associated risks of such an endeavour would be too high (Stachowiak 1973).

Transferring a system from reality into a mental, manipulable depiction is known as modelling. A model is a system that has originated by purposeful, abstract depiction of another system (Krallmann et al. 2002). Models always fulfil the characteristics of depiction, abbreviation and pragmatism, i.e. models of something are built for someone, for a certain time interval, at a given level of abstraction and for a defined purpose (Stachowiak 1973; Krallmann et al. 2002; Kuhn and Wenzel 2004).

16.3.2 Modelling Methods in Supply Chain Design

Several modelling methods and corresponding software are available for supply chain design. Their differentiation can be made according to their consideration of system characteristics. System analysis fundamentally distinguishes between the structure and behaviour of systems. Structure comprises a static view mapped onto the system, while behaviour depicts its dynamics, i.e. the processes running within the existing structures (Scheer 2002).

Structures can be perceived as static, i.e. invariant over the course of time. Structures of value creation networks, too, are invariant on short- and mid-term time horizons. Taking a longer term perspective, structural changes are understood to be the means or goal of network design. As these changes are of a long-term, predictable or plannable nature, their static consideration in modelling is adequate (Weber 2002).

Structures feature deterministic characteristics. As they are either an invariant or an actively changeable design parameter, their behaviour is foreseeable and is not characterised by stochastic peculiarities. Structural design is of a long-term nature and often connected to far-reaching investment decisions. The goals of modelling and evaluation are to ensure cost-effectiveness and efficiency of structural alternatives.

System behaviour is dynamic, i.e. consecutive time periods influence each other. Processes are, for example, affected by waiting cues and sequence turbulence that span several time periods. The design of value creation and logistics processes has to consider process dynamics in order to assure their feasibility, adequacy and efficiency. Simultaneously, processes are shaped by stochastic system behaviour. Status cannot be predicted deterministically. Their characteristics and, consequently, their appearance, can only be described by probability distributions. Process design is short-term in character in comparison to structural design with regard to the goal of process feasibility. At the same time, it pursues adequate processes to fulfil the performance goals posed.

It becomes clear that different modelling methods are essential for the design of structures and processes respectively. While structural decisions are of long-term character and are analysed using static-deterministic system characteristics, process questions focus on system behaviour and achieving performance goals with a background of dynamic-stochastic influences.

In this way modelling methods can be distinguished. We present an evaluation of "static scenario comparison" and "mathematic optimisation", representative of structure modelling methods. We also assess "dynamic simulation" and "system dynamics" as methods of process assessment. Assessment is made according to the requirements for the supply chain design process, i.e. efficiency, speed, granularity and level of abstraction.

16.3.2.1 Static Scenario Comparison

Static scenario comparison is the automated calculation of KPIs for analysing supply chains based on cause-and-effect relationships describing system behaviour and analysing the effects of changing these relationships. Temporal behaviour is depicted statically; probabilistic behaviour is represented deterministically. Static scenario comparison has its origins in planning practice and has grown popular by the increased availability of spread sheet applications. Software applications have taken this approach and integrated it into standard software environments. Its focus of application is the rapid and efficient generation of static-deterministic KPIs for network design.

Essential for the application of this method is the comparison of different scenarios. An as-is status of a network is depicted and validated and alternative solutions for the planning case are created and documented in additional scenarios. These scenarios are then compared and best cases identified. The potential for improved solutions is heavily influenced by the experience and knowledge the user is able to apply when generating the alternative scenarios.

Static scenario comparison is limited to static and deterministic system characteristics in modelling for supply chain design. This directly impacts the degree of fulfilment for the requirements posed. The modelling effort is modest due to the limited consideration paid to system characteristics; efficiency is therefore high. This fact also supports the rapid generation of scenarios and leads to the overall rapidity of the method. It is characterised by a medium level of granularity. While product/network structures and processes can be finely modelled, demand values are considered as averages, often for monthly or yearly periods. Static modelling, in many cases, does not make use of finer granularity. The method provides a mid level of abstraction. It neglects dynamic and stochastic influences and so the KPIs generated have to be considered with these limitations in mind, especially when used for predictions of system behaviour, e.g. the level of inventory over the course of time.

16.3.2.2 Mathematic Optimisation

Mathematic optimisation also focuses on modelling and analysing network structures. While static scenario comparison is a descriptive method, mathematic optimisation is prescriptive, i.e. the former describes the effects of parameter changes, while the latter provides additional default values on parameter characteristics in order to achieve a certain target. Mathematic optimisation utilises decision or optimisation models. In addition to cause-and-effect relationships, these comprise target equations for assessing and selecting alternatives. Thus, an optimisation model is the formal depiction of a decision problem, where the most suitable, i.e. optimal, alternative regarding the targeted goals is to be selected (Scholl 2004). In mathematic models all aspects depicted are described by cardinally measurable metrics. Elements of the real system are represented by parameters and variables and are interrelated to each other by equations. Assessing such quantitative models is the task of operations research. Due to the long-term character of the changes in network structures, temporal behaviour is most often represented statically (Weber 2002).

In summary, mathematic optimisation extends the description of system behaviour by target equations in order to find optimal network parameters. System modelling is static and deterministic. Stochastic optimisation models are often not efficiently solvable. The relative efficiency of mathematic optimisation can be considered "medium" in the context of the supply chain design process. The effort required to model parameterisation and overcome structural model defects is significantly higher than for static scenario comparison. Similarly, rapidity is considerably lower. Parameterisation and solving structural defects has to be performed to generate a current state network, as well as for modelling and analysing further alternatives. This delays the assessment of multiple scenarios. The results produced can also be considered to be of low granularity. To optimise models of a realistic size, it is necessary to reduce problem complexity as this will ensure solvability and provide efficiency. This is achieved by taking the depiction of reality to a higher lever of aggregation and abstraction. Mathematic optimisation is of "medium" abstraction. When compared with static scenario comparison and dynamic and stochastic systems we find characteristics are neglected. The KPIs generated have to be considered under these limitations. In addition, many optimisation problems cannot be solved – only acceptable solutions generated by heuristics. Furthermore, the solution found might not always be the global optimum for the problem, but rather a local one. Both of these aspects limit the original target of optimality of the method.

16.3.2.3 Dynamic Simulation

Dynamic simulation has developed from material flow simulation to the simulation of complex value creation networks. Simulation is defined as the depiction of a system with its dynamic processes in an experimental model in order to bring insights that are transferable to reality (VDI 2000). By systematically varying model parameters in repeated simulation runs, a pointedly quantitative investigation of model behaviour is undertaken. The basis of modelling is the description of cause-and-effect relationships, extended by temporal system development achieved through dynamic modelling. Dynamic simulation obtains its validity through the consideration of stochastic characteristics (Kuhn and Wenzel 2004). While deterministic consideration is supported, the user gains a greater level of realism through stochastic influences. Dynamic simulation does not strive for optimality, but is a descriptive method. Dynamic simulation models the course of time discretely, and here it differs from system dynamics, which depicts the course of time continuously.

Dynamic simulation comprises cause-and-effect relationships as well as their behaviour over the course of time. Additionally, these relationships are differentiated by the probability of their appearance. Thus, dynamic simulation depicts systems dynamically and stochastically. This ensures a high degree of representative reality and relates to the fulfilment of supply chain design process requirements. However, the efficiency of dynamic simulation is rather low. The additional effort required for formulating and placing parameters upon system behaviour and stochastic interrelations is very high, especially compared with the aforementioned methods (Kaczmarek 2002; Liebl 1995; Rabelo et al. 2003; Rall 1998). It is usually necessary to expend this effort for the base scenario and also for each alternative. This causes significant time consumption and slows the process. As it presents a high degree of reality, dynamic simulation achieves a very high level of granularity. Some sample applications include single-object tracking and capacity and demand fluctuation over the course of time (Rabelo et al. 2003). Dynamic simulation also presents reality with a low level of abstraction.

Dynamic simulation of system behaviour allows for close-to-reality modelling and provides detailed KPIs. All in all, dynamic simulation provides highly granular and realistic system modelling, while requiring a great deal of modelling effort and long modelling times.

16.3.2.4 System Dynamics

Ever since the fundamental works on urban dynamics by Forrester (1958) system dynamics has been applied to modelling interactions and flows between different elements in complex systems (Sterman 2006). Differential equations depict the interactions between different subsystems and the influence of delays. Therefore, system dynamics models are aggregated models that predominantly describe holistic rates of change instead of specific events within the system modelled (Rabelo et al. 2003). The system dynamics method focuses on the modelling and evaluation of processes. The essential difference with dynamic simulation is the continuous depiction of time. The passage of time is not based on events or discrete time steps, but instead appears continuous through the application of differential equations. Furthermore, the underlying cause-and-effect relationships are represented deterministically (Buchholz 2006). Modelling stochastic effects is rather difficult (Abelev 2000) and may be seen as currently not sufficiently matured (Ridalls et al. 2000). System dynamics is a method that provides understanding of the dynamics of systems on an aggregated level and offers design guidelines in supply chain management.

For logistics systems this approach is unsuitable as the identification of individual objects and their effect on the system is not possible (Kuhn and Hellingrath 2002; Ridalls et al. 2000; Holweg and Frits 2005). By the deterministic depiction no stochastic interrelations can be considered. Quantifying these interrelations requires a great deal of effort, especially with increasing numbers of system elements and their mutual influences. Thus, system dynamics models are aggregated models with interrelations that are simplified when compared with reality. The efficiency of system dynamics modelling is consequently considered as low, as detailed models demand a great deal of effort to determine parameters and for validation. This also affects the speed of modelling. With reduced complexity, modelling might be faster, but for models with the level of detail required in supply chain design, the speed is lower. Only low levels of granularity can be achieved. The requirement for the modelling to be undertaken at an aggregated level does not allow for more detail, e.g. in the depiction of product structures and differentiation between standard and optional components (Holweg and Frits 2005). System dynamics delivers a medium-level degree of detail. While stochastic effects are neglected, the consideration of dynamics is a big asset. However, system dynamics seems less suitable for modelling and evaluating processes in supply chain design due to the aggregation and simplifications required, as well as the low degree of granularity and the lack of representation of stochastic effects.

16.3.3 Assessing the Modelling Methods

Analysing the modelling methods has shown that each of them is characterised by a focus on certain system characteristics and therefore bears distinct advantages

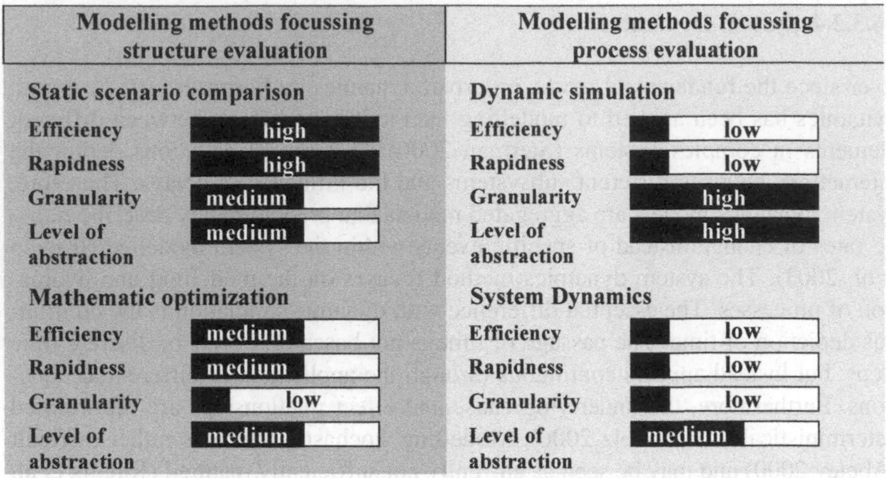

Modelling methods focussing structure evaluation		Modelling methods focussing process evaluation	
Static scenario comparison		**Dynamic simulation**	
Efficiency	high	Efficiency	low
Rapidness	high	Rapidness	low
Granularity	medium	Granularity	high
Level of abstraction	medium	Level of abstraction	high
Mathematic optimization		**System Dynamics**	
Efficiency	medium	Efficiency	low
Rapidness	medium	Rapidness	low
Granularity	low	Granularity	low
Level of abstraction	medium	Level of abstraction	medium

Fig. 16.5 Degree of performance of modelling methods

and disadvantages. The qualitative comparison in Fig. 16.5 is presented in the context of method application in supply chain design and with the reasoning that only the separate comparison of structural methods and process methods is sensible, as they are very different in their purpose and modelling characteristics. At the same time statements within these groups clearly define the characteristics to be expected when these methods are applied in the supply chain design process.

Static scenario comparison presents higher efficiency and rapidity compared with mathematic optimisation. At a similar level of abstractions its granularity is higher. Thus, static scenario comparison will be pursued as a method of structural evaluation in supply chain design.

Dynamic simulation demands a great deal of effort and requires longer modelling times to deliver models with the degree of reality required. While system dynamics remains on a lower level of granularity, dynamic simulation allows for very high granular modelling. Considering stochastic system characteristics in dynamic simulation also results in lower levels of abstraction. Therefore, dynamic simulation will be applied for the evaluation of processes following static comparison.

This comparison also determines the integration of both methods into the supply chain design process. Static scenario comparison allows for modelling and rapidly evaluating a large number of scenarios based on KPIs of medium granularity and levels of abstraction. Subsequently, the resulting low number of feasible scenarios is analysed at a high level of granularity and a lower level of abstraction by dynamic simulation.

This section presents the integrated method of static scenario comparison and dynamic simulation for application in supply chain design and describes the process of applying this method.

16.4 An Integrated Method for Supply Chain Design

The challenge of supply chain design can briefly be summarised as the rapid and efficient answer at the degree of detail required to the questions stemming from the design of value creation networks. The integrated method of static scenario comparison and dynamic simulation will address this challenge.

Initially, a large number of design alternatives is possible in principle and therefore have to be evaluated. The large number of alternatives leads to a time-consuming evaluation process to adequately investigate them all. Effort for this evaluation is to be minimised in order to come to decisions efficiently. This goal is opposed by the large number of potential alternatives and the effort required in assessing them. The KPIs for evaluation and selection of alternatives have to be realistic and reliable.

The integrated method leverages the advantages and avoids the disadvantages of the respective individual modelling methods. Static scenario comparison allows for the rapid and efficient assessment of a large number of scenarios and provides KPIs with a medium level of detail. Dynamic simulation analyses the remaining scenarios in detail with higher detail KPIs, of lower abstraction, as presented in Fig. 16.6.

This method is acceptable, as the degree of detail of KPIs from the static scenario comparison is sufficient to provide a reliable selection from among the large number of scenarios possible. The approach makes it unnecessary for all the alternatives to undergo the detailed assessment of dynamic simulation, as the static scenario comparison allows for the differentiation between feasible scenarios and identifies the scenarios to be excluded. The integrated method is advantageous as it better fulfils the requirements posed compared with the application of individual

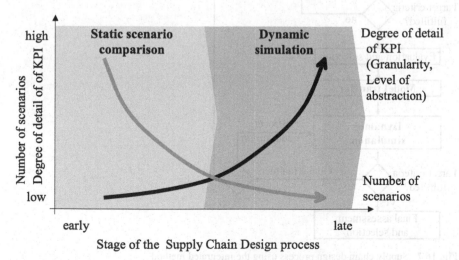

Fig. 16.6 Integrated method of static scenario comparison and dynamic simulation

methods. The rapidity of the supply chain design process is higher, as the greater numbers of possible scenarios are assessed by static scenario comparison, which is relatively fast. Consequently, only a few feasible scenarios are selected and analysed using the more time-consuming dynamic simulation. This step is accelerated further as large parts of the model are transferred via an interface from static scenario comparison to dynamic simulation and therefore are quickly and effortlessly available to help complete the dynamic simulation model. The lion's share of model generation in dynamic simulation is therefore omitted. In this latter stage, remaining scenarios are analysed by KPIs of high granularity and low abstraction. Thus, the alternative scenarios analysed at the end of the supply chain design process have been assessed using KPIs with the greatest degree of realism and detail possible.

The integrated method promises to deliver a rapid and efficient supply chain design process with final scenario assessments made using the most detailed KPIs.

16.4.1 Supply Chain Design Process for the Integrated Method

The supply chain design process for the integrated method orientates itself in its basic form using the elements and procedures provided by VDI and Wolff and Nieters (Wolff and Nieters 2002; VDI 2000). It is depicted in Fig. 16.7.

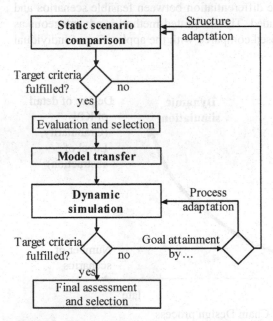

Fig. 16.7 Supply chain design process using the integrated method

16.4.2 Implementing the Integrated Method

Next, we will describe an actual industrial implementation of the integrated method. After a brief introduction of the underlying software products, the interface developed and its application are presented.

16.4.2.1 4flow vista

4flow vista is software for supply chain design that is based on the static scenario comparison method. It has been developed and is marketed by 4flow AG. Its industrial origins lie in the design of networks for the automotive industry. The software is built up from modules that allow case specific application.

4flow vista allows for comprehensive and detailed modelling of value creation networks, restricted by the underlying modelling method's static-deterministic system characteristics. The strengths of 4flow vista are its ability with static structure-focussed KPIs for costs, inventory, lead time or capacities. Its weakness is that dynamic process KPIs cannot be assessed. The analyses can be applied to arbitrary sections and elements of a network. Modelling is efficient and rapid due to the moderate data requirements.

16.4.2.2 OTD-NET

OTD-NET is software for network design based on the dynamic simulation method. It has been developed by Fraunhofer IML and is used for process-orientated network modelling and evaluation, specifically for the analysis of the order-to-delivery process, i.e. from customer order to delivery to the customer. OTD-NET originated in the automotive industry and is applied mainly by OEMs. The software consists of three modules, the Graphical Modelling Environment, the OTD-NET simulation core and the OTD Analyser.

OTD-NET allows for very detailed and realistic modelling of value creation networks based on the underlying dynamic simulation methodology. It is especially useful in the modelling of complex information flow, material flow and planning processes. The effort expended in outlining the model's scope and producing the required quality for input data is very high. However, the user-friendly graphical modelling environment reduces this effort. OTD-NET generates detailed KPIs for the dynamic behaviour of the system, which determines the logistics performance of the value creation network, e.g. delivery lead times and delivery reliability as well as inventory, utilisation and cost figures over time.

16.4.2.3 The Interface Between 4flow vista and OTD-NET

Congruent model constituents are transferred between 4flow vista and OTD-NET via an interface between the two software applications and an executable simulation model is generated automatically.

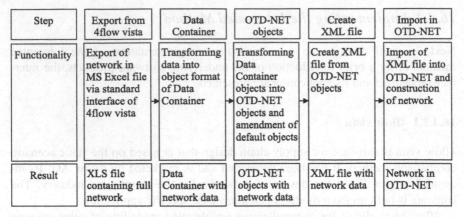

Step	Export from 4flow vista	Data Container	OTD-NET objects	Create XML file	Import in OTD-NET
Functionality	Export of network in MS Excel file via standard interface of 4flow vista	Transforming data into object format of Data Container	Transforming Data Container objects into OTD-NET objects and amendment of default objects	Create XML file from OTD-NET objects	Import of XML file into OTD-NET and construction of network
Result	XLS file containing full network	Data Container with network data	OTD-NET objects with network data	XML file with network data	Network in OTD-NET

Fig. 16.8 Operation of the interface between 4flow vista and OTD-NET

The interface transfers data for locations, product structures, demands, transport relations and containers as well as material flow processes modelled from these elements. Missing data on information flow processes are added using a number of default parameters. The simulation model is instantly executable. It has to be amended by manipulation of the relevant simulation parameters and subsequently validated before it can be applied and sound analyses produced.

To operate the interface a consistent process was developed, applying the standard interfaces based on MS Excel in 4flow vista and XML in OTD-NET. All conversion process steps from one format to the other are fully integrated and implemented in a Java application. This process and its single steps are presented in Fig. 16.8.

16.5 Application Cases

The integrated method has been applied in a number of industrial case studies in order to validate it, prove its applicability and quantify its potential in fulfilling the requirements posed by the supply chain design process. A short overview of three case studies is provided here.

16.5.1 Global Networks Facing Increased Market Demand

This case study investigated structural questions for the development of the global production and distribution network of a motorcycle manufacturer. Currently, the manufacturer operates a production plant in central Europe, has most of his suppliers in central Europe and ships the final products to ten globally distributed distribution centres. Against a background of growing demand, with growth of different intensity in the various regions, the question of whether and where to locate additional production facilities was posed. The question was addressed by assessing a total of eight scenarios. Within these, the number of production facilities was varied from one to three, with production location choice made between Europe, North America and Asia, and the supplier structure adapted depending on the location of the production facilities. The scenario analysis resulted in KPIs for the total cost of logistics, the total cost, taking into account price differences for local sourcing, and average expected delivery times to the custumer.

The scenario with a total of three production sites and local sourcing strategy provided the best solution. Compared with the current network, the logistics costs dropped by 15%, total costs decreased by 17% and a decrease in the average delivery time to the customer of 70% was achieved.

The interface transformed approximately 140 model objects in 4flow vista into 1,300 model objects in OTD-NET, proving the efficiency of the integrated method.

16.5.2 Product Design for Logistics

The supply network for the future modular product structure of a vehicle cockpit was compared with that of a current cockpit. The scenario comparison showed that cost penalties were not incurred in the supply network due to a move from the current state to the modular product structure. Simultaneously, the dynamic simulation showed that the delivery lead time and reliability goals aimed for could be achieved.

The interface allowed for the efficient transfer of both scenarios. For each scenario approximately 250 model objects in 4flow vista were transformed into 4,300 model objects in OTD-NET.

16.5.3 Strategic Network Design for Future Automotive Production

In a strategic case study for the European automotive industry, future modular and conventional products, centralised and decentralised network structures, and associated logistics processes were assessed. Key results gave indications for the effects of different design measures and their advantages and disadvantages regarding logistics costs, logistics performance and pollutant emissions. The detailed results are presented in Chap. 19.

For the scenarios assessed, 2,200 model objects from 4flow vista were automatically transferred and transformed into 14,000 model objects in OTD-NET.

16.6 Conclusion

Supply chain design is an essential constituent of supply chain management as it influences the consecutive phases in supply chain management and determines a significant share of the supply chain costs incurred during subsequent operation. This chapter identified the demand for a supply chain design process that allows for rapid, efficient and realistic modelling and evaluation. Existing modelling methods have been found lacking in these requirements as their individual characteristics lead to limitations in fulfilling these performance criteria.

An integrated method of static scenario comparison and dynamic simulation was proposed and elaborated to overcome the shortcomings of the individual methods. The integrated method is efficient, rapid and provides KPIs of the granularity and levels of abstraction requested in the course of the supply chain design process. The advantages of the approach have been proven in several industry case studies and it can be readily applied in further supply chain design projects, as this work has utilised commercially available software products.

References

Abelev L (2000) Ereignisbasierte und kontinuierliche Simulation. Lehrstuhl für Softwaretechnologie, University of Dortmund

Beckmann H (2004) Supply chain management: Grundlagen, Konzept und Strategien. In: Beckmann H (ed) Supply-Chain-Management – Strategien und Entwicklungstendenzen in Spitzenunternehmen. Springer, Berlin, pp 1–97

Bovet D, Martha J (2000) Value nets. Breaking the supply chain to unlock hidden profits. Wiley, New York

Buchholz P (2006) Modellgestützte Analyse und Optimierung. Systeme und Modelle. Lehrstuhl für Praktische Informatik, University of Dortmund. Available via http://www4.cs.uni-dortmund.de/Lehre/0641127/Folien/MAO_1.pdf. 10.07.2006

Chopra S, Meindl P (2004) Supply chain management – strategy, planning and operations, 2nd edn, Pearson Prentice Hall, Upper Saddle River

Christopher M (1998) Logistics and supply chain management, 2nd edn. Financial Times Prentice Hall, London

Corsten D, Gabriel C (2002) Supply Chain Management erfolgreich umsetzen. Grundlagen, Realisierung, Fallstudien. Springer, Heidelberg

Forrester JW (1958) Industrial dynamics. Productivity Press, Portland

Handfield R et al (1991) Involving suppliers in new product development. Calif Manag Rev 42(1):59–82

Hellingrath B, Mehicic Eberhardt S (2006) Werkzeuge für die Gestaltung der Supply Chain. In: Software in der Logistik – Marktspiegel. HUSS, Munich, pp 94–97

Hellingrath B, Laakmann F, Nayabi K (2004) Auswahl und Einführung von SCM-Softwaresystemen. In: Beckmann H (ed) Supply-Chain-Management – Strategien und Entwicklungstendenzen in Spitzenunternehmen. Springer, Berlin, pp 99–122

Holthöfer N, Lessing H (2004) Grundlagen neuerer SCM-, APS-Ansätze. http://www.competence-site.de/pps.nsf/95EFD7D2306B023DC1256949007717C8/$File/4-grundlagen-scm-aps.pdf. 27.10.2004

Holweg M, Frits KP (2005) Second century. Reconnecting customer and value chain through Build-to-Order. MIT Press, Cambridge

Kaczmarek M (2002) Definition von Anforderungen an die Modellierung und Analyse der Supply Chain. SFB599 Modellierung großer Netze in der Logistik. Technical Report 02007, University of Dortmund, Dortmund

Konrad G (2005) Theorie, Anwendbarkeit und strategische Potenziale des Supply Chain Management. Deutscher Universitätsverlag, Wiesbaden

Krallmann H et al. (2002) Systemanalyse im Unternehmen. Vorgehensmodelle, Modellierungsverfahren und Gestaltungsoptionen, 4th edn. Oldenbourg Wissenschaftsverlag, Munich

Kuhn A, Hellingrath B (2002) Supply chain management. Optimierte Zusammenarbeit in der Wertschöpfungskette. Springer, Berlin

Kuhn A, Wenzel S (2004) Simulation logistischer Systeme. In: Arnold D et al (eds) Handbuch Logistik, 2nd edn. Springer, Berlin pp A2-41–A2-60

Levy DL (1997) Lean production in an international supply chain. Sloan Manag Rev 38(4):94–102

Liebl F (1995) Simulation: problemorientierte Einführung, 2nd edn. Oldenbourg, Munich

Persson F, Olhager J (2002) Performance simulation of supply chain designs. Int J Prod Econ 77(3):231–245

Rabelo L et al (2003) A hybrid approach to manufacturing enterprise simulation. In: Chick S et al (2003) Proceedings of the 2003 Winter Simulation Conference. Society for Computer Simulation International, San Diego, pp 1125–1133

Rall B (1998) Analyse und Dimensionierung von Materialflußsystemen mittels geschlossener Warteschlangennetze. Institut für Fördertechnik und Logistiksysteme (Wissenschaftliche Berichte des Institutes für Fördertechnik und Logistiksysteme der Universität Karlsruhe (TH), Karlsruhe

Ridalls CE et al (2000) Modelling the dynamics of supply chains. Int J Syst Sci 31(8):969–976

Scheer A-W (2002) ARIS – vom Geschäftsprozess zum Anwendungssystem. Springer, Berlin

Scholl A (2004) Grundlagen der modellgestützten Planung. In: Arnold D et al (eds) Handbuch Logistik, 2nd edn. Springer, Berlin A2-1–A2-9

SCM Competence and Transfer Center (2003) SCM Marktstudie 2003: Supply chain management software – Planungssysteme im Überblick. Fraunhofer IRB, Stuttgart

Seidel T (2006) Integrating static scenario comparison and dynamic simulation: a combined methodology for supply chain design. In: Thoben K-D et al (2006) Proceedings of the 12th International Conference on Concurrent Enterprising. Centre for Concurrent Enterprise, Nottingham, pp 123–130

Stachowiak H (1973) Allgemeine Modelltheorie. Springer, Vienna

Sterman JD (2006) Business dynamics – systems thinking and modelling for a complex world. McGraw Hill, New York

Stock JR, Lambert DM (2001) Strategic logistics management. McGraw-Hill, Boston

VDI (2000) VDI-Richtlinie 3633: Blatt 1: Simulation von Logistik-, Materialfluss- und Produktionssystemen. Grundlagen. VDI-Handbuch Materialfluss und Fördertechnik, vol. 8. VDI-Verlag, Düsseldorf

Weber H (2002) Asset optimization im Fokus der strategischen Netzwerkoptimierung. Supply Chain Manag2:1–10

Wolff S, Nieters C (2002) Supply chain design – Gestaltung und Planung von Logistiknetzwerken. In: Pradel U, Süssenguth W (eds) Praxishandbuch Logistik. Deutscher Wirtschaftsdienst, Cologne, sect. 3.5

Chapter 17
Moving Towards BTO – An Engine Case Study

Michael Toth[1], Thomas Seidel[2], Katja Klingebiel[3]

[1] Fraunhofer-Institut für Materialfluss und Logistik, Dortmund, Germany
[2] 4flow AG, Berlin, Germany
[3] ebp-consulting GmbH, Stuttgart, Germany

Abstract. This chapter introduces a case study that focuses on a first-tier supplier delivering engines to several OEM vehicle plants. Assessing the global product strategy of the engine plant analysed and its OEMs, it could be foreseen that the current build-to-stock (BTS)-orientated production system will change to a more flexible, customer-orientated production system with fewer safety stocks and shorter lead times. To avoid large engine inventory and enable more flexible, cost-efficient production the supplier is urged to switch to a stockless BTO production and a just-in-sequence (JIS) supply of engines within 4 days of call-off. Consequently, the case study covers the challenges in implementing BTO at this first-tier supplier, i.e. the development and validation of a new logistics concept, including planning and material flow processes for production, supply concepts for engine modules, late configuration and JIS delivery concepts for engines. The visionary state for the engine assembly provides remarkable progress towards the BTO paradigm and demonstrates a real-world application of the BTO paradigm. Furthermore, it demonstrates the successful application of the guidelines and methods described in the chapters before.

17.1 Scope of the Case Study

The main objective of the case study is to support the implementation of BTO strategies at a first-tier engine supplier and to evaluate the potential and risks of applying such a customer order-orientated production system within a small time window between call-off and delivery. The second objective of the case study is to identify the requirements for the implementation of just-in-time (JIT)/just-in-sequence (JIS) strategies at eligible second-tier suppliers. The third objective of this case study is to analyse the potential of a combined static analysis and dynamic simulation approach for the evaluation of BTO scenarios.

The current processes at the engine plant and its OEMs, as well as the future BTO concept, will be described before the results of the case study are illustrated in detail.

17.1.1 Current State Processes

The European automotive OEM, supplied by the engine manufacturer under study, has BTO processes that are highly customer-orientated with regard to both their sales and production. The OEM's objective is to ensure that the customer receives his individual vehicle on an agreed date, ideally a date requested by the customer. The minimum cycle time between vehicle order receipt and customer delivery is currently 10 working days. The customer should have the option to change certain specifications of his order up until a few working days before completion of the vehicle. One hundred percent delivery reliability to the customers is obligatory.

The process consists of three major sub-projects: online ordering, tracking and tracing within distribution and a new production system (the latter being the driver for the case study). The new production system builds upon the concepts of late order tagging, a frozen stable horizon and order-decoupled pre-fabricated painted vehicle bodies. The assembly sequence stability is guaranteed by a sufficient stock of painted bodies. This is achieved by producing a small number of bodies in each of the specific variants.

Thus, a longer stable horizon for planning and production is implemented within the process by freezing the sequence of orders from the point of order dispatching. Furthermore, this supports greater flexibility on the horizon before order dispatch. This benefits first-tier suppliers who supply just-in-time to the standard production request. This production request incorporates not only the final order, but also the requested date of arrival for a component and the sequence in which the components should be supplied. By setting the stable planning horizon, production and delivery lead time is extended considerably by this information enhancement – from a few hours in the past to 4 working days now.

The current logistics process at the engine supplier is split into two sections. The assembly of "base" engines with low numbers of variants follows a BTS paradigm. The late configuration process configures the base engines resulting in a high number of variants utilising a BTO process responding to the 4-day call-off from the vehicle plants. These processes are decoupled using a base engine storage centre, with an average inventory of up to 3 working days. The call-off of base engines from the storage utilises a standard production request. The late configuration process in the engine plant takes place in the sequence required by the vehicle plants. In the current logistics process, the late configuration is the order penetration point.

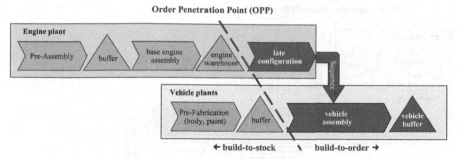

Fig. 17.1 Current state logistics process at the engine plant

The inbound network of the engine plant is organised using standard BTS strategies with consumption-driven safety stock levels. The supply chain processes, upstream from this point, are triggered without any reference to the final customer order, instead using forecasts. Thus, the second-tier suppliers produce according to the forecast and store the finished parts in internal warehouses and at the high-bay warehouse of the engine plant.

The current process guarantees a constant assembly workload for the base engine production lines and a stable availability of parts. It also provides the flexibility required for JIS supply to the European vehicle plants. However, this is achieved by holding large inventories of engines and incurring high capital commitment costs, due to the BTS strategies applied. Hence, the challenges are to remove these cost centres and enable reliable BTO production with decreased inventory stocks, reliable processes and short lead times. Figure 17.1 illustrates the current state process of the engine supply chain. The order penetration point is located before the late configuration process at the engine plant.

17.1.2 To-Be BTO Processes

With an expected growth in the number of base engine variants and new IT solutions to speed up production planning, the objective for the future is to shift the logistics and production paradigm for base engines from BTS to pure BTO. In future, the production schedule for base engines and the late configuration process will be triggered by the standard production request, removing the 3 days' worth of base engine storage. The pull from the vehicle plant will tighten internal processes at the engine supplier plant and reduce the finished goods storage volumes to less than 1 day of demand in future. The engine buffer's future role would be that of a sequencing facility and a minimum buffer to level off disturbances in the supply or assembly processes – comparable to a body store in a vehicle plant.

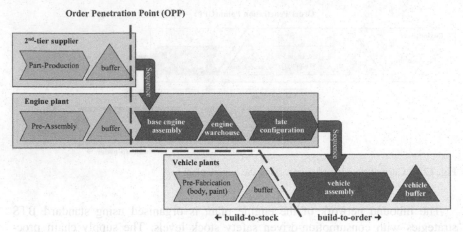

Fig. 17.2 New logistics process based on the BTO paradigms

In order to cope with short-term demand fluctuations, long-term demands that originate from overseas plants can be applied to level production. In addition, it would be necessary to negotiate greater production flexibility, for example, working time, in order to cope with short-term fluctuations in the production process. These new lead time requirements and the inventory reduction for in- and outbound buffers will mean second-tier suppliers have increased requirements placed upon them with regard to supply flexibility and reliability. High variant engine components could be JIT and/or JIS, with an available supply lead time of as little as 1–2 days depending on the shift model in the engine plant and the distance to the supplier site. Physical proximity, as well as reliable production and transportation processes, will be the key to an undisrupted supply. Only standard parts should be delivered to stock.

Figure 17.2 shows the new logistics process based on the requirements of the OEM vehicle plant's new BTO strategies. The image shows the order penetration point at the OEM level, for the engine plant and the integrated second-tier suppliers.

The case study evaluates the new to-be BTO processes. The next section will describe the measurable objectives of the case study, the model conception and the application of the combined static and dynamic approach leading to tangible results and requirements for the implementation of the specified concepts.

17.2 Application of the Combined Approach and Model Scope

The research was conducted using a procedure model for case study-based evaluation. The procedure model identifies real industry needs based on their current state and improves their processes using the BTO concepts developed. The evaluation of the derived concepts is done by using computer models. Both static and dynamic software prototypes were developed and employed where most appropriate (see Fig. 17.3). The static model environment was provided by 4flow vista® and the dynamic one by OTD-NET.

17.2.1 Objectives of the Case Study

The case study was divided into two phases. In the first phase the production system, inbound logistics costs and the outbound network of the engine plant were analysed using a combined static and dynamic modelling approach to identify potentially successful process scenarios.

The first step was to identify the model scope and detailed objectives. The initial task was to analyse the logistics costs expected when switching to near-stockless BTO production and JIT/JIS supply systems. The cost question was answered using the static modelling approach using the 4flow vista® software prototype. A large number of different scenarios based on average inventory levels and cost factors were analysed and reduced to a very few acceptable scenarios. Then, dynamic simulation of the identified, acceptable scenarios was carried out to ensure that the requirements for production planning and flexibility could be achieved. The questions that had to be answered were:

- How much flexibility is needed internally concerning long-term demand fluctuation and short-term programme deviation and adjustments?
- Is programme levelling with long-term orders possible or even necessary?
- Is it possible to achieve 100% supply reliability to the OEMs?

The detailed description of the scope of the OTD-NET model and the tangible results of the first phase will be given later.

During the second phase the case study was extended to the inbound network of the engine plant, aiming at the identification of requirements for the integration of second-tier suppliers in a multi-tier BTO process. The parameters analysed included lead time, safety stock levels and of course order and delivery reliability. The results of the second phase produced detailed requirements for the BTO processes and optimal safety stock levels to provide reliable supply, taking into account demand and transportation fluctuations. Figure 17.3 summarises the industry objectives, concepts used and shows the applied procedure model for case study-based evaluation.

17.2.2 Model Scope

The model scope has been defined based on the objectives set out. For the evaluation it is necessary to have a model of the real world that is complex enough for the analysis of the given questions and simple enough to keep it understandable.

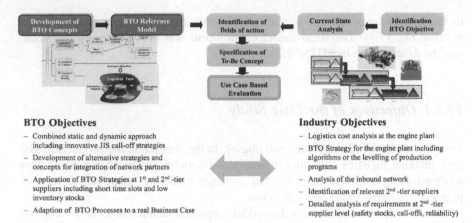

BTO Objectives

- Combined static and dynamic approach including innovative JIS call-off strategies
- Development of alternative strategies and concepts for integration of network partners
- Application of BTO Strategies at 1st and 2nd -tier suppliers including short time slots and low inventory stocks
- Adaption of BTO Processes to a real Business Case

Industry Objectives

- Logistics cost analysis at the engine plant
- BTO Strategy for the engine plant including algorithms or the levelling of production programs
- Analysis of the inbound network
- Identification of relevant 2nd -tier suppliers
- Detailed analysis of requirements at 2nd -tier supplier level (safety stocks, call-offs, reliability)

Fig. 17.3 Industry objectives and applied concepts

Achieving a reasonable level of abstraction meets the dual challenge of providing reliable results that can be transferred into reality and allowing for rapid modelling and assessment of the case study.

Therefore, the model scope can be described as follows. Focussing on a single engine type with 125 variants and its associated supply, planning and assembly, as well as distribution processes leads to a model with 20 second-tier suppliers, three of which have been analysed in detail for second-tier integration into a multi-tier BTO approach.

The engine plant was modelled with its internal production network and processes including five representative customers, i.e. four European vehicle plants and one overseas plant. The case study started by iteratively collecting and improving planning data, which were integrated into a robust model of the network, using 4flow vista® planning and logistics modelling software. Subsequently, logistics and assembly processes were detailed with cost and lead time measures. Recent demand data for the engine, covering a 5-month long sample, were transferred from the OEM's operational systems.

There are various questions that cannot be answered to a satisfactory degree by static analysis. Limitations result from time-consuming activities and changes during a time period. Static analysis lacks the ability to determine possible bottlenecks in capacity, stock-outs and flexibility needs based on the system's dynamic behaviour. Also, it cannot assess the performance of different planning process approaches because it does not replicate them.

Those dynamic effects can be identified and analysed by dynamic simulation. Simulation will provide highly granular KPIs and allow the assessment of the dynamic behaviour of supply networks. However, dynamic modelling often demands more data and greater modelling effort than equivalent static models.

An automated transfer process was created, called vista2OTD, that allowed for the efficient exchange of structure and product information from the static 4flow vista environment to the dynamic OTD-NET simulation environment (see Fig. 17.4).

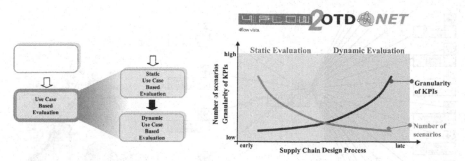

Fig. 17.4 Efficient model exchange using the vista2OTD interface

For dynamic simulation the model scope needed to be expanded by adding specific processes and data, such as the planning logic for the order and delivery processes, probability distributions for production and transportation processes and fluctuating demands based on realistic demands and daily call-offs. Furthermore the exact JIT and sequencing strategies were modelled (for the OEM plants during the first phase and for eligible second-tier suppliers in the second phase of the case study) as well as the production planning algorithm at the engine supplier site. Thereby, the realistic behaviour of the supply chain was reproduced and the to-be processes were integrated into a dynamic simulation model. Validating the model in OTD-NET paved the way for answering the questions posed.

17.3 Results of the Case Study

This section describes the results of the case study based on the identified industry objectives and the defined model scope. The sub-sections cover the two phases of the case study as well as the benefits of the combined static and dynamic approach.

17.3.1 Results of the Static Analysis

The static assessment comprises a number of different analyses. The first step is the visualisation of the inbound supply and plant level network structure and processes. Next is the computation of inventory levels and logistics costs. Finally, a throughput analysis identifies high-volume second-tier suppliers who appear to have the capability to be part of the BTO approach.

Visualisation of logistics networks and processes is an important means in network design as it facilitates comprehension and discussion of the scope of the matter addressed. For the case study at hand the logistics network was modelled in a schematic view, including the suppliers delivering to the engine plant, the engine

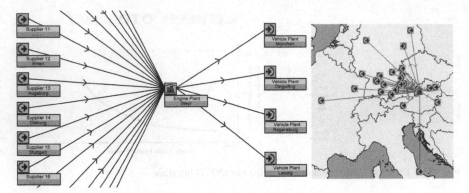

Fig. 17.5 Schematic and geographic view of the logistics network

plant itself and the vehicle plants. The corresponding geographic view was reduced to cover only the inbound network. Both views are depicted in Fig. 17.5.

In order to answer the detailed questions posed for BTO feasibility it is necessary to cover the network structure and processes inside the engine plant itself. Therefore, the model has been extended to an integrated plant level that depicts all relevant internal locations and processes from goods in to goods out, including JIS and JIT supply processes, assembly, engine storage and late configuration. The schematic high-level overview of this internal network is presented in Fig. 17.6.

One of the major goals of the BTO paradigm is the reduction of inventory levels and/or the relocation of inventory to lower cost stages upstream in the supply chain. The first analysis therefore was to identify process inventory levels within the engine plant. The analysis supports two conclusions. First, the processes considered are run very "lean", with minimal inventory levels in the processes before and after the engine storage. Second, engine storage accounts for more than 97% of the total inventory. Therefore, this is the prime focus of further assessment.

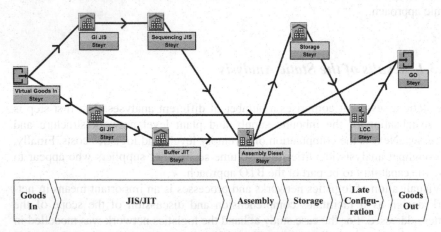

Fig. 17.6 Schematic view of the internal network of the engine plant

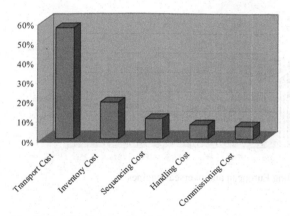

Fig. 17.7 Distribution of logistics costs across cost categories

In order to comprehensively cover the possible impact of BTO adoption, an assessment of logistics costs for the defined case scope was conducted. The logistics costs for various cost categories were analysed for the engine production and corresponding supply processes. It was found that the major cost contributors were transport costs at 57% and inventory costs at 19% of total logistics costs. The distribution of logistics costs is depicted in Fig. 17.7.

Of additional interest was the daily supply volume of the second-tier suppliers delivering to the engine plant. This information formed the basis for selecting suppliers that were assessed for an extension of the BTO paradigm beyond the engine plant and towards its suppliers.

The static network assessment has provided us with a comprehensive overview of the network and processes considered. Visualisation facilitated understanding and discussion among the people involved. Both inventory and logistics cost analysis provided robust performance indicators, providing direction for the areas requiring further assessment. The engine storage area was identified as the prime target for evaluating the impact of implementing stockless production. Transport and inventory costs are the main cost contributors that have to be taken into account in that context. Finally, a number of key suppliers have been identified for use when extending the BTO paradigm into the second tier.

17.3.2 Dynamic Evaluation of the New Process Concept

The objective of the dynamic evaluation was to demonstrate that it is possible to deliver engines within a 4-day order and supply process with very low levels of safety stocks. In close discussion with the engine plant's planners a possible planning process was designed for the closer integration of the engine supplier into the global processes.

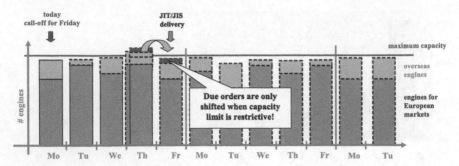

Fig. 17.8 Planning algorithm including European and overseas engines

Based on forecast-based, rolling, mid-term production planning of all vehicle plants, a general production programme is generated at the engine supplier site to level capacities and plan shifts. On the short-term planning horizon, this production programme is adapted to the actual demands of the OEMs, which have to be delivered JIS according to the short-term call-off. These demands have a lead time of 4 days and do not change according to the quantities and sequence information submitted. Based on this stable planning horizon the engine supplier is able to produce all required engines in a BTO process in this time slot. However, due to fluctuating demand quantities and the aim of achieving a constant and levelled production programme with few deviations, it was necessary to derive an algorithm to level the production programme close to the maximum capacity, whilst taking all production and planning restrictions into account. Therefore, overseas engine orders were used as a form of buffer to level the production programme. These engines have a forerun of several weeks and can therefore be rescheduled with a larger degree of freedom (see Fig. 17.8).

After implementing the rolling daily planning and levelling algorithm with a stable 4-day call-off in the simulation environment, it was possible to analyse the impact of demand fluctuations. This included demand peaks and bottleneck situations on the production programme, in order to identify how much flexibility is required to compensate for those situations.

The planning process implemented was tested utilising realistic assembly processes and capacity bottlenecks, whilst supply, production and transport were also integrated. Furthermore, realistic lead time fluctuations were anticipated. The simulation showed that despite a volatile vehicle plant demand with regard to vehicle volume and mix, 100% reliable delivery and low-volume fluctuation could be achieved. With the use of long-term orders as a production programme levelling mechanism, a highly stable assembly process could be implemented. Long-term orders were still built before the due date, resulting in a specifiable storage need.

The results also showed that a lower than expected level of flexibility was required. Due to an adequate long-term order quota, the production programme could be levelled (see Fig. 17.9). The work showed that only minor flexibility, less than 4% of production, was necessary to support the switch to a BTO production

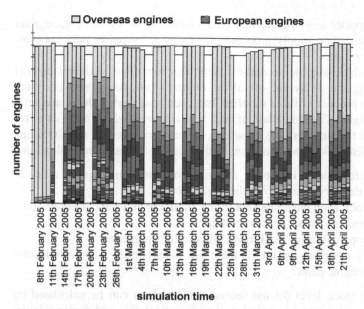

Fig. 17.9 Extract from the levelled production programme

system. Short-term fluctuations resulting from deviations between forecasts and call-offs could be levelled by reallocation of long-term orders. Furthermore, a required safety stock level for engines could be calculated.

The capacity prerequisites for the second-tier of the supply chain have been assessed through call-off volume analyses, with the aim of extending the BTO concept through the value stream. A number of critical, high-volume and high-variant suppliers were identified. This was the starting point for the second phase of the case study.

17.3.3 Integration of Second-Tier Suppliers into BTO Concepts

The first phase of the case study evaluated the outbound network of the engine plant, logistics costs of different scenarios and the feasibility of BTO concepts within a 4-day JIS call-off process from the European OEM plants. The second phase of the case study analysed the inbound supply chain of the engine plant and examined how eligible second-tier suppliers might be integrated into the call-off processes with lower levels of safety stocks, JIS supply and reliable processes. Furthermore, we have to analyse the key requirements to help the second-tier suppliers move towards BTO.

Today, the inbound network of the engine plant operates with BTS strategies. The sequencing of high-variant parts is realised in the engine plant in dedicated sequencing areas. Parts are available in a high-bay warehouse, which is either operated by a supplier (e.g. vendor-managed inventory) or by the engine plant itself. Other current state supply processes are based on supplier park concepts with short delivery times.

Based on the three eligible second-tier suppliers identified, different call-off strategies, utilising the 4-day call-off at the engine plant, have been analysed. There, the supply process of these suppliers has been specified in detail and integrated into the model. The simulation of such a model enabled the analysis of resulting demand at the supplier sites (based on the required engines, the BOM and the derived part demands). Questions remained that are essential to an evaluation:

- How much safety stock is needed at the second-tier suppliers in order to achieve 100% delivery reliability to the engine plant?
- What safety stock level has to be kept in the engine plant to avoid out-of-stock situations if suppliers do not deliver on time?
- How do the supply concepts react to volatility in transport times?
- Are the supply concepts feasible if there are disturbances in the production process in the engine plant?

A suitable safety stock level for the second-tier suppliers can be calculated by simulating different demand scenarios based on historical data and demand deviation assumptions. The results enable the identification of possible stock-outs and the identification of a suitable inventory level to guarantee 100% delivery reliability. Based on a calculated inventory level, it is also necessary to specify a disposition strategy to manage the inventory. Due to the lead time of a call-off and the distance of the supplier from the customer plant, it is either possible to react to – for example, higher demand – directly and produce the absent parts before delivery, or the absent parts can be taken from the safety stock and the supplier has to refill the inventory as soon as possible. Both strategies have been analysed and suitable inventory levels could be identified.

However, the applicability of such a call-off strategy depends on realistic demand deviations, capacity restrictions and production philosophies at the second-tier suppliers. Additional restrictions like batch production, high lead times or transportation strategies have to be taken into account. Nevertheless, the dynamic simulation enables the detailed analysis of such scenarios and the identification of inventory requirements and possible stock-outs.

Based on the demand scenarios and calculated safety stock levels at the second-tier supplier sites, transport volatility and failures can be integrated into the supply chain. These measures have a significant impact on the availability of parts at the engine plant. Based on volatile transport times and failures it is possible to identify an appropriate safety stock level for the engine plant. In particular cases, this means up to 30% less inventory than today, a figure that takes into account increasing and decreasing stock levels due to product mix changes. Figure 17.10 shows the impact of volatile transport times and failures in reliability (this was the basis for the calculation of safety stock levels at the engine plant).

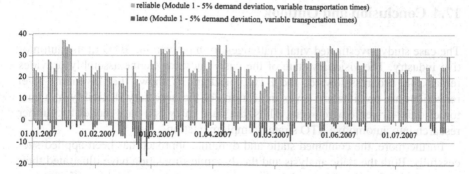

Fig. 17.10 Reliability of the engine production based on second-tier integration and without safety stocks

The second phase of the case study identified factors influencing the integration of second-tier suppliers into a short-term JIS call-off process. Scenarios using generic demand fluctuation assumptions showed the applicability of the approach. These effects can be compensated for by suitable safety stocks at the supplier and engine plant sites. A continuous planning process has to be taken into account to guarantee sustainable results due to changing demand behaviours.

17.3.4 Key Results

Leveraging the two methodologies' strengths the project team focussed on answering cost and supply or distribution capacity questions with a static model based on the 4flow vista® environment, while lead time, assembly flexibility and delivery reliability issues were analysed dynamically with OTD-NET.

Major cost drivers of the engine production and distribution network have been identified as the transportation cost for continental and intercontinental distribution, as well as capital commitment and warehousing costs for the high-bay engine warehouse.

Two measures allow the implementation of the BTO paradigm in the current supply chain. First, the engines with overseas destinations are used to level the production plan for smooth, undisturbed assembly. Second, some flexibility in shift working is necessary in order for staff employed to follow assembly volume fluctuations without assembly sequence disruption. Implementing this approach will lead to significant reductions in inventory in the high-bay engine warehouse.

The second phase of the case study identified factors influencing the integration of the second-tier supplier into a 4-day call-off process. The integration of eligible second-tier suppliers also requires the analysis of restrictions at the supplier sites.

17.4 Conclusion and Outlook

The case study investigated vital challenges in implementing BTO in the automotive industry. The visionary state of the case study for engine assembly provides remarkable progress towards the BTO paradigm. It shows that inventory level reductions are possible and gives indications of the necessary flexibility in a customer order-driven environment. Thus, the case study is a base model for further research into automotive BTO strategy implementation.

Furthermore, the combined static and dynamic approach has been applied successfully. Both the static analysis and the dynamic simulation have illustrated their range of useful applications and the integrated approach presented an efficient way of analysing large-scale supply chain scenarios, identifying feasible scenarios, taking cost and average capacity, as well as lead time measures, into account. The subsequent dynamic simulation enabled the detailed evaluation of a few reasonable scenarios focussing on the dynamic system behaviour and appropriate KPI. The analysis provided senior management with sufficient evidence to pursue BTO strategies in series production – thus fulfilling the aim of the study.

Chapter 18
How the Electro-Mechanical Valve Train Accelerates Logistics and Reduces Costs

Thomas Seidel[1] and Thomas Huth[2]

[1] 4flow AG, Berlin, Germany
[2] FEV Motorentechnik, Aachen, Germany

Abstract. Engines using a mechanical variable valve train (MVVT) have improved torque, emissions and efficiency as valve timing on both the intake and outlet camshafts can be adjusted to the power required from the engine as a function of gas pedal position and engine speed. This valve train system has been in mass production since 2002 and is produced by BMW under the name "VANOS." The electro-mechanical valve train (EMVT) goes beyond mechanically manipulating valve positions and enables the independent control of single valves by separate electronic operation. From a performance point of view, this enables a further reduction in fuel consumption and exhaust emissions, whilst increasing engine output. From a production and logistics perspective a number of advantages result from the EMVT. While the MVVT consists of 33 components with a total of 160 single parts, the EMVT is built from only four components with 85 single parts. This simplification of the product directly affects assembly and logistics. The EMVT will result in a significant shortening of the assembly process as approximately one-third of the assembly steps for the MVVT will be omitted without replacement. Consequently, assembly costs, material supply and handling costs, as well as assembly lead time are greatly reduced. The EMVT has fewer, higher density parts. Whilst this causes an increase in weight, it decreases the storage volumes required by an estimated 80%. The EMVT further benefits the inbound supply chain, as 18 suppliers have to be coordinated for production of the MVVT, but as few as three suppliers can deliver the EMVT components. The consequence is a drastic reduction in inbound logistics complexity, and an assessment has shown a transport cost reduction of up to 40%. It has become apparent that the EMVT is a key example of how product design can improve production and logistics processes. It can also provide significant cost advantages and a considerable reduction in production lead time.

18.1 Introduction

The key demands on future cars are compiled from stringent legislation and manufacturer's commitments to reducing pollutant emissions and customer-driven demand for higher performance, with increased comfort and enhanced safety provision. European legislation on hazardous emissions will be tightened up in 2008. EURO 5 limits are aiming to reduce HC, NO_X and particulates emissions under even more stringent test cycles. Even more demanding legislation is discussed for the future. Concern about the green house effect has led most European car manufacturers to commit to the introduction of cars that "on average will emit less than 140 g/km CO_2 by 2008" (ACEA 2007; SEC 2007). This corresponds to 25% lower fuel consumption compared with the 1990 level. Simultaneously, customers will require even more safety and comfort, which in most cases increases weight, conflicting with the demand for lower energy consumption. In addition, customers wish for even higher vehicle performance; however, costs of ownership should not rise.

Accepting that much work is required to comply with future emission targets, the main goal for the development of advanced engines is the improvement in fuel economy in order to reduce CO_2 emissions. The technical solutions available include: turbo- or super-charging, fully variable valves and direct injection. An obvious strategy for substantial improvements in fuel economy is to downsize engine displacement while retaining the original torque curve by charging (Pischinger et al. 2003). So far, this concept has been limited due to the fact that at low engine speeds turbo charging is not efficient, so high torque at low engine speeds in a small engine is not available.

Conventional valve train (CVT) systems for combustion engines consist of several mechanical parts, i.e. camshaft, camshaft bearings, valves, springs, rocker arms, valve shaft caps etc. Each part of the valve train has to be manufactured specifically for an engine and requires complicated logistics support at high cost. The current selection of "state of the art" valve trains, such as the mechanical variable valve train (MVVT), have improved torque, emissions and efficiency as valve timing on both the intake and outlet camshafts can be adjusted to the power required from the engine as a function of gas pedal position and engine speed. However, these will not be competitive in the near future. The current solutions lead to complex logistics at high cost. This is because they are deficient in the essential criteria for a competitive product, including low production complexity, reduced lead time, customer demand for variety, cost-effectiveness, eco-friendliness and technical requirements. The optimal future valve train would consist of a module that will simply bolt onto the engine block and plug into the power supply. Modularisation is, therefore, a key enabler for rapid BTO with simplified logistics.

To simplify the logistics chain and lower costs, a valve system module using a single piece of hardware and a variety of software applications has been developed.

Power variety can be achieved at a very late stage of the production process by software configuration. This system is able to fulfil the differing technical requirements of combustion engines, such as the torque curve, engine maximum speed, engine power, internal exhaust gas recirculation, engine exhaust emission and fuel consumption. This technological solution is called the electro-mechanical valve train (EMVT).

In this chapter, the differences in the manufacturing and logistic structure of a MVVT and an EMVT have been analysed based on data evaluation of the current and future situations.

Before further analysis of the technical and logistical benefits of the EMVT is undertaken, it is necessary to examine the markets to ensure that customers will be purchasing cars that require this form of engine technology. Figure 18.1 depicts the prediction of market share and penetration of the different engine concepts in Western Europe up to the year 2015. The diesel market share has risen considerably recently due to rapid improvements in engine characteristics. Indeed, sales of diesels will soon reach 50% of total car sales in Europe (Trampert 2006). Concerns have been raised that there will be a slow-down in direct injected diesel engine sales and this may be further intensified by the cost impact that increasingly stringent regulations have on their production and logistic system. However, this slow-down is technically driven and replacing conventional diesel technology with a new combustion process like homogeneous charge compressed ignition (HCCI) may mitigate these effects. It is predicted that the market share of turbo charged gasoline engines will grow considerably. Many new gasoline engines will use direct injection including turbo charging, although most will operate with homogeneous combustion. These engines use fully variable valve systems so these will have increasing market penetration.

Since 2005, combinations of advanced technologies have become available. The interest in natural gas as an alternative fuel is rising throughout Europe. In

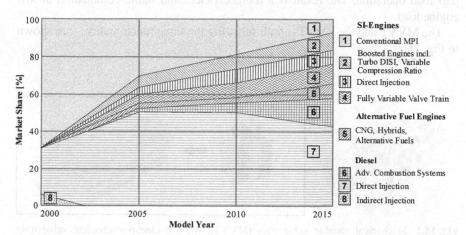

Fig. 18.1 Share of engine technologies in Western Europe over time (FEV 2006)

Germany, the number of refilling stations doubles each year. Let us consider hybrid technology. Initially, hybrids with advanced SI engines (spark ignition) were introduced in Japan and the US. However, so far their acceptance in Europe has been limited. Fuel cell technology in cars is not expected to become available in this decade due to unsolved issues of complexity and hydrogen infrastructure.

As a stand-alone technology, the EMVT system has firm prospects of achieving a 7% market share. However, in the future this system will be combined with advanced techniques like DI (direct injection) and boosted engines and thus the potential of the EMVT system allied with other technologies may reach beyond 25% of the total market share. This is enough to justify investment of considerable effort in the development of both the product and the supporting logistics.

18.2 Technical Description of the MVVT Compared with the EMVT

This chapter provides an overview of the working principles of MVVT and EMVT.

The MVVT system is based on a CVT with additional mechanics to achieve valve train variability. The variable intake valve motion is created by the fully mechanic "controlled relay lever unit" shown in Fig. 18.1. The MVVT has a conventional intake cam, but it also uses a secondary eccentric shaft with a series of levers and roller followers. The eccentric shaft is turned by a servomotor. Based on torque request signals taken from an electronic gas pedal, the servo motor changes the phase of the eccentric shaft, modifying the lift of the intake valve. This is controlled by a valve control unit. The entire MVVT system is pre-assembled and inserted as a module into its position in the cylinder head. A further positive effect of the smaller valve opening gap is a very high charge motion in part load operation. The result is a more efficient and stable combustion at low engine load.

The MVVT and the EMVT – both fulfilling the same functionality – are shown in Fig. 18.2.

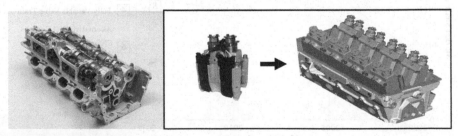

Fig. 18.2 Mechanical variable valve train (MVVT; *left*) and electro-mechanical valve train (EMVT; *right*)

A fully variable solution allows the use of a dethrottled spark-ignition engine. Thus, a throttle for load control (power output) is no longer necessary, so that the engine can aspirate with higher efficiency. In particular, it satisfies the requirements for compact design, reliability, weight and cost-efficiency. Due to the entirely independent and demand-orientated control of the intake and exhaust valves with regard to valve timing and lift, this fully variable control system allows each valve to generate its own drive. Besides the major impact on reduced pumping losses, the use of a driving device with the fully variable ability of free valve timing also influences the combustion process and different engine strategies. This provides user benefits such as improved cold start and warm-up operation and also reduces idle speed, cylinder and valve deactivation and optimised internal residual gas recirculation.

The technical advantages of the fully variable valve train provide some discernable improvements in driving performance. The driver benefits from an optimised speed range with high low-end torque, which is increased by 25% compared with conventional valve trains. In addition, future iterations of the fully variable valve train will benefit from increased demand for low fuel consumption vehicles. The introduction of fully variable valve trains will result in fuel savings of up to 25%. That means that the CO_2 emission that is coupled with fuel consumption will decrease. The exhaust-gas emission before the catalytic converter (NO_x and HC) is lowered by the possibility of internal exhaust gas recirculation, whereby the constantly reduced emission limits can be adhered to. Therefore, a fully variable valve train system offers increased driving pleasure to the customer along with lower emission values and fuel consumption.

This chapter explains an important working principle of the EMVT. For example, the linear spring-mass oscillator is explained. This new system for operating the valves of an internal combustion engine uses magnetic forces to open and to close the engine valves. The major EMVT system components are a magnetic armature, two electromagnets and two springs where one magnet and one spring act in the valve-closing direction and the other magnet and the other spring in the valve-opening direction. In addition, other components for fixing, guiding and adjustment are assembled with the major components into one actuator housing (Fig. 18.3).

The moving parts of the spring mass system consist of components of the actuator and of the engine valve including the valve spring, and the valve spring retainer with its valve keys and the valve shim. The mass of further parts must also be considered as they are also moved by the actuator. Actuator systems with electromagnets as operating sources and with opposite acting springs can store potential energy at the end positions of movement (valve open/valve closed) when the magnets keep the armature using holding current. During transition (closed to open/open to closed) the potential energy is transformed into kinetic energy where the magnets are used to overcome the losses (e.g. friction) during movement. When "switched off" (the magnets have no current) the system is held in a defined balanced position (neutral position/middle position) by the two springs.

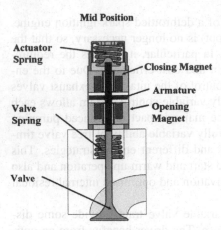

Fig. 18.3   Spring-mass oscillator system principle of an EMVT

During the starting process the armature is set into oscillation from its middle position by alternating magnetic currents in the upper and lower magnet until the amplitude results in the armature reaching one of the magnet's pole surfaces. The potential energy of the spring-mass system is greatest when the armature is at the end positions. The maximum value of kinetic energy is reached at the maximum speed of the armature travel, normally at the half valve travel point. At each end position a force caused by the deflected springs acts on the spring-mass system, which returns it in the direction of the middle position. The magnets can hold the armature at the end positions for a given period of time. The holding time can be chosen independently by switching the power to the appropriate magnet on or off. Due to the coupling between the valve and the armature, the valve timing can be optimised with regard to engine speed and load points.

The EMVT is not currently in mainstream use. However, in order to provide a useful comparison it can be compared with MVVT, a technology that has been in mass production since 2002 and has already shown advantages regarding fuel consumption and exhaust emissions. The mechanical system has fewer degrees of freedom regarding valve and cylinder deactivation. Therefore, the reduction of fuel consumption is only 20% compared with the 25% possible with the EMVT.

18.3 Production and Logistics Benefits of the EMVT

Knowledge of the basic elements of EMVT technology is widely held. However, the technology is at a very early stage of development with respect to mass production application. At present, the logistical benefits resulting from the application of the EMVT in high-volume engine production can be derived from a comparison with the MVVT. We must also consider the legal, economic, ecological and functional requirements, as well as the technical maturity of an assumed

EMVT production process that is expected to be developed and implemented in the next 3–5 years.

The first investigation determined the different construction characteristics of the MVVT and EMVT, using their BOMs to analyse the potential impact on logistics. We focus on three distinctive aspects: the assembly process, the storage process and the logistics network.

18.3.1 Comparing the Assembly Process

The specification of valve train systems is determined by the five most important elements of an internal combustion engine (crankshaft, crankcase, piston, camshaft, and cylinder head). Each engine is matched to a particular system. In a BTO engine plant, to meet the required customer-orientated adaptability as well as short lead time, all variants of these elements and their sub-components would have to be kept in storage. A fully variable valve train system cannot be economically or rationally managed, so a reduction in variants and simultaneous simplification of the main elements is required.

The impact of an EMVT on the main engine elements design, manufacture and assembly (cylinder block and head, cover etc.) has also been considered. A simplified and intelligent assembly and logistics concept can be fashioned based upon an all-purpose and technically matured EMVT, differentiated by software.

One strategy to control such a production system is to reduce the number of individual parts of an assembly. The fewer parts to be manufactured, the less logistic effort is required. Module components are a key to solving the dilemma. In the case of the EMVT, the functions of the relay lever unit and the chain drive module are substituted and compared with the MVVT.

In the case of different valve train concepts the main changes in terms of the manufacturing structure will occur on the engine assembly line. The baseline engine assembly steps with a MVVT system are shown in Fig. 18.4. The complete machined cylinder head with preassembled plugs, valve guides and seats will be supplied from the machining facility, likewise the EMVT engine block.

The most obvious difference of the cam-less EMVT engine manifests itself in the simplicity of the machined cylinder head. The EMVT cylinder head does not require the bulky bearings of the camshaft, the oil pipes, a variety of fixing threads and numerous drilling operations. This reduces the total machining time by up to 20% compared with the MVVT cylinder head.

Each box in Fig. 18.4 represents one assembly step. The cylinder head and engine block are "supplied" in the upper left corner and run simultaneously through the line. The block assembly is influenced little by changing the valve train; therefore, the focus will mainly be on the head assembly. The process starts with someone putting the head on to an assembly pallet. The next step is to identify each head by reading in the individual head bar code. This allows the tracking of each head so that for different versions the right parts can be delivered just-in-sequence on the

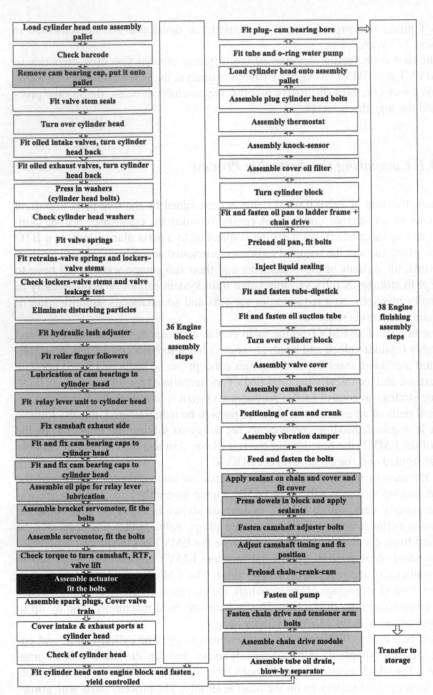

Fig. 18.4 Engine assembly process (stages with *light shading* are superfluous/stages with *dark shading* are additional)

same assembly line. Some assembly steps are marked to demonstrate the changes in the process that switching from a MVVT to an EMVT would bring. In the case of an EMVT, 22 assembly stations are redundant and only one new assembly step has to be added. The unnecessary and new assembly steps are highlighted.

Camshafts are not necessary for an EMVT engine so the first step eliminated is camshaft bearing cap removal. One can assume that the valve assembly will not change in any essential way, so either head type could run through this section.

Major changes occur in the valve train assembly. While the EMVT head is forwarded to the actuator module assembly station the MVVT engine has to be equipped with the camshafts and the valve train parts, such as hydraulic lash adjusters and finger followers. In addition, preassembly of the camshaft adjusters with the camshafts can be bypassed. After mounting the cylinder head to the block, the chain drive is built up; another set of processes not required for the EMVT.

Using an EMVT engine clearly simplifies engine assembly. In total, the machining time of the cylinder head can be reduced by 20% and it is possible to remove 22 assembly steps. This reduces the engine assembly time by 20 min.

18.3.2 Comparing the Storage Process

A calculation was made for the storage requirements of the two different valve train components and the results were compared. For this case, the following base data have been applied:

- Mass production in 2006 >330,000 engines per year
- Three hundred working days per year
- Six hundred and sixty engines per shift
- Two shifts per day
- Storage volume magazine assembly line, 660 engines
- Main storage assembly line with a 3-day capacity
- The required electrical power for the EMVT is delivered by an upgraded generator

The result of this analysis shows a reduction in the necessary storage capacity from 202 m^3 to 27 m^3, which is a volume decrease of 87%. The total mass of the stored material increased from 42 tonnes to 48 tonnes, which is 14%.

18.3.3 Comparing the Logistics Network

The EMVT, with its reduced BOM, has a direct impact on the corresponding inbound logistics network. With fewer components and parts necessary, the number of suppliers can be reduced significantly. In this case, as few as three suppliers have been selected to deliver the EMVT components to the engine plant. The

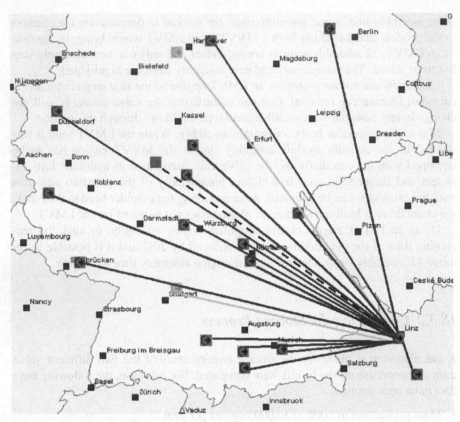

Fig. 18.5 Supplier structure – *dark lines* for MVVT suppliers, *bright lines* for EMVT suppliers, *dashed line* indicates supplier of both the EMVT and MVVT component

selection was based on the current product portfolio, which qualifies them as suppliers capable of producing the EMVT components. In juxtaposition, the inbound network of the MVVT constitutes 18 suppliers, which were included in the model. In our study, we found that one of the suppliers qualifies for delivering components for both the EMVT and the MVVT. This supplier is modelled identically in both inbound networks. Both of the inbound networks are depicted in Fig. 18.5.

The assessment of the impact that the EMVT has on the inbound logistics network is based on a number of quantitative planning parameters for handling, storage and transport costs. Handling costs are represented by the effort and associated cost of supplying and moving the containers from the storage site at the engine plant to the assembly line. Storage costs are computed by applying standard cost factors for the utilisation of storage space in the engine plant's storage facilities. Average inventory ranges for the current MVVT component spectrum have been translated and applied to the equivalent EMVT component sets. Transport costs are founded on a two-dimensional tariff matrix. The first dimension is the volume or loading metres of the truck utilised. The second dimension is the distance travelled.

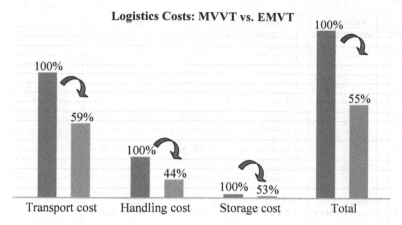

Fig. 18.6 Logistics costs of MVVT and EMVT

This approach allows for non-linear cost progression to be represented. The cost values themselves are based on comparable tariffs quoted by logistics service providers. Transport frequencies have been modelled, achieving a cost optimum for transport, storage and inventory cost.

The results of the logistics cost comparison between the MVVT and the EMVT show significant potential for savings in all logistics cost categories. Transport costs for the EMVT are more than 40% lower than for the MVVT. Although the increased weight of the EMVT components more quickly meets the weight limits of the trucks, the number of journeys is reduced as there are fewer suppliers. Handling costs are reduced by 56% as fewer parts need to be supplied to the assembly line. Storage costs are reduced by more than 45% because of the low number of EMVT components and their higher packaging density. In total, 45% of the logistics costs for the MVVT can be saved by switching to the EMVT. These statistics are displayed in Fig. 18.6.

18.4 Conclusion

The EMVT shows how design and technology can be improved to enable BTO, whilst simultaneously improving fuel consumption, meeting emissions legislation and lowering costs. Figure 18.7 summarises its advantages. The change from a MVVT to an EMVT makes 21 assembly steps superfluous. The number of first-tier suppliers can be reduced from 18 to three. This reduction is possible thanks to a decreased number of variants and components for the valve train, i.e. four components instead of 33. The required storage volume is also significantly decreased, by 87%.

The engine family variants can utilise one single valve train. The different variants will be catered for through software changes in the valve control unit. With

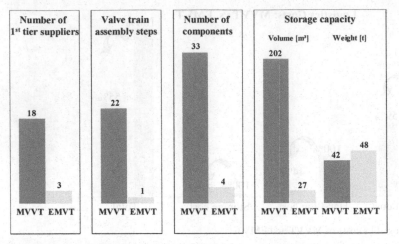

Fig. 18.7 Comparison of MVVT and EMVT in mass production

these systems the number of engine modules and variants can be decreased significantly. The goal of reducing variants will result in the need for a single base engine for several power classes. With only one base engine the storage capacity decreases and the flexibility increases further. The final engine power class is produced using supercharging with adapted fuel injection and the fully variable valve train. By combining these three hardware modules, the final engine power output can be adjusted using software changes. This approach results in optimised exhaust emissions and efficiency for every operating point and power class. In addition, the customisation of the power class can be implemented late in the production process, facilitating the transition to BTO.

References

ACEA European Automobile Manufacturers' Association (2007) Car industry support s reducing CO2 emissions. Available via
   http://www.acea.be/index.php/news/news_detail/reducing_co2_emissions_working_together _to_achieve_better_results/. Accessed 28 Nov 2007
FEV (2006) Internal source, FEV Motorentechnik GmbH
Pischinger S et al (2003) Der Weg zum konsequenten Downsizing. MTZ 5/2003, p 398
SEC (2007) 60 COMMISSION OF THE EUROPEAN COMMUNITIES (2007) Results of the review of the Community Strategy to reduce CO2 emissions from passenger cars and light-commercial vehicles, Brussels, 7.2.2007. Available via
   http://ec.europa.eu/environment/co2/pdf/sec_2007_60_ia.pdf. Accessed 28 Nov 2007
Trampert S (2006) Entwicklungstrends von Motoren und Getrieben vor dem Hintergrund der ACEA-Selbstverplfichtung. ATZ 6/2006, p 466

Chapter 19
Network Design for Build-to-Order Automotive Production

Kati Brauer and Thomas Seidel

4flow AG, Berlin, Germany

Abstract. Establishing a pure BTO system in the European automotive industry demands the investigation of new concepts concerning product structures, planning and execution processes and supply chain design. This chapter describes possible supply chain designs and compares scenarios for future network structures in the European automotive industry. The qualitative comparison of alternatives in component supply, vehicle assembly and distribution is supported by a quantitative model-based analysis. In this respect a network containing close to 200 first-tier and second-tier suppliers, four and 15 vehicle final assembly plants, a number of distribution centres and 500 dealer locations all over Europe has been modelled. The scenarios investigated differ with regard to modular vs conventional sourcing, centralised vs decentralised final assembly and a number of different distribution concepts. The evaluation delivers key performance indicators regarding logistics costs as well as transport times. Apart from these economic measures, overall transport mileage and pollutant emissions are included in the evaluation to reflect the environmental impact. In addition, a dynamic evaluation has been applied to estimate lead times and reliability, and study the dynamic behaviour of the networks considered. As a result, implications for the design of future BTO automotive networks in Europe are derived.

19.1 Introduction to Automotive Value Creation Networks and Build-to-Order Supply Chain Design

Value creation networks in the automotive industry have grown over years and consist of a large number of related companies (see Fig. 19.1). Supply networks are traditionally structured in tiers with raw material suppliers at the lowest level and first-tier component, module, and systems suppliers directly connected to the OEM. A recent development is the evolution of big "tier 0.5" suppliers and the agglomeration of suppliers close to vehicle plants in so-called supplier parks, which are discussed in a different chapter within this book. After the final assembly at the OEM plant most of the vehicles are stored close to the plant, in distribution centres or at local dealer compounds until the final customer picks them up. This leads to an average of close to 2 months' new vehicle stock coverage in Europe (Miemczyk and Holweg 2002).

In order to overcome the high costs of large finished vehicle stock and establish a pure BTO system in the European automotive industry, product structures, planning and execution processes and supply chain design have to be examined. Regarding supply chain design, a large number of publications have evaluated differences between BTO and BTS network structures (e.g. Christopher 2000; Fine 2000; Fisher 1997; Gabriel 2003; Reeve and Srinivasan 2005). The main results are that BTO supply chains have to be agile and responsive, with a focus on low lead times to the final customer. On the contrary, BTS network structures focus on leanness and efficiency in production and component supply. Lead time and responsiveness are less important in this field as customers can be served from the finished product inventory.

Even though lead times have been identified as a critical factor in BTO networks, costs are still an important criterion. This leads to the fact that BTO automotive networks have to be lean and responsive at the same time. Apart from time and cost measures environmental impact will be a third dimension for the evaluation of automotive value creation networks. In the context of rising political and social interest in environmental issues, this dimension has to be kept in mind when alternatives for the design of automotive value creation networks are developed.

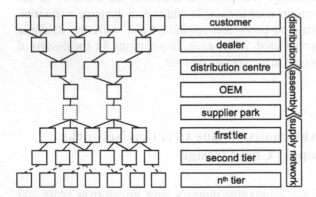

Fig. 19.1 Automotive value creation network

19.2 Alternative Designs for Automotive Value Creation Networks

In the course of this study, the automotive value creation network was divided into three areas: component supply, vehicle assembly and distribution. For each area, a number of alternative designs will be introduced in the following sections.

19.2.1 Component Supply

Regarding component supply, two alternatives are compared in this study. On the one hand, there is the conventional strategy of sourcing of parts, components and minor subassemblies, resulting in a rather large number of suppliers directly connected to the OEM. In Fig. 19.2 it becomes clear that the conventional sourcing strategy leads to relatively greater effort for the OEM, as many suppliers and parts have to be coordinated at the vehicle plant.

The second option is to decrease effort in final assembly and source large preassembled modules from a smaller number of module suppliers. This role could be taken up by former first-tier suppliers or supplier parks. The applicability of modular sourcing concepts depends on the product structure of the car that is to be built. Imperatives for modularity are agreed architecture, detailed interfaces and standardisation (Baldwin and Clark 1997). The modularisation approach investigated in this study is based on the innovative product structure that was introduced in earlier chapters on modular architecture. The final product is divided into front end, greenhouse front, greenhouse rear, rear end, exhaust system and outer panels.

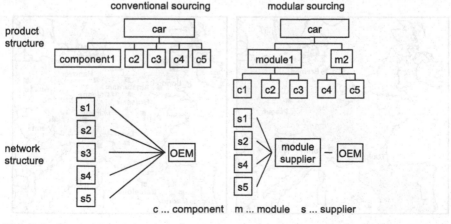

Fig. 19.2 Alternative sourcing strategies

Each module can be designed and produced independently and is joined with the others in the final assembly.

The possibility of independently developing, designing and producing the modules is one of the great benefits of this approach, along with the decrease in internal variety and complexity. Additional advantages are time-saving and reduced effort in final assembly, as well as the outsourcing potential that is created (Piller and Waringer 1999).

While these facts support the application of modular sourcing in the automotive industry, the issue of whether or not the transport and handling of those modules lead to a disproportionate increase in overall logistics costs has to be examined.

19.2.2 Vehicle Assembly

Two concepts, the centralised and decentralised location of OEM plants, are considered concerning the final vehicle assembly. Figure 19.3 illustrates a centralised structure with four plants and an example of a decentralised network structure with 15 plants and their respective supply networks in Europe.

Benefits and shortcomings of centralisation and decentralisation have been covered in a number of publications mainly focussing on stock-keeping and distribution (e.g. Alicke 2005). Decentralised stock-keeping allows short delivery times. If customers' orders are served from centralised stock, the transport costs and times are higher and longer, but the same service levels can be achieved with lower levels of inventory, as demand fluctuations across plants may compensate for one another. Other aspects to be considered are fixed costs and overheads, which are lower with fewer locations. Economies of scale are a further argument in favour of centralisation.

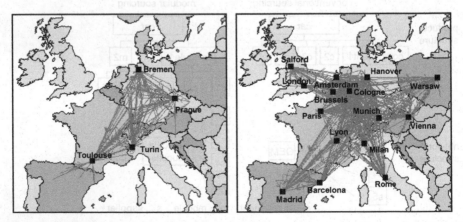

Fig. 19.3 Centralised and decentralised final assembly

Most of these findings can also be applied to the assessment of centralised and decentralised locations of production facilities. When customers require short lead times, decentralisation of final assembly saves outbound transport time and costs. While finished product inventory is abolished in BTO production, stock-keeping of bought-in parts and components is still a relevant question. The main arguments in favour of centralised production in the automotive industry are the high fixed costs and economies of scale in body production (according to Nieuwenhuis and Wells 2003). With the use of space frames instead of monocoques, the shortcomings of lower production volumes in decentralised scenarios can be reduced (see Chap. 7). The body parts can be produced centrally and then supplied to the various vehicle plants.

19.2.3 Distribution Network

Vehicle distribution includes transporting the vehicle from the assembly plant to the final customer. The distribution lead time is an important component of the total order-to-delivery time and influences the responsiveness of the whole network. While the supply network and the final assembly have fundamentally changed under the influence of lean production initiatives, the distribution system has remained essentially unchanged since the 1980s (Urban and Hoffer 2003). Currently, most vehicles are shipped via national or regional distribution centres to dealers and then delivered to the customer (Holweg and Miemczyk 2003). As lead times to the customer are becoming an important success factor in automotive production, the assessment of alternative distribution holds remarkable potential for the redesign of automotive value creation networks. The alternatives considered in this analysis are depicted in Fig. 19.4.

The first option, which shows considerable potential for short transport times, is *direct delivery* from the vehicle plant to the local dealer, or all the way to the final customer. There is no intermediate transhipment point necessary and the routes are easy to coordinate. However, direct transport tends to cause higher logistics costs. Either the vehicles are stored temporarily until an economic lot can be shipped or transports leave the OEM plant with low capacity utilisation. Since the first option increases lead time to the final customer, the second option seems more viable in a BTO environment. The consequences are higher transport costs, which are a drawback of this solution.

The use of *delivery runs* partly compensates for the low capacity usage of direct deliveries. In a delivery run, vehicles are delivered directly from the OEM plant to multiple dealer locations. This provides the advantage that there are no intermediate locations needed. The capacity usage increases, especially at the beginning of the tour and transport costs can be lower, if small volumes are distributed to dealers in geographical proximity on a regular basis (Chopra and Meindl 2004). The coordination of such tours, however, tends to be more complicated. Tours have to be redesigned on a daily basis, particularly in the case of fluctuating demand.

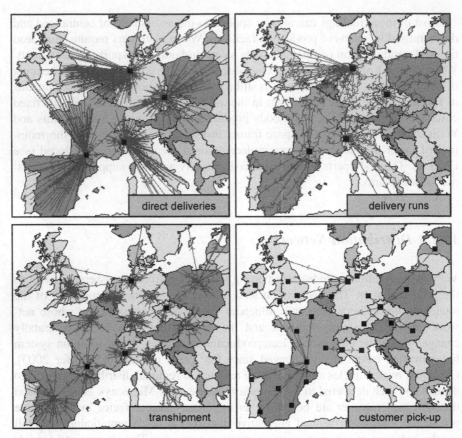

Fig. 19.4 Distribution concepts

A further disadvantage is the increase in transport time for the last dealers to be served on a tour due to intermediate stops and detours on the route.

The use of *transhipment points* reduces planning complexity, as routes are split up into consolidated transport from the vehicle plant to the transhipment point and from there to the local dealer. This is especially beneficial if long distances between plants and dealers in specific regions have to be covered. Routes with high capacity utilisation bear economies of scale and could be run by alternative means of transport (e.g. trains), which would further decrease transport costs. In contrast to today's distribution centres, transhipment points would not hold any stock, but just forward the vehicles to the local dealers. As well as the use of direct deliveries, local delivery runs could also serve at these distances. However, compared with direct delivery from the OEM plant, the lead time to the customer increases due to the additional handling effort and detours.

The last concept to be considered is the *customer-pick-up* solution, where the end-customer picks up their vehicle from a local pick-up station instead of having it delivered to their home address. Customer pick-up is already offered by a number of OEMs (e.g. Volkswagen in Wolfsburg, Audi in Neckarsulm). In order to apply this concept on a larger scale, the distances between the final customers and the pick-up station would have to be kept at a reasonable level. In this study a total of 30 pick-up stations all over Europe have been modelled with more than 90% of all customers being within a 250-km radius. While this option radically cuts down transport costs and time, the effort required to organise the customer pick-up has to be kept in mind. Capital investments in pick-up stations, refunds of travelling expenses etc. become necessary in order to reduce the inconvenience to the customer.

19.3 Model-Based Evaluation of Design Alternatives

In addition to the qualitative assessment of the supply, assembly and distribution concepts introduced, a model-based quantitative analysis was performed. The next section summarises the characteristics of the modelling approach applied. The following comparison of alternative scenarios delivers KPIs for logistics costs and performance, as well as the environmental impact.

19.3.1 Characteristics of the Model Employed

The concepts introduced in the last section have been integrated into a number of scenarios (see Fig. 19.5) and then modelled within standard software for supply chain design.

The scope of the model included close to 200 first-tier and second-tier suppliers, four and 15 vehicle plants, 4–30 transhipment points and pick-up stations, and 500 dealer locations all over Europe. More than 1,000 transport relations per scenario have been modelled in order to reflect the tight integration in the automotive industry.

The scenarios were modelled within the integrated static and dynamic modelling approach introduced in Chap. 16. The key performance indicators, which

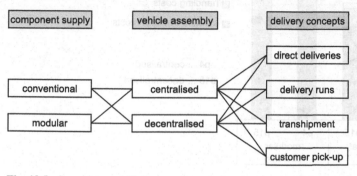

Fig. 19.5 Scenarios modelled

have been evaluated in the static analyses, are logistics costs, transport time and environmental impact. For one potentially feasible scenario, a dynamic evaluation of lead times and reliabilities was performed.

19.3.2 Results

The *comparison of modular and conventional sourcing* in centralised and decentralised scenarios shows few differences in logistics costs (see Fig. 19.6). When modules are assembled close to the OEM plant, as assumed here, no significant increase in logistics costs occurs. Nevertheless, handling costs increase by about 20%, because a new tier is added to the value creation network. This cost penalty should be measured against the savings in the final assembly process.

The second finding is that in decentralised scenarios modules should be assembled as close as possible to the OEM plant as the transport of modules over long distances increases transport costs as they are voluminous and heavy. In decentralised scenarios, however, it has been shown that plants in the same region can share supplier parks without an increase in costs, as the consolidation of inbound transport to the supplier parks more than compensates for the additional costs that occur in delivering the modules to the OEM plant.

In general, inbound logistics costs are lower in centralised scenarios because of lower transport costs and reduced inventory. The lower transport costs are a consequence of the higher capacity utilisation when only four plants are served.

A cost comparison of the different *distribution concepts* is provided in Fig. 19.7. Direct delivery has the highest transport costs for both centralised and decentralised scenarios. In decentralised scenarios, the vehicles are assembled closer to the

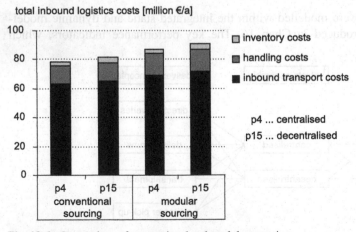

Fig. 19.6 Comparison of conventional and modular sourcing

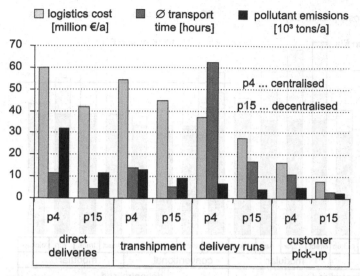

Fig. 19.7 Comparison of distribution concepts

final customer, thus leading to cost savings of about 30% compared with centralised scenarios. The introduction of distribution centres shows significant benefits for the centralised scenarios, as the distances to be covered are larger. Switching from direct deliveries to delivery runs yields the highest potential cost savings. About 35% of outbound transport costs can be saved regardless of the location of the vehicle plants. The introduction of customer pick-up stations cuts transport costs even further. Yet, it has to be kept in mind that part of the savings potential of about €200 per car is needed to organise the pick-up.

Transport times and pollutant emissions also show the lowest values in scenarios with customer pick-ups. However, those figures have to be read carefully, as the customer has to cover the distance from and to his home address, which consumes time and adds further emissions. Besides the customer pick-up solution, delivery runs also cause relatively low pollutant emissions. In contrast, for the centralised scenarios the average transport time shows a large increase and reaches more than 60 h. The transhipment option combines reasonable transport times with moderate levels of pollutant emissions, even though it does not yield very high cost savings. Direct deliveries cause the highest environmental impact, especially in centralised scenarios when the pollutant emissions of direct deliveries become about three times higher than in all other scenarios.

Rapidity and responsiveness in distribution networks have their price. Achieving very low transport times leads to pollutant emissions and increases delivery costs, which make up about half of the total logistics costs when direct deliveries are used to distribute the vehicles. Transhipment points and delivery runs appear to be the environmentally friendlier solutions. The use of alternative means of transport on the consolidated routes to the transhipment point or customer pick-up station can further decrease emissions.

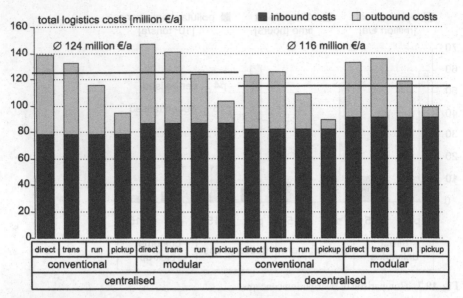

Fig. 19.8 Comparison of centralised and decentralised final assembly

In order to evaluate the location of the *vehicle assembly*, both inbound and outbound costs have to be considered (see Fig. 19.8). Scenarios with centralised final assembly have slightly lower inbound costs, while scenarios with decentralised assembly induce significantly lower outbound costs. In total, decentralised scenarios have been shown to be more cost-efficient if very short lead times and direct deliveries are needed. Comparing all scenarios, the average benefit of decentralisation was about 7% of total logistics costs. If alternative distribution concepts are permitted, the cost benefit of decentralisation decreases and the higher inbound costs become more relevant.

The results of the *dynamic lead time and reliability evaluation* are illustrated in Fig. 19.9. The analyses were performed with the scenario using a conventional sourcing strategy, decentralised final assembly and direct deliveries as the main distribution concept. In the dynamic model, demand peaks and tight capacity constraints have been integrated in order to reflect realistic conditions for scheduling and production in a BTO environment.

The first finding is that average total lead times from order to delivery range between 5 and 7 days. The fluctuations are caused by varying distances between different final customers and vehicle plants. Capacity constraints at supplier plants and in the vehicle assembly can necessitate re-scheduling and the postponement of operations. Weekends and holidays are another source of disturbance that has been taken into account. The results show that even under the realistic conditions of a dynamic environment short order-to-delivery times are achievable within the modelled scenario. The evaluation of the delivery reliability supports this assumption, and shows that 92% of all customer orders could be delivered within the promised lead time.

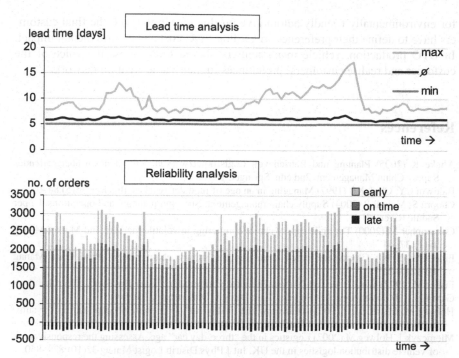

Fig. 19.9 Dynamic evaluation results

19.4 Conclusion

The objective of this chapter was to present the results of a study on BTO supply chain design. A number of alternative concepts for the European automotive industry have been assessed. With regard to the supply network, modular and conventional sourcing strategies have been compared, with the result that the cost penalty for sourcing large preassembled modules can be kept at a reasonable level, if module suppliers are located close to vehicle plants. With regard to vehicle assembly, the decentralised assembly concept with 15 plants spread across Europe has shown to be beneficial with respect to logistics costs. As for distribution concepts, delivery runs, transhipment and customer pick-up are cost-wise sensible solutions. The shortest delivery times can be achieved with customer pick-ups or direct deliveries, the latter being associated with the highest pollutant emissions.

All in all, there certainly is no one best solution for BTO automotive production in Europe. This research has shown that there are very sensitive trade-offs among time, cost and environmental impact in the design of automotive value creation networks. Depending on the future development of those three dimensions, different concepts will be suited best. While decentralised assembly and fast delivery concepts are instruments that react to demands for shorter lead times, delivery runs and consolidated shipment with alternative means of transport are viable solutions for environmentally friendly automotive production. In the end, the final customers have to define their preferences regarding time, cost and environmental impact. In BTO production, vehicle manufacturers should then follow the voice of the customer and realise customers' preferences in their value creation network.

References

Alicke K (2005) Planung und Betrieb von Logistiknetzwerken. Unternehmensübergreifendes Supply Chain Management, 2nd edn. Springer, Berlin

Baldwin CY, Clark KB (1997) Managing in an age of modularity. Harv Bus Rev 75(5):84–93

Chopra S, Meindl P (2004) Supply chain management. Strategy, planning and operations. Upper Saddle River, NJ

Christopher M (2000) The agile supply chain: competing in volatile markets. Ind Market Manag 29(1):37–44

Fine CH (2000) Clockspeed based strategies for supply chain design. Prod Oper Manag 9(3):213–221

Fisher M (1997) What is the right supply chain for your product? Harv Bus Rev 2:105–116

Gabriel C (2003) Strategisches Supply Chain Design. Dissertation, Bamberg

Holweg M, Miemczyk J (2003) Delivering the '3-day-car' – the strategic implications for automotive logistics operations. J Purch Supply Manag 9(2):63–71

Miemczyk J, Holweg M (2002) Logistics in the "three-day car" age. Assessing the responsiveness of vehicle distribution logistics in the UK. Int J Phys Distrib Logist Manag 32(10):829–850

Nieuwenhuis P, Wells P (2003) The automotive industry and the environment. A technical, business and social future. Woodhead, Cambridge

Piller FT, Waringer D (1999) Modularisierung in der Automobilindustrie – neue Formen und Prinzipien. Modular Sourcing, Plattformkonzept und Fertigungssegmentierung als Mittel des Komplexitätsmanagements. Shaker, Aachen

Reeve JM, Srinivasan MM (2005) Which supply chain design is right for you? Supply Chain Manag Rev 9(4):50–57

Urban DJ, Hoffer GE (2003) The virtual automotive dealership revisited. J Consum Market 30(6):570–578

Part V
Implementation

Chapter 20
Automotive e-hubs: Exploring Motivations and Barriers to Collaboration and Interaction*

Mickey Howard, Richard Vidgen and Philip Powell

School of Management, University of Bath, Bath, UK

Abstract. Business-to-business electronic marketplaces or 'e-hubs' are adopted by organisations seeking to achieve dramatic reductions in cost. While initially heralded in such industries as the automotive sector as the key to restructuring old economy firms, the claims for e-hubs now appear optimistic. This chapter explores collaboration and interaction by examining four cases of e-hub adoption by vehicle manufacturers and suppliers. A conceptual framework emerges from this examination that helps to assess the real benefits of electronic applications – not the hyperbole – by revealing firm and industry level motivations and barriers. The framework explains the dissonance between expected and realised benefits, and extends thinking on information system barriers. It concludes with recommendations for how best to adopt e-hubs in terms of supply topology, buyer–supplier relationships, leadership and how to avoid disbenefits.

20.1 Introduction

The rise of the internet and the rapid spread of electronic business across world markets have left few industries unchanged. Healthcare, gaming software, private utilities, aerospace, and automotive are all examples of sectors exploiting the digital economy to address supply chain visibility, supplier relationships, distribution and pricing, customisation, and real-time decision-making (Swaminathan and Tayur 2003). Since its inception in the 1990s, e-procurement, enabled by business-to-business (B2B) electronic marketplaces, or 'e-hubs', has been feted by organisations seeking to implement new business models and achieve dramatic reductions in transaction costs (Bakos 1998; Timmers 1998; Min and Galle 2003). While heralded in the auto sector as the solution to restructuring so-called

---

* Note: this chapter has been reprinted from Howard et al. (2006), with permission from Elsevier.

'old economy' firms with estimated annual savings of $100–200 billion in North America alone, the initial claims for e-hubs appear to be optimistic (Bauer et al. 2001). e-hubs have not only endured the ignominy of a dotcom crash, but are now criticised over difficulties in adoption, indeterminate lifespan and failure to create value (Connelly 2001; Counsell 2002; Alves de Quieroz et al. 2002; Daniel et al. 2003; ANE 2004; Arbin and Essler 2005).

Despite considerable investment and high expectations of savings from e-procurement, there is little evidence of benefits realised beyond indirect goods such as office stationery. Further, the dawn of the second century of vehicle manufacture finds the global car industry in crisis: automakers beset by 40–80 days of unsold inventory, one-fifth of European customers driving home new cars that are not what they intended to buy, and 85% of total waiting time attributable to bottlenecks in information flow before an order even reaches production (3DayCar 2002; Holweg and Pil 2004). This chapter develops a framework that highlights the dissonance between expected and realised benefits from e-hubs. It extends thinking on sharing information between vehicle manufacturers (VM) and suppliers, helping to explain the context of both firm and inter-firm barriers to information exchange. While the importance of stakeholder dynamics is already understood in the industry (Howard et al. 2003), this research focuses on B2B e-procurement across four firms and investigates why benefits from e-hubs have not materialised. Individual cases of e-business adoption abound in information systems (IS), but industry studies of multiple stakeholders and collaborative electronic platforms are rare. Hence, the unit of analysis in this study is collaborative platforms – e-hubs – and the phenomenon under investigation is the inter-firm adoption of e-hubs.

20.2 Reviewing the e-hub: Theory and Practice

This review draws from automotive and other industries to build a picture of e-hub development. The first section synthesises the classification of e-hubs within the context of B2B e-procurement. The second section assesses the current state of automotive e-hubs and presents a map of industry structure. The third section reviews the concept of motivations and barriers to collaboration and interaction during inter-organisational system (IOS) adoption. It concludes with an outline of a preliminary framework that is extended after the analysis.

20.2.1 e-hub Classification

The term 'e-hub' covers portal, trade exchange, an internet-driven electronic marketplace, and a cyber-purchasing system (Kaplan and Sawhney 2000; Skjøtt-Larsen et al. 2003). e-hubs can be defined as "web-based systems that link multiple businesses together for the purposes of trading or collaboration" (Daniel et al. 2003, p. 39). The appeal of e-hubs is clear: by aggregating a large number of buyers and sellers and automating transactions, they expand the choice available to buyers, give sellers access to new customers and reduce transaction costs (Bakos 1998). e-hubs provide a marketplace or collaborative platform where operators earn revenue by extracting fees for the transactions. Prior to the dotcom/technology stock market crash in April 2000, members could expect benefits from cost reductions using internet auctions and founders could expect additional revenue arising from the subsequent flotation of the exchange on international financial markets. Whereas business-to-customer (B2C) on-line trading has dominated attention in the media (e.g., Amazon, eToys, and Lastminute.com), B2B trade through e-hubs remains in its "infancy" (Skjøtt-Larsen et al. 2003, p. 208) and is "still in an early growth phase" (Barratt and Rosdahl 2002, p. 111).

e-hubs tend to be classified in terms of what they exchange, e.g. products and services, and how they exchange it, for example, either systematic sourcing and negotiated contracts, or spot sourcing and commodity trading (Kaplan and Sawhney 2000). Systematic sourcing involves contracts negotiated with qualified suppliers, involving long-term contracts and close relationships between buyers and sellers. In spot sourcing, the buyer's goal is to fulfil an immediate need at the lowest cost, such as commodity trading for steel by NewView – formerly the 'e-Steel' hub. In their description of the governance structures of e-markets, Baldi and Borgman (2001) consider two dimensions to be of particular importance. First, the role of the owner: the owner of the market can be an active market participant or an independent third party. Second, the competitive relation of the owners: where the firms owning and operating the market can be direct competitors outside this venture. This results in different ownership structures, presented below with examples from the automotive industry:

- A *private trade exchange* is owned and operated by a single firm (e.g. 'Supply-Power' owned and operated by General Motors).
- A *third party exchange* is operated by a group of non-competing firms or one that is not considered a trading partner (e.g. 'SupplyOn' operated by Bosch, Continental, Ina, SiemensVDO, SAP, and ZF Friedrichshafen).
- A *consortium exchange* is where the ownership is shared between competing firms (e.g. 'Covisint' owned by Ford, General Motors, DaimlerChrysler, Renault-Nissan and Peugeot Citroen – until sold in January 2004).

e-hubs are often synonymous with terms such as e-commerce and e-procurement; yet, there are significant distinctions. e-commerce involves the "electronic trading of physical goods and of intangibles such as information" (Timmers 1998, p. 3). Hence, some forms of e-commerce have existed for over 20 years, for instance: electronic data interchange (EDI), computer assisted lifecycle support (CALS), and computer aided design (CAD) (Webster 1995; Kuroiwa 1999; Croom 2001). e-procurement is defined more specifically as "using Internet technology in the purchasing process" (De Boer et al. 2002, p. 26). Today, e-procurement is concerned with buy-side,

B2B electronic markets, using inter-organisational platforms or hubs to connect purchasing, material planning and logistics, and product development functions with online tools such as catalogues, auctions, requests for quotations, invoicing, and collaborative engineering.

Traditional EDI originated from firms wishing to automate the exchange of data internally and with partners, as a secure link between firms offering a reliable means of communicating purchase orders, build schedules and forecasts (Swatman and Swatman 1992). However, EDI integration faced a number of difficulties in the automotive industry: high entry costs, a proliferation of standards, and coercive pressures from powerful vehicle manufacturers. Firms using EDI found themselves tied into a technology that replicated the hierarchical nature of traditional, adversarial customer–supplier relationships (Sako 1992; Lamming 1993). For instance, the Ford Motor Company in the 1980s had a basic objective in developing its EDI network 'Fordnet' as a competitive weapon: gaining competitive advantage by locking suppliers into its systems and keeping competitors out (Webster 1995).

Today, e-commerce differs from EDI as it provides an inter-organisational information system that fosters market-based exchanges between agents in all transaction phases (Bakos 1998). It overcomes the technological barrier of a bespoke system through web-enabled technology that uses the internet for 'many-to-many' information exchange among multiple firms. The emergence of WebEDI in the late 1990s, using eXtensible Markup Language (XML) and personal computers, offers a low-cost solution for suppliers seeking connection to their business partners via the internet (Vidgen and Goodwin 2000). e-commerce also offers on-line development using 'virtual spaces' in which manufacturers and suppliers can collaborate on joint engineering projects. Therefore, the potential benefits are not only reductions in transaction costs, but also the enabling of buyer–supplier partnerships and new product innovation. This explosion in worldwide connectivity has led to a proliferation of e-hubs, resulting in profound implications for business relationships. A most significant impact is the change in dynamics of increasingly friction-free markets, which are not attractive for suppliers that had previously "depended on geography or customer ignorance to insulate them from low-cost sellers" (Bakos 1998, p. 41).

20.2.2 Current State of the Automotive Market

In 2000, the launch of 'Covisint', the biggest and most powerful automotive e-hub, was announced as the beginning of a new era in industry purchasing and supply chain management. The founder members, Ford, General Motors and DaimlerChrysler (later joined by RenaultNissan and PeugeotCitroen) anticipated significant component price reductions and customer responsiveness by combining purchasing economies of scale and internet technology. However, rival manufacturers and component suppliers were already developing their own solutions and were reluctant to join Covisint over fears of accepting a subordinate role. As private trade exchanges proliferated, Covisint's vision of offering collaborative procurement, lower transaction costs, and the introduction of a universal system standard began to diminish (Helper and MacDuffie 2003). In January 2004, the e-hub that was supposed to transform the auto industry was disassembled and put up for sale, having already sold its on-line auction service and shut down its electronic parts catalogue (ANE 2004). Covisint was bought by Freemarkets, which, in turn, was purchased by Ariba and is now being used in the US healthcare market.

An overview of automotive industry structure shows the integration of e-hubs into vertical and horizontal buy-side electronic markets (Fig. 20.1). In the race to re-engineer the industry from its old economy origins, current structures appear closer to a loose arrangement of spokes than a hub. Overlapping networks compete for limited membership across the industry, resulting in isolated pockets of collaboration and irregular information flow. Figure 20.1 presents the structure and flow across all tiers in the supply chain by classifying e-hubs in terms of membership and network relationships. Membership is represented by the affilia-

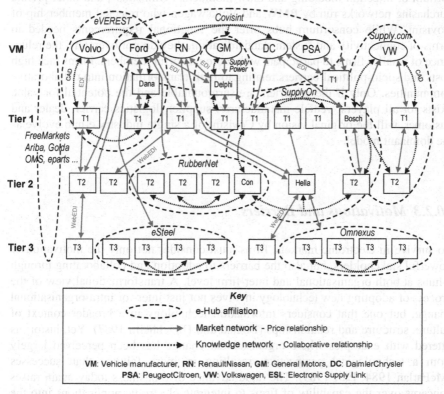

Fig. 20.1   Automotive industry e-hub structure in 2003

tion to e-hubs by VMs and suppliers. The relationship between industry stakeholders is represented by the flows within and between e-hubs – shown as solid and dotted arrows in Fig. 20.1 – driven by price/markets and collaboration/knowledge. Price relationships are short-term, based on considerations of market forces (Sako 1992). Collaborative relationships involve long-term exchanges and include information or knowledge sharing that supports new product and service development (Lamming 1993). All types of transactions associated with the flows depicted in Fig. 20.1 were observed during the research.

e-hubs represent the focus or point of exchange for the aggregate supply and demand of goods and services in electronic markets. Yet, Fig. 20.1 exposes the difficulty experienced by automotive firms setting up hubs to serve both price-based market networks and collaboration-based knowledge networks. The consortium hub (e.g. Covisint) represents a horizontal model of e-procurement that excludes suppliers from participating as an equal partner. Suppliers are not being offered shares in the venture because of the combined purchasing power of VMs when they get together as a club. Suppliers are sceptical of win–win promises because they fear their profit margins will be slashed through internet auctions and their product knowledge commoditised. Moreover, other exchanges have emerged in the industry, such as a consortium of tyre manufacturers, 'RubberNet', a consortium of injection moulding and blow moulders, "Omnexus", and other private purchasing networks run by BMW and Volkswagen who rejected membership of Covisint. While consortium hubs offer the functional requirements needed in terms of connectivity and system standardisation, they do not overcome the reluctance of smaller firms to participate with more powerful members. Firms face high costs and socio-political barriers relating to their transformation into collaborative communities. Only if these barriers are surmountable and the potential for value exists for all players: vehicle manufacturer, supplier, logistics carrier, dealer and customer, will the original vision by Covisint of a single industry e-hub evolve as the dominant model.

20.2.3 Motivations and Barriers

To survive in electronic markets, firms must consider not only the motivation of powerful stakeholders, but also the barriers to adoption and collaborating through e-hubs at both organisational and inter-firm level. A transformational view of the process of adopting new technology involves not just inter- or intra-organisational change, but one that considers information technology in a broader context of culture, structure and relationships (Boland and Hirschheim 1987). Yet, history is littered with examples where organisational change has been perceived largely from a technological perspective, resulting in more failures than successes (McFarlan 1984; Earl 1989). The rapid advance of e-business today again raises concerns over the capability of firms to integrate electronic applications into the business environment (Galliers 1999). This is importance for stakeholders motivated by the cost reductions associated with shifting from paper-based purchasing to e-purchasing, but obstructed by barriers ranging from lack of sharing communication standards to buyer–supplier relationships (Barratt and Rosdhal 2002; Min and Galle 2003; Skjøtt-Larsen et al. 2003). In the automotive industry, even top executive Jac Nasser, ex-CEO of Ford Motor Company, was dismissed for poor performance related to e-business initiatives. Accused that he "fell under the spell of the Internet", the Ford vision of web technology as the 21st century equivalent of the moving assembly line has so far failed to materialise (Connelly 2001, p. 42).

Motivation is viewed as an external standard of organisational effectiveness, where the ability to create acceptable outcomes and actions is judged in terms of how well it meets the demands of the various groups concerned with its activities (Pfeffer and Salancik 1978). Motivation means not only the incentive of adding value from information systems, but involves some element of risk for firms. A sense of "expected" and "realised" benefit describes the potential and eventual outcomes that drive stakeholders to pursue a goal or set of objectives. In the emerging e-business world the expected benefits from e-hubs are dramatic reductions in transaction cost, improvements in service quality, faster product delivery, and other more opportunistic factors, such as leveraging component price reductions through electronic auctions (De Boer et al. 2002). Hence, in this context stakeholder motivation forms an integral part of information system planning in industry and part of the "sociology of technology", where the whole process of adoption and change "... is not simply a technical-rational process of 'solving problems'; it also involves economic and political processes in articulating interests, building alliances and struggling over outcomes" (Webster 1995, p. 31).

Barriers research can be traced to early literature on inter-organisational systems, which considers the interchange of information as the basis of all activity (Barret and Konsynski 1982, p. 101). In their assessment of the impact of IOS, Barret and Konsynski describe "factors of concern" as how new technologies are introduced into the firm and the resultant effects on organisational structure, users, and the IT department. Early attempts to unify the "fragmented models" of IS implementation reveal five categories of factor: individual, structural, technological, task-related and environmental (Kwon and Zmud 1987). Argyris (1990) describes organisational defences as being one of the most critical barriers to learning where organisational change is dependent on an ability to learn, requiring people to collaborate through a shared vision. Hence, planning for IS-related change requires clear strategic objectives and a systematic means of identifying project constraints, barriers or features (Earl 1989).

Today, the planning, implementation, and maintenance of an e-procurement programme to create value across supply chains represent a considerable challenge (Presutti 2003). While the value of e-hubs is consistently reported as comprising cost reduction, market reach and knowledge transfer (Bakos 1998; Kaplan and Sawhney 2000; Amit and Zott 2001; Barratt and Rosdahl 2002), there is little consensus over the supply chain factors that constrain e-hub adoption (Skjøtt-Larsen et al. 2003). Further, an assumption persists in firms that simply acquiring technology such as e-hubs somehow represents a solution in itself (Hagel 2002). In reality, the first obstacle that inhibits e-hub development is shortage of capital, often resulting in a quick rethink of strategy where buyers and suppliers can take an equity share. Some companies also experience supplier reluctance to adopt e-hubs, which stems from the suppliers' unease about the trend towards "commoditisation and the squeeze on supplier's profit margin" (Baratt and Rosdahl 2002, p. 118). This is particularly significant for the auto industry, faced with rebuilding confidence after events such as the dotcom crash, the departure of ardent supporters of e-commerce and the ongoing burden of replacing IT legacy systems.

Several models exist today that are based on IS-related barriers, including King and Thompson's Facilitators and Inhibitors for Strategic IT (1996), Kirvennummi et al.'s Barriers Information Framework (1998), and Heeks and Davies's e-Government Barriers (1999). Typically, these present key organisational factors that facilitate or inhibit the adoption of technology. The studies are conducted as firm-level analysis where IT is used as a competitive weapon to gain advantage over rivals. This chapter argues that such studies of firm level inhibitors and enablers to IT adoption are insufficient to allow full understanding in cases where e-hubs are adopted by multiple firms. The integration of web technology requires considerable attention to both the intra- and inter-organisational context to achieve the levels of collaboration and interaction needed between partners. Hence, the terms 'motivations' and 'barriers' are adopted here to focus greater attention on issues of power, expected benefit, and level of analysis. This is presented as a preliminary framework to act as a starting point from which to guide conceptual thinking and structure the findings (Fig. 20.2).

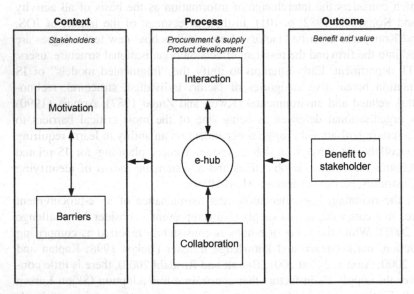

Fig. 20.2 Preliminary framework

20.3 Approach to Developing the Framework

To understand the interaction between e-hubs and the environment, the framework must capture enough detail to get close to the problem, yet remain sufficiently flexible to enable exploratory research. Hence, a three-part model is used in the framework (Fig. 20.2) to explore stakeholder motivation and barriers (context), collaboration and information sharing through electronic applications (process), and benefit to the stakeholder (outcome). Developing a contextual perspective and breaking it down to its "three essential dimensions" enables understanding of organisational and industry change (Pettigrew and Whipp 1991, p. 26). Context here means the circumstances that form the setting for the event, or "over-arching structures and systems...that facilitate or inhibit" an individual, group or organisation's impact (Denison et al. 1996, p. 1006). Investigating context provides a rich description of events and is particularly useful in research situations involving change at all levels, such as the change to world economies brought about by globalisation and the diffusion of technology.

This investigation builds on work by the 3DayCar programme, launched in 1999, to study customer order fulfilment in the UK automotive industry. It entered a new phase in 2004 that focussed on the technologies required to enable BTO in short lead times across Europe. A key finding of the original study is of the 40 days it takes to deliver a new car in the UK, around 85% of the delay is derived from information processing – order entry, order processing and vehicle scheduling – and only 15% from manufacture and distribution (3DayCar 2002). The framework in Fig. 20.2 provides the foundation for this research, based on an initial study during the 3DayCar programme as e-hubs emerged in the industry (Howard et al. 2002). Meetings held with the founding firms of Covisint in Dearborn, US, and in Europe established that e-hubs represent a radical departure from earlier business practices: they require high levels of collaboration between multiple partners involving interdependency, interaction, and information exchange, which presents a significant opportunity for organisational research. Two vehicle manufacturers and two suppliers were chosen on the basis of their experience in purchasing and supply, and their contribution in terms of helping to understand the challenge of adopting e-hubs within and between their organisations.

Over a 20-month period in 2001–2003, 30 semi-structured interviews were conducted with managers and directors from procurement, supply, product development and IT, using the framework as the basis for questioning. Selected interviews with staff from IT consultancies and e-hubs were also included in order to gain additional insight and gauge the level of new technology being accepted by the industry. At the end of the fieldwork the investigators facilitated a wash-up session attended by all staff to agree and discuss what were the key issues arising from the interviews. The interview transcripts were codified and presented in tabular form as a focus for analysis by logically leading the investigator to reflect on the gap between expected and realised benefits, and by considering the type and level of barrier responsible. This enabled the investigators to re-examine the preliminary framework and develop it further from the study data.

20.4 Automotive Case Analysis

The case interviews identify the perceived benefits of e-hubs together with industry and organisational barriers. The four cases (identity concealed for commercial confidentiality) and their associated e-hubs are now considered in depth and summarised. The analysis concludes by examining the motivations and barriers for interaction through e-hubs across all cases.

20.4.1 Case 1: "Motorco"

Motorco is undergoing major restructuring following the announcement of a $4.5 billion loss and 17,000 job cuts in 2002. The CEO and executive board in Detroit are seeking significant cost reductions across global operations driven by material price, transaction efficiency and supplier rationalisation. As the firm struggles to regain profitability world-wide, it aims to reduce an annual purchasing spend of $100 billion by "$1 billion" through taking advantage of the economies of scale to leverage material costs and increased visibility to source cheaper suppliers (Launch manager). Central to this strategy is a massive e-procurement initiative involving the introduction of a standard application to all purchasing departments: "... it's the third largest investment this company's got at the moment" (Purchaser, e-business). Covisint is perceived as an opportunity to develop the electronic marketplace as a means of utilising the reverse auction on commodity and production materials. The expected benefits in order of priority are material price reduction, minimisation of paper transactions, and electronic audit capability.

However, despite an e-business team of 300 people at the firm, realised benefits in the form of lower material costs, increased transaction efficiency and control over maverick spending, have only partly fulfilled expectations. Covisint was delayed by federal and EU authorities concerned over possible breaches of anti-trust legislation, and by the resignation of a succession of European CEOs due to the e-hub failing to meet its growth targets. Internal IS-related problems also persist in adopting e-procurement, where Motorco's approach is to "institutionalise" the new system to comply with the needs of the central purchasing commission who approve all spending (Purchaser, e-business). This has encountered resistance by managers who already have well-established purchasing relationships with suppliers. In addition, system operators who were used to having bespoke software must use only standard e-procurement systems. Thus, Motorco has some way to go before reaching its goal of "moving the buying community away from the transaction [and] giving them the tools which will help them in negotiation, strategic thinking and disseminating information from a lower level" (IT Manager).

Despite initial success in price reductions of non-production, commodity catalogue items, a key industry barrier for the company remains the reluctance by suppliers to subscribe to Covisint over fears of the effect of reverse auctions on component prices. Motorco is renowned for its cost focus, and this affects the willingness of partners to accept e-procurement: "their perception and [level of] readiness is not good" (Launch Manager). The e-hub is already facing delays from conflicting organisational structures in North America and Europe, cultural mismatch between the head office in Detroit and associated European brands, and supplier suspicions over component price cuts. Shifting from vertical integration towards a flatter, open market-based system represents a major challenge for Motorco whose own employees acknowledge that it is "still a hierarchical company" (Purchaser, e-business). Further, in preparation for switching over to Covisint, little consideration was given to the purchasing and supply legacy systems still used by the company's manufacturing sites. "The tricky thing is how do we decommission those systems? How do we convert the data and switch off?" (Purchaser, e-business). It is now recognised that during the adoption of the e-hub "limited buy-in by stakeholders [during] registration has been a nightmare. In hindsight our track record hasn't been good with system implementation – in future we need to take bite size chunks" (Change Manager).

20.4.2 Case 2: "Carco"

Carco was one of the first manufacturers in Europe to become seriously committed to slimming its distribution pipeline and reducing inventory costs over a decade ago. Its purchase in 1999 means it has become the benchmark for BTO capability for its parent organisation Motorco. In order to continue its policy of customer-ordered production, Carco must integrate its sequenced in-line supply of components further upstream to second-tier suppliers, while implementing Motorco's latest global purchasing programme for partners using the e-hub "eProcure". Yet, the structural differences between the two organisations became evident from the outset: "[Carco] is an agile and open organisation. When we are part of this global project the differences between us and our colleagues are quite visible" (Procurement Director). Despite such differences, an early and unexpected benefit has been the opportunity for Carco executives to learn from Motorco, particularly concerning the need to achieve "efficiency benefits from web-based e-procurement [and] a more strategic and integrated approach to procurement" (Procurement Director).

While managers at Carco at first believed that the key to increasing supply chain efficiency was through total visibility (originally promised by Covisint), they now believe that the key is the synchronisation of material and orders. Carco is obliged to integrate eProcure into its own business as well as first-tier suppliers. Yet, there are serious concerns that Carco's participation in the e-hub will be perceived by suppliers as evidence of Tayloristic business practices and a strong cost reduction focus, leading to an erosion of trust between the vehicle manufacturer and supply partners. eProcure may become "both an internal and external threat for suppliers [by introducing] new competitors...they are quite scared we are going to show prices that [Carco] has and are different to [Motorco]" (IT Manager).

Carco's culture of autonomy and corporate citizenship is substantially different from those of its Detroit parent organisation. However, as a partner they are obliged to adopt the e-hub despite its design based on American organisational processes, which are very different to their own. Internally, Carco staff expect to achieve process efficiency benefits, to develop more strategic buyers using centralised data, and to cope with an expanding new vehicle development programme through fully cross-functional system integration. Yet, it is generally acknowledged that current processes are "based on paper and personality", involving time-intensive transactions via multiple legacy IT systems (Purchasing Director). The adoption of eProcure aims to change the Swedish carmaker's current ad hoc approach to purchasing and supply irrevocably, although "people are worried about

Table 20.1 Vehicle manufacturer benefits and barriers

	Motorco / *Covisint*	**Carco / *eProcure***
Benefits		
Expected	• Indirect price reduction (P) • Direct price reduction (N) • Minimise paper transactions (P) • Electronic audit capability (P) • Transaction efficiency (P) • Control maverick spending (P)	• Reduce inventory costs (N) • Increase supply chain efficiency (N) • Create more strategic buyers (P) • Cross-functional integration (N) • Opportunity to learn from Motorco (U)
	F = Fully realised P = Partly realised N = Not realised U = Unexpected	
Industry barriers	• Supplier resistance to subscribe to reverse auctions and VM-owned e-hubs • Anti-trust legislation delaying launch	• Fears that suppliers will perceive e-hub as adoption of 'arms-length' business practices - leading to erosion of trust
Firm barriers		
Structural	• New systems must be institutionalised to align with the needs of the centralised purchasing commission • Conflict from US & EU firm structures	• Concern over eProcure limited choice of suppliers • High cost of subscribing to electronic catalogues
Cultural	• Cultural mismatch between US and European offices • Employees acknowledge that Motorco is still a hierarchical-style company	• Concern over Carco's unique independent Swedish origins being affected by American process and culture
Managerial	• Resistance by managers who already have established purchasing relationships • CEO leadership difficulties over the management of e-commerce	• Current processes managed by paper and personality • Managers fear redundancy through the supplier e-database
User	• Staged implementation of Covisint means it is difficult to identify the system as a definitive product for users • Operators used to bespoke system design means difficulties in adapting to a standard package	• User fears over the effect of increased transparency by eProcure restricting the individual's autonomy • People worried about losing their own way of working
Technical	• Decommissioning legacy systems	• Multiple IT legacy systems

[losing] their own way of working" (Procurement Director). The next step for eProcure is the most critical – production parts – as this directly affects the manufacturing plants, goods distribution and the link with dealerships. However, a common concern in Carco is over the level of resources needed to implement eProcure effectively. This has been underestimated because "we have a totally different process from [Motorco]...you affect everything with one tool and process. The tool and process we are trying to implement now is adapted for [Motorco]. So now we have a culture and system problem together!" (Materials Planning and Logistics).

The outcomes of the case research with the two vehicle manufacturers are summarised in Table 20.1. This shows the expected, unexpected and realised benefits, in addition to industry and firm level barriers. Its purpose is to force investigators to look beyond initial impressions and see the evidence through multiple lenses in order to sharpen research definition and validity (Eisenhardt 1989). The data illustrate significant differences between the vehicle manufacturers in their approach towards e-hubs. Motorco perceives Covisint as an exclusive club to reduce supplier component prices and thus lower costs. Carco perceives eProcure in terms of the opportunity for process improvements, but also as a potential threat to its organisational culture and supplier relations. Overall, both are motivated by the expected gains to be achieved in lower costs and efficiency. Yet, a common industry level barrier to the adoption of e-hubs for the vehicle manufacturer is supplier reluctance to participate in a system that may force price cuts, disrupt stocking and labour agreements, and offer little benefit to suppliers in return. The supplier approach to e-hubs is examined next.

20.4.3 Case 3: "Partco"

Partco is a major shareholder in the supplier e-hub SupplyOn, founded in 2000 to host on-line procurement for first- and second-tier component suppliers. The organisation is one of the biggest first-tier suppliers in Europe and specialises in delivering technology solutions such as direct fuel injection systems for the expanding diesel engine market. SupplyOn is one of the few automotive e-hubs that has survived the dotcom crash and emerged with a strong business model, presented as "a marketplace from suppliers for suppliers and a strong partner for vehicle manufacturers" (IT Manager). The e-hub now serves 2,700 partners by supporting an electronic connection and training service for activities that span the entire product life-cycle. It focuses on collaborative engineering, electronic commodity catalogues, WebEDI service provision, and offers links to other e-hubs such as the tyre consortium Rubbernet. In contrast to Covisint, only 3% of SupplyOn's business involves auctions.

The expected benefits for Partco in developing SupplyOn were reductions in manual processes, cost transparency, and more responsive order-to-delivery lead times. Yet, the company has now strengthened its competence in building electronic supply chains and realised a number of benefits through fully integrated e-procurement, simultaneous engineering, and leading negotiations in industry standards. Partco is renowned for innovative engineering, but was concerned it should develop its on-line capabilities in engineering and purchasing. Hence, the supplier's motivation for developing an e-hub differs significantly from vehicle manufacturers such as Motorco: "I hope reduced price will be one of the secondary benefits, but it's not our focus" (Purchasing Manager).

Introducing an on-line materials purchasing and development system into Partco meant it initially encountered reservations from managers and users. Some senior managers wanted to wait until other suppliers had implemented their systems first. However, Partco was eager to gain first-mover advantage, particularly over system standards. At first some internal users became reluctant after the reality of using an internet system did not live up to expectations. Moreover, the lack of a common standard during system trials meant suppliers were not involved until Partco had achieved full back-end integration of its own mainframes. Yet, the time and resources dedicated to achieving a swift launch of SupplyOn resulted in an unexpected benefit: a shift towards standard operating procedures and practices. This complements the vision for SupplyOn to become an integrated application that serves all supply partners, not just an electronic directory or on-line bidding site for the prime manufacturer.

Partco's goal to develop SupplyOn as a collaborative engineering hub for building relationships in the supply chain and not simply focussed on component price, means that it takes a relatively ambivalent view towards Covisint as a potential competitor: "There is conflict, but also a complementary approach" (Consultant Engineer). The first-tier supplier has now floated SupplyOn as an independent organisation, although it remains the majority shareholder, and aims to strengthen the e-hub's position in the market as a service provider for automotive supply partners. Partco's current objective for SupplyOn is to "optimise and accelerate the business processes between suppliers and their partners...to drive industry growth through efficient communication and transaction development" (Purchasing Manager).

20.4.4 Case 4: "Lampco"

Lampco is a specialist lighting systems supplier faced with increasing demand for commercial and passenger vehicle headlights. It has recently been requested by Motorco to use Covisint's purchasing and stock control function after initial training last year. As a manufacturer of a specialised product, Lampco is more fortunate than other suppliers in its business relationships with vehicle manufacturers, because "there is less commercial pressure to make it cheap. They're usually after technology, or a certain quality" (Kaizen Manager). Yet the introduction of Covisint represents a radical change for the second-tier supplier: "You don't know whom you're dealing with any more...I can see a problem of the faceless world" (Supply Manager).

Covisint provides the internet platform for the electronic component release system to be used by Motorco on all new product introduction projects. Lampco can see little benefit from Covisint, which has so far charged the supplier after completing its basic training: "They tell us how to use their system and charge us at the same time" (Supply Manager). There is also anger over Motorco forcing the supplier to adopt new technology and being penalised by the performance metrics system if they refuse. Lampco recently attempted its first online auction and the experience has done little to boost confidence in e-hubs: "not many bidders and not compatible [with the business process]" (Purchasing Manager). Despite weeks of preparation, the auction soon dropped below minimum pricing levels and gave no opportunity for direct negotiation with the vehicle manufacturer. Lampco had hoped for cost and efficiency savings from the increasing availability of e-procurement tools in the industry, but the only benefit to emerge so far is a faster response to invoice queries.

The success of Covisint depends largely on membership levels; yet, many of the lighting manufacturer's suppliers are retailers from outside the industry where vehicle parts do not make up the bulk of their trade; hence, subscription to the e-hub is unlikely. "They could easily turn round and make things difficult for us" (Systems Manager). Yet, despite the criticisms of the e-hub, managers at Lampco admit that the potential to improve transaction efficiency within its own organisation is high: "We are very manual here and not very integrated. Too many Excel spreadsheets [and] paper everywhere" (Purchasing Manager). However, in other parts of the supply chain there are complaints over the burgeoning electronic communication network representing a threat to suppliers who average six bespoke EDI systems per plant with no common functionality. The proliferation of e-hubs across all levels of the industry means: "We won't be in a position to support everything they throw at us" (Systems Manager). If the rate of introduction of new systems across the industry continues, Lampco will not be able to cope with the demand to adopt these applications: "I just wish vehicle manufacturers would get together and come up with a common standard" (Purchasing Manager).

In an unprecedented move, the Board of directors in Germany has supported their regional plants in a complaint over the inefficient operation of Covisint and accepted plans to outsource the task of searching for relevant data fields in the e-hub to a third party service provider. Yet, concern exists internally amongst staff over the Board's lack of awareness regarding the broader opportunities and threats of e-commerce. New technology is often considered as best left to the experts and there is "very little involvement from managers" (Systems Manager). There is also the attitude that inter-organisational communication via hubs is a backward step: "It's password this, password that, it's just a minefield of repetitive actions" (Purchasing Manager). Although information is shared freely within Lampco, this level of trust does not extend to external information systems: "Covisint is outside [Lampco]...this facelessness, you can't trust someone you don't know" (Supply Manager).

The outcome of the research with the suppliers is summarised in Table 20.2. It illustrates two contrasting approaches to e-hubs: the aspirations of the first-tier supplier Partco to gain first-mover advantage by launching SupplyOn as a collaborative electronic community, and the misgivings reinforced by the lack of realised benefits by the second-tier supplier Lampco over their interaction with Covisint.

While Lampco has achieved minor benefits in transparency and audit capability, the overall increase in process complexity from Covisint has increased costs, resulting in a net disbenefit to the firm. Partco's development of SupplyOn has resulted in reductions in manual waste, improved process transparency and enhanced collaborative product development capability. This case also illustrates an instance of unexpected benefit, where the preparation of a common business language in readiness for SupplyOn's launch had the effect of standardisation of procedures between supply partners. Considering both cases, common barriers to the adoption of e-hubs are competitive conflicts of interests, internal organisational resistance, lack of adherence to universal industry standards and network complexity.

Table 20.2 Supplier benefits and barriers

	Partco / *SupplyOn*	Lampco / *Covisint*
Benefits		
Expected	• Reduced time-to-market (P)	• Cost reduction (N)
	• Minimise manual processes (F)	• Process efficiency savings (N)
	• Cost transparency (P)	• Delivery - faster response to invoice
	• Online product development (F)	queries (P)
	• Competence in e-supply (F)	
	• To lead in electronic standards (F)	
	• Standardised procedures (U)	
	F = *Fully realised* P = *Partly realised* N = *Not realised* U = *Unexpected*	
Industry barriers		
	• Some competitive conflicts of interest	• Increasing complexity from a burgeon-
	with Covisint	ing industry communications network
	• Need for greater recognition between	• Difficulties fostering trust in e-hubs:
	supply partners for industry IT standards	view of Internet as a 'faceless world'
		• Little incentive for Tier 2 suppliers
Firm barriers		
Structural	• Considerable commitment required in	• Hampered by manual and poorly inte-
	terms of time and resources to adopt or	grated processes
	develop e-hubs	• Lacking in sufficient internal resources
		to support e-hubs
Cultural	• Reservations by staff against e-business	• A general attitude that e-hubs are
	in general	a backward step
Managerial	• Desire by some senior managers to wait	• Concerns over lack of management
	and see how other firms conducted their	awareness over the opportunities and
	e-business strategy	threats of e-hubs
User	• User expectations of big benefits not	• Too much paperwork and manual
	realised in the short-term	processes within the firm
		• Logging-on process involving pass-
		words seen as bureaucratic
Technical	• Lack of a universal industry system	• Lack of adherence to common EDI
	standard: suppliers were involved only	standards
	after Partco had achieved its back-end	• e-hub is slow during portal navigation
	integration	

The four cases highlight the expected benefits and barriers during the process of adoption and interaction with e-hubs. They reveal the most common expected benefit from e-hubs as the reduction of cost for vehicle manufacturers and suppliers collaborating in electronic catalogues, quotation requests, auctions, component delivery, and product development. However, whereas the motivation for Motorco is to achieve economy of scale by combining total purchasing spend through Covisint to leverage supplier price reductions, the motivation for Partco is economy of scope by tackling cost across several core areas of the business, using SupplyOn for product development and order fulfilment, not just procurement. Motorco adopted a 'big bang' approach to Covisint by investing heavily in the project and using its power in the supply chain to persuade suppliers to participate. Motorco's top-down approach is also related to its affiliate company's difficulties over assimilating the eProcure e-hub into its own organisation, difficulties arising from striking differences in structure and culture between Motorco and Carco. The relationship between Motorco and partners Carco and Hella reveals the significance of governance in the auto industry: in this case the potential for powerful buyers to exercise coercion over suppliers. Here, the motivation to interact through e-hubs reflects earlier approaches to collaboration typically in terms of arms-length/ adversarial as opposed to obligational/contractual relationships (Sako 1992). This research shows Motorco, Carco and Lampco with gaps where expected benefits have not materialised, while only Partco has realised several expected (and unexpected) benefits by adopting a strategy that includes *all* firms, based on reducing supply costs across all areas of the business. The next section applies the cases to the conceptual framework.

20.5 Discussion: Revisiting the Framework

The framework is revisited by synthesising the data from the vehicle manufacturer and supplier cases (Fig. 20.3). The context of the study reflects not only the range of expected benefits, which extend beyond conventional measures of cost, quality, and delivery, but also the scope of firm and industry level IS-related barriers. Industry dynamics in terms of stakeholder salience involving power, legitimacy, and urgency, mediates between motivations and barriers, and is described in detail in Howard et al. (2003). The barriers, which concern the industry as a whole, highlight the issue of resistance to e-hubs due to the fear of price cuts, lack of trust in more powerful stakeholder groups, and the increasing burden of network complexity brought about by the proliferation of e-hubs. The competitive conflict between consortium, private, and third-party e-hubs to increase membership does not address long-term concerns over adherence to universal industry standards and protocols.

The significance of organisational, firm level barriers is also represented by the framework in Fig. 20.3. Structural alignment reflects the difficulties for firms to align e-hubs with current organisational hierarchies and the associated problems of

interfacing with internal purchasing and supply systems. The cost of e-hub sub-
scriptions and the internal human resource issue rises in proportion to the number
of external networks used by the firm. Cultural differences reflect the mismatch in
outlook between US and European businesses brought together as an attempt to
create closer relations by interacting via e-hubs. This means that suppliers or
manufacturers with strong associations with brand or national identity often feel
that their independence is threatened by automated links to other organisations.
The lack of leadership by senior management is illustrated by a generally poor
awareness of the potential impact of the internet and the difficulties over e-hub
adoption. User buy-in reflects the fears of people who associate the transparency
of e-hubs with a perceived threat to individual autonomy. Procurement and materials planning and logistics system users are often reluctant to abandon their bespoke IT systems because of the implications of retraining and the uncertainty
over long-term benefits from e-hubs, and the stability offered by the status quo.
'Accessibility' implies the difficulties of logging on and navigating through pass-
word-protected sites, which are often slow, complex, and provide little support for
the user. Legacy IT system infrastructure represents a barrier to e-hub adoption
because of the difficulties of interfacing bespoke technology to a common business standard across many organisations. These barriers are derived from the case
summaries in Tables 20.1 and 20.2, and are developed into conceptual categories
to create a summary of barriers (Table 20.3).

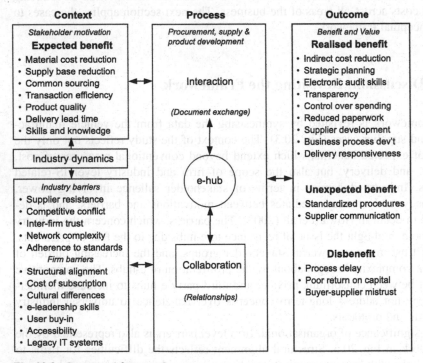

Fig. 20.3 Conceptual framework

The process space in the framework highlights the difference between e-hub interactions, involving issues related to document exchange, and the vastly greater difficulties associated with collaboration and the management of inter-firm relationships (Fig. 20.3). Interaction here addresses the mechanics of an e-hub-assisted exchange between remote databases, while collaboration addresses a broader spectrum of relationship concerns over power, control and the management of inter-organisational systems by stakeholders. The outcome space in Fig. 20.3 reveals that, as with any enabling technology, adopting an e-hub does not guarantee benefits to the firm, may realise unexpected benefits, or even result in 'disbenefit'. This means that the cost of e-hubs may never be recovered (e.g. Covisint), buyer–supplier relationships may be damaged and electronic applications may adversely affect product and service quality. The net effect is that there is no stakeholder benefit and no value created for the customer.

Table 20.3 Summary of barriers

	Summary	Barriers from cases
Industry barriers		
Relational	Buyer-supplier agreement	Supplier resistance to auctions
	Competitive conflict	Erosion of inter-firm trust
		Competitive conflict between e-hubs
		Conflict from firm performance transparency
Topological	Industry standards	Increase in individual system standards
	Legislation	Network complexity
		Anti-trust legislation against consortia e-hubs
Organisational barriers		
Structural	Firm-technology alignment	Alignment of e-hubs with existing processes
	Financial resources	Conflict caused by differences in firm structure
		High costs of subscription
		Lack of resources to cope with e-hubs
Cultural	Autonomy	Cultural mismatch between US and EU offices
	Attitude to risk	Concern over erosion of affiliate firm origins
		Choice of common business language
		Widespread reservations over e-business
Managerial	Control	Resistance from well-established managers
	Leadership	Current manual processes support status quo
		CEO difficulties leading electronic initiatives
		Inertia and lack of awareness over 'e'
User	Fear of change	Individual user fears over effects of transparency
	Accessibility	Difficult to shift from bespoke to standard system
		Resistance to e-hubs often from older users
		Logging-on process perceived as bureaucratic
Technical	Infrastructure	Multiple IT legacy systems
	Delay	Lack of standards impede e-hub integration
		e-Hubs slow during portal navigation

20.6 Contribution to Theory and Practice

The contribution to theory is reflected in the value of the framework showing 'benefit dissonance' as well as extensions to the literature on information systems barriers. Benefit dissonance identifies the differences between expected benefit and realised benefits from adopting e-hubs where, particularly in the case of Covisint, the difficulties of integrating a direct component purchasing system across supply partners was significantly more difficult than anticipated. While isolated cases of unexpected benefits occurred, stakeholders such as Motorco were unprepared for the eventuality of disbenefit as a result of e-hub adoption *increasing* process delay and buyer–supplier mistrust of the system. Barriers in IOS do not just exist within individual firms (Barret and Konsynski 1982; Kwon and Zmud 1987), but occur at industry level including supplier resistance, inter-firm trust, and network complexity, in addition to e-leadership, structural, and cultural difficulties at firm level.

In terms of contribution to practice there are considerable differences in the motivation to adopt e-hubs from one tier to another. Vehicle manufacturers such as Motorco are motivated by the use of inter-organisational systems as a tool to implement short-term price reductions. Others, such as Carco, perceive e-hubs as a potential enabler of more long-term relationships; hence, there is a conflict between the price-based perspective of Motorco and the relational approach of Carco. First-tier suppliers such as Partco aim to develop a knowledge-sharing platform for all partners with supply cost reductions as a secondary objective. Second-tier suppliers such as Lampco lack the resources to cope fully with the demands placed on them to adopt IOS from more powerful players, which leave them vulnerable to coercive, top-down approaches to e-hub adoption. Thus, four recommendations for the auto industry emerge:

1. *Topology* – electronic applications are at a nascent stage of development across the industry and considerable work remains in building a common IT infrastructure or "e-supply topology". Topology is critical to the way in which constituent parts such as e-hubs are interrelated, and includes control of industry standards, provision of translation services, and legislation that connects Europe to other regions in the world.
2. *Buyer–supplier relationships* during the adoption and use of e-hubs are key to collaboration and creating benefit; yet, considerable conflict is revealed between vehicle manufacturers and suppliers. Vehicle manufacturer expectations of short-term gains are counterbalanced by suspicions from suppliers of exclusive VM ownership and enforced collaboration where partners must adopt new technology regardless of the consequences. A potentially more profitable model that rejects the Covisint approach is the open membership and knowledge-sharing approach by SupplyOn, offering B2B collaborative tools in return for a standard fee.

3. *Leadership* is revealed here to be lacking in the auto industry. This includes a general lack of e-business awareness outside the IT department, an inability to offer adequate support and guidance, and senior managers setting business objectives for e-hubs that are simply too ambitious.
4. *Disbenefit* occurs during the process of adopting electronic applications. Yet, none of the cases paid adequate attention to long-range planning that mitigates the risk of delay or downsizing of e-hubs, with budgets running to billions of euros. Applying the concept of expected, unexpected, and realised benefits at regular intervals during project development enables the practitioner to track changes to e-hub performance, and crucially, help to highlight the barrier responsible.

20.7 Conclusion

This investigation helps to dispel the notion of rapid change in the transition from supply hierarchy to electronic markets, where the nature of inter-firm relationships during IOS adoption often reflect characteristics of the industry's traditional origins as a vertically integrated mass production paradigm. Despite optimistic predictions over e-commerce diffusion and value creation worldwide (Presutti 2003; Min and Galle 2003) the investigation reveals benefit dissonance from a range of barriers across multiple firms operating in a less information-intensive, manufacturing-based sector. The study enables understanding of the real outcomes from e-hubs – not the hyperbole – by revealing motivations and barriers in the context of buyer–supplier collaboration. Yet, some surprises are revealed in the framework; for instance, there are no concerns highlighted over organisational security issues. This study reveals an opposite view, highlighted by the current concerns of e-hub users over system accessibility. An issue for further investigation is the apparent lack of a flattening of industry structure from improved information flow through e-hubs, with at present only improvements in transactional automation. Is it just a matter of time before structural changes occur in the industry, or are other forces at work? Counterintuitive to the argument of disintermediation, this study shows how improved information flows seem to have reinforced the roles of some first-tier suppliers in relation to the vehicle manufacturers.

Given that the outcome of e-hubs in terms of benefits is somewhat different from industry expectations, further work is planned to extend the case analysis to include second- and third-tier suppliers, examine the use of e-hubs in the component aftermarket, and focus on other networks, not only in manufacturing sectors such as aerospace, but also service sectors such as government and retail. While this research reveals one approach adopted by vehicle manufacturers, i.e. Motorco and the launch of Covisint, there may be other supply relationship strategies to explore, for instance, in terms of VM-to-supplier or supplier-to-supplier alliances. The goal is to explore further insights into why e-hubs fail, explain conditions for success, and compare the role of electronic applications across multiple sectors in terms of collaboration and interaction.

References

3DayCar (2002) Towards a customer driven system – a summary of the 3DayCar Research Programme. Report by the 3DayCar Research Team. Available via www.3daycar.com

Alves de Queiroz I, Stucky W, Hertweck D (2002) The e-business battlefields of the automotive industry. International Conference of Electronic Business, Beijing

ANE – Automotive News Europe (2004). A decimated Covisint is put up for sale. 26 Jan, p 17

Amit R, Zott C (2001) Value creation in e-business. Strat Manag J 22:493–520

Arbin K, Essler U (2005) Covisint in Europe: analysing the B2B auto e-marketplace. Int J Automot Tech Manag 5(1):31–45

Argyris C (1990) Overcoming organizational defenses: facilitating organizational learning. Allyn and Bacon, Boston

Bakos Y (1998) The emerging role of electronic marketplaces on the internet. Communications of the ACM 41(8):35–42

Baldi S, Borgman H (2001) Consortium-based B2B e-marketplaces – a case study of the automotive industry. 14th Bled Electronic Commerce Conference, Slovenia, pp 629–645

Barratt M, Rosdhal K (2002) Exploring business-to-business marketsites. Eur J Purch Supply Manag 8:111–122

Barret S, Konsynski B (1982) Inter-organisational information sharing systems. MIS Q Special Issue:93–105

Bauer M, Poirier C, Lapide L, Bermudez J (2001) E-business: the strategic impact on supply chain and logistics. Council of Logistics Management

Boland RJ, Hirschheim RA (1987) Critical issues in information systems research. Wiley, Chichester

Connelly M (2001) Where Jacques Nasser went wrong. Automot News 15 October:42

Counsell A (2002) How dinosaurs are cleaning up amid the rubble of the dotcoms. Financial Times Information Technology 5 June:1

Croom SR (2001) The dyadic capabilities concept: examining the processes of key supplier involvement in collaborative product development. Eur J Purch Supply Manag 7:29–37

Daniel E, White A, Harrison A, Ward J (2003) The future of e-hubs: findings of an International Delphi Study. Information Systems Research Centre, Cranfield Centre for Logistics and Supply Chain Management

De Boer L, Harink J, Heijboer G (2002) A conceptual model for assessing the impact of electronic procurement. Eur J Purch Supply Manag 8:25–33

Denison D, Hart S, Kahn J (1996) From chimneys to cross functional teams: developing and validating a diagnostic model. Acad Manag J 39(4):1005–1024

Earl M (1989) Management strategies for information technology. Prentice Hall, London

Eisenhardt K (1989) Building theories from case study research. Acad Manag Rev 14(4):532–550

Galliers R (1993) Research issues in information systems. J Inform Tech 8:92–98

Hagel J (2002) Out of the box: strategies for achieving profits today and growth tomorrow through web services. Harvard Business School Press, Boston

Heeks R, Davies A (1999) Different approaches to information age reform. In: Heeks R (ed) Reinventing government in the information age – international practice in IT-enabled public sector reform. Routledge, London

Helper S, MacDuffie JP (2003) B2B and modes of exchange: evolutionary and transformative effects. In: Kogut B (ed) The global internet economy. MIT Press, Cambridge

Holweg M, Pil F (2004) The second century – reconnecting customer and value chain through build-to-order. MIT Press, Cambridge

Howard M, Vidgen R, Powell P, Graves A (2002) Are hubs the centre of things? Case studies of e-procurement in the automotive industry. In: Wrycza S, Kautz K, De Marco M, Galliers R (eds) Proceedings of 10th European Conference on IS, Gdansk, pp 1517–1526

Howard M, Vidgen R, Powell P (2003) Overcoming stakeholder barriers in the automotive industry: building to order with extra-organizational systems. J Inform Tech 18:27–43

Howard M, Vidgen R, Powell P (2006) Automotive e-hubs: exploring motivations and barriers to collaboration and interaction. J Strat Inform Syst 15:51–75

Kaplan S, Sawhney M (2000) E-hubs: the new B2B marketplaces. Harv Bus Rev May/June:97–103

King WR, Thompson T (1996) Key dimensions of facilitators and inhibitors for the strategic use of information technology. J Manag Inform Syst 12(4):34–53

Kirvennummi M, Hirvo H, Eriksson I (1998) Framework for barriers to IS-related change: development and evaluation of a theoretical model. In: De Gross J et al (eds) Proceedings of Joint IFIP Working Conference, Finland, pp 509–528

Kuroiwa S (1999) Growing an open business environment from CALS, JIT and supply chain. Logistics in the Information Age, 4th Conference Proceedings of Information Systems Logistics, Florence, 353–358

Kwon K, Zmud R (1987) Unifying the fragmented models of information systems implementation. In: Boland RJ, Hirschheim RA (eds) Critical issues in information systems research. Wiley, Chichester

Lamming R (1993) Beyond partnership – strategies for innovation and lean supply. Prentice Hall, London`

McFarlan WF (1984) Information technology changes the way you compete. Harv Bus Rev May/June:142–150

Min M, Galle W (2003) E-purchasing: profiles of adopters and nonadopters. Indust Market Manag 32:227–233

Pettigrew A, Whipp R (1991) Managing change for competitive success. Blackwell, Oxford

Pfeffer J, Salancik G (1978) The external control of organizations – a resource dependent perspective. Harper and Row, New York

Presutti WD (2003) Supply management and e-procurement: creating value added in the supply chain. Ind Market Manag 32:219–226

Sako M (1992) Prices, quality, and trust. Inter-firm relations in Britain and Japan. Cambridge University Press, Cambridge

Skjøtt-Larsen T, Kotzab H, Grieger M (2003) Electronic marketplaces and supply chain relationships. Industrial Marketing Management 32:199–210

Swaminathan J, Tayur S (2003) Models for supply chains in E-business. Manag Sci 49(10):1387–1406

Swatman PM, Swatman PA (1992) EDI system integration: a definition and literature survey. Inform Soc 8:169–205

Timmers P (1998) Business models for electronic markets. Electron Market 8(2):3–8

Vidgen R, Goodwin S (2000) XML: what is it good for? Comput Contr Eng J 11(3):119–124

Webster J (1995) Networks of collaboration or conflict? Electronic data interchange and power in the supply chain. J Strat Inform Syst 4(1):31–42

Chapter 21
Automotive Supplier Park Strategies Supporting Build-to-Order

Joe Miemczyk[1] and Mickey Howard[2]

[1] Audencia School of Management, Nantes, France
[2] School of Management, University of Bath, Bath, UK

Abstract. Build-to-order (BTO) is a production strategy that aligns to the demands of the 21st century where the industry is challenged to achieve flexibility from elongated supply chains that cross the globe and yet rely on inaccurate demand forecasts. A study of European manufacturers explores whether supplier parks are an essential part of BTO using a conceptual framework developed from the literature. The findings question the idea that simply locating suppliers close by to vehicle assembly plants reduces delivery lead time and inventory overall. The study finds that not all types of supplier parks are designed for BTO, where the cases reveal a wide variety of types, ranging in scale and proximity. The originality of the research is a unique study that redefines both automotive supplier park terminology and the relationships with BTO.

21.1 Introduction

Build-to-order (BTO) has been described as a production strategy that fits the demands of the 21st century, fulfilling customer orders in short lead times through responsive manufacturing and information exchange (Gunasekaran 2005; Holweg and Pil 2004; Howard et al. 2003; Holweg and Miemczyk 2002). Yet, a considerable challenge is how to achieve flexibility from extended supply chains that retain elements of the destructive cycle of make-to-forecast (Holweg and Pil 2001). Today, automotive supply chains hold weeks of component stocks, driven by a combination of vehicle manufacturer forecasts and supplier concerns over "stock-out" arising from quality or delivery issues. Globalisation of the industry has meant that low value vehicle parts are now shipped from all corners of the world. For instance, to complete an engine assembly in the United Kingdom, the oil pump takes 8 weeks to arrive from South Korea, represents 26 days worth of inventory, and travels over 8,000 nautical miles. One way to achieve the increased level of flexibility demanded by BTO in recent years is through clusters of suppliers located in close proximity to production; we define this as a supplier park. Supplier parks are emerging as increasingly common in automotive and other industries; yet, supplier parks and the implications for responsiveness are loosely defined in operations literature (Chew 2003; Cullen 2002).

Current descriptions of supplier parks in the automotive industry include "decentralised production in local assembly units ... are located close to the car assembly plant" (Millington et al. 1998, p. 180), and "a confined area in proximity to the assembly plant" (Larsson 2002, p. 769). Given that supplier parks have been used for over a decade – for instance SEAT in Barcelona – it is surprising that the links with BTO have not been explored sooner. Hence, the aim of this research is to identify the role of supplier parks in BTO, and to ask if they can be considered imperative for BTO. Eight cases of supplier parks are examined across Europe, where the phenomenon under investigation is BTO and the unit of analysis is the supplier park. The chapter develops a conceptual framework derived from the literature with which to explain the data and provide a clearer understanding of the drivers, moderating factors and outcomes. The current gap in knowledge over how supplier parks may facilitate BTO justifies the use of an exploratory multiple case study. In this context case studies are considered one of the most powerful methods in operations management, particularly in the development of theory (Stuart et al. 2002; Voss et al. 2002).

21.2 Background

This section describes the objectives and requirements of BTO and the supporting role of supplier parks in the automotive industry. The literature review is structured by examining the drivers and factors of supplier parks to support BTO. Several questions emerge from the literature that underpin the conceptual framework used to structure the enquiry and focus our analysis.

21.2.1 The Objectives of Build-to-Order

Build-to-order requires different performance objectives from traditional mass production approaches, such as make-to-stock. Hence, before devising a supply chain, Fisher (1997) recommends considering factors such as the predictability of demand for the product and the need for physical efficiency and market responsiveness. Fisher argues that it is the drive for efficiency in the process of supplying innovative high-variety products in industries such as automobiles, personal computers, and other consumer goods that account for so many broken or unresponsive supply chains. In their seminal paper, Holweg and Pil (2001) use the ideas of Slack (1991) and Upton (1994) to develop core objectives for BTO: processes, product, and volume flexibility. Flexibility is defined by Slack (1991, p. 77) as the "...ability to change, to do something different", whose framework includes aspects of flexibility not only across the total operation or system, but also the supply network. First, suppliers need to be integrated so that they can see orders based on real demand from customers, allowing process flexibility in the supply chain. Second, customisation needs to be brought closer to the customer instead of relying on finished goods, hence enabling product flexibility (Ward and Duray 2000). The third objective – volume flexibility – requires negotiation with workers and suppliers to reduce the dependence on full-capacity utilisation (Slack 1991).

One of the main requirements from a supply chain perspective is to closely tie supplier production schedules into customer production schedules (Holweg and Pil 2001). Geographic distance can be a major constraint to this level of integration between suppliers and customers for the following reason. The daily assembly schedule and vehicle assembly sequence is of little use where suppliers are located hundreds or thousands of miles away with commensurately long delivery lead times. Hence, suppliers and customers hold stocks to cope with the issues of lead time and schedule variability. Strategies are needed to control the cost of flexibility for BTO where suppliers are continually under pressure from the original equipment manufacturer (OEM) to reduce the time needed to deal with variations between planned production and actual orders. One such strategy is the development of the supplier park.

21.2.2 What Are Supplier Parks?

The co-location of supplier facilities close to vehicle assembly plants involves individual suppliers setting up dedicated facilities only for that customer (Millington et al. 1998). Previous authors have used terms such as "local assembly units" or "local dedicated units", which refer to geographically close individual supplier ties (Millington et al. 1998; Larsson 2002). The broader concentration of production sites is commonly known as an "industry cluster" and may be thought of as including supplier parks within that definition (Saxenian 1994). We have developed our own definition of supplier parks because to date they have been described only superficially, in the broadest sense of the term. Thus, a supplier park is defined here as:

"A concentration of dedicated production, assembly, sequencing or warehousing facilities run by suppliers or a third party in close proximity – i.e. within 3 km – to the OEM plant."

The number of automotive supplier parks has grown over the past decade, especially in Europe and currently totals 23 sites. Most OEMs have implemented some kind of supplier park, including Ford, GM, Fiat, Peugeot, Renault, BMW and Volkswagen. Typical activities carried out in automotive supplier parks include warehouse and inventory management, sequencing, manual assembly and late configuration, and range in size consisting of between seven and 24 suppliers (Kochan 2002).

The motivating principles – or drivers – for developing supplier parks appear to vary across the descriptions of supplier parks. That supplier parks themselves vary widely (i.e. size, location, activity), suggests there is no simple relationship between drivers and characteristics of supplier parks. This chapter draws from the contingency theory to develop internal and external moderating factors (Kast and Rosenzweig 1981) that may intervene before supplier parks enable BTO. Drivers and factors are discussed in the next section.

21.2.3 Motivations for Supplier Parks

One of the key trends in the automotive sector is the increase in variant numbers of individual models of cars (Holweg and Greenwood 2001). This trend has led to an increase in the part numbers required by assembly plants and thus has had an impact on the inventory policies of vehicle manufacturers (VMs) and the general need to maintain mix flexibility to remain competitive (Berry and Cooper 1999). In this case, assembly plants either hold a greater amount of inventory to ensure supply of the correct parts or install more responsive supply chain processes, such as sequenced in-line supply (SILS). Where SILS has been implemented, the time between a car starting final assembly and the fit point of the particular part (such as a seat) is given to a supplier to deliver the part exactly as specified (Doran 2001). Where this short order cycle time is only a matter of minutes, the supplier is often located close to the OEM plant (Larsson 2002).

Another trend related to increasing product variety in the automotive industry is the move towards simplifying production by introducing modules (Hsuan 1999; Fredriksson 2002; Sako and Murray 1999). Arguments for modular supply include cost reduction through lower supplier wages and overheads, and inventory reduction, increased space and simpler transactions (Baldwin and Clark 1997; Doran 2003; Von Corswant and Fredriksson 2002). Firms can mitigate the negative impact of product variety on operational performance by using modularity in the design of product family architectures (Salvador et al. 2002). Taking modular supply to the extreme leads to the idea of "modular consortia" where each module supplier locates next to the OEM plant, and has responsibility for all suppliers in the module, investing in the facility with the OEM and even assembling the module directly into the vehicle in some places (Collins et al. 1997). In theory, the practice of configuring complex product architecture through modular design with standard interfaces (Sanchez and Mahoney 1996) enables greater flexibility when considering supply chain strategies such as outsourcing (Hsuan 1999). Yet, current practice shows that supplier parks are also represented by suppliers of commodity components (e.g. nuts and bolts), parts that are bulky, and high variety parts that can be late-configured just before delivery to the vehicle assembly line.

Volume flexibility is seen as a further key to obtaining competitive advantage and there are a number of methods by which manufacturing firms can achieve this (Jack and Raturi 2002). The decision to co-locate a supplier facility near the OEM assembly plant can also be driven by a need for volume flexibility, for example, where capacity is taken by an additional assembly line. The cost of holding this inventory may be shifted to the supplier instead of making use of an OEM-controlled warehouse.

A significant driver for setting up a co-located supplier facility is the opportunity for funding development of local production sites. Regional and local development agencies often have funds to establish production sites, especially in areas identified as economically disadvantaged, for example, where European structural funds are made available (Larsson 2002). Regional development agencies may then approach large production facilities to offer them a subsidised infrastructure for further development of production facilities to encourage economic growth.

21.2.4 Other Factors That Affect Supplier Parks

While the previous section describes drivers, this section identifies factors that moderate how supplier parks support the objectives of BTO.

21.2.4.1 Start-up Costs

The common objective for government funding for supplier parks tends to centre round assisting the vehicle manufacturer to remain competitive (Larsson 2002). If the investment cost of increasing flexibility is high, then assistance from local authorities is likely to be sought. However, the extent to which these agencies will fund the development may affect the ability to achieve the proposed objectives, and more specifically enable BTO. If the funding covers the development of all production facilities, then the cost of initial start-up may be significantly lower than if only basic infrastructure such as road links are included. Therefore, start-up costs have a significant impact on the overall cost of increasing flexibility.

21.2.4.2 Choice of Supplier

The choice of supplier brought onto a supplier park will depend on the type of component or module being supplied. Co-location is likely where the part is specific to a particular vehicle, such as seats, cockpits, or external structures, such as bumpers. This is especially the case where there are a number of variants of the part per model. Bulky parts such as front- and rear-end modules that are costly to ship are also likely to be brought within close proximity to the final assembly line. Typically, these product sub-systems (or modules) have been integrated into manufacturers' operations, but the trend in recent years has been to outsource more of these major "chunks" of the product architecture, thus increasing distance between assembly operations (Sako and Warburton 1999). Such transactions can be thought of as having "high asset specificity" – an attribute with a number of associated problems.

Transaction cost economics (TCE) argues that assets specific to a transaction are more likely to be internalised than non-transaction-specific assets (Williamson 1979). Thus, if a supplier delivers parts that are specific to one vehicle, then the OEM is more likely to seek hierarchical control to reduce opportunism. Asset specificity is a key concept to understanding the benefit of specialised supplier networks according to Dyer (1996). For example, site specificity has been described as where "successive production stages are located in close proximity to one another to improve coordination and economize on inventory and transportation costs" (Dyer 1996, p. 273).

A particular problem that can occur is opportunistic re-contracting, where either the buyer or supplier can act opportunistically when contracts are renewed (by increasing prices or decreasing service levels, for example). Klein et al. (1986) describe the dealings that culminated in a vertical merger in the 1920s between General Motors (GM) and Fisher Body, a leading supplier of the new style of closed auto bodies. An exclusive dealing arrangement significantly reduced the possibility of GM acting opportunistically by demanding a lower price for the bodies after Fisher made the specific investment in production capacity. Unfortunately, these pricing provisions did not work out in practice. The shift in demand from open- towards closed-style bodies meant that GM was unhappy with the price it was being charged by its now very important supplier. In addition, Fisher refused to locate their body plants adjacent to GM's assembly plants, a move GM claimed was necessary for production efficiency, but which required a large and very specific investment on the part of Fisher. Finding the contractual relationship intolerable, GM began negotiations for purchasing the stock of Fisher Body, culminating in a final merger agreement in 1926.

The degree to which post-contractual opportunistic behaviour occurs is dependent on how specific the assets are to the transaction, and therefore how difficult it is to write contracts accounting for all contingencies. If supplier facilities at supplier parks have highly specific assets (i.e. physical, human and site-related), then the risks of opportunistic re-contracting are higher (Millington et al. 1998). Specific assets can also lead to strategic inflexibility, as the OEM is dependent on the co-located supplier. In terms of supplier parks, this issue was summarised by one automotive supplier as "while the set-up fosters a long-term partnership, it reduces flexibility in quality or cost disputes" (Cullen 2002). Both these issues lead to a lack of supply chain flexibility, arguably an undesirable attribute for BTO.

21.2.4.3 Other External Pressures

Outsourcing capacity also influences the development of a supplier park. Capacity at the OEM plant, such as assembly of modules, can be outsourced to a supplier to increase the flexibility required for BTO. Considerable barriers to the successful achievement of this are the institutional norms that develop in firms in order to build legitimacy. One type of institutional norm that has a particular influence over manufacturing firms is the presence of strong unionisation – established to protect the interests of the workforce. OEM trade union representatives may not agree that efficiency benefits will be gained from outsourcing production operations and that the interests of the work force will be downgraded; hence, the union is likely to resist such a move. Such resistance to outsourcing operations has been well documented, for example at the General Motors Lansing assembly plant over the outsourcing of module assembly (Marinin and Davis 2002). Lean strategies are neither wholly supported nor resisted by unions, and questions still arise over the effect of these institutional norms (Shah and Ward 2003).

21.2.4.4 Just-in-Time Activities

Just-in-time (JIT) refers to the movement of material to the right place at the right time. Elements essential to its success concern the capability of suppliers to participate through information technology, thereby enabling frequent communication (Wafa et al. 1996). JIT supply into vehicle manufacturers is expected to increase in the future, with more suppliers having to cope with its associated demands (Von Corswant and Fredriksson 2002). Schonberger and Gilbert (1983) propose that the success of JIT practised by firms implementing lean principles is associated with geographically proximate suppliers. However, research has also shown that this is not always the case (Wafa et al. 1996). Specifically, information and communication technologies are able to mitigate the effects of distance on successful JIT defined as reductions in inventory, component rejects and delivery lead time. Thus, if JIT capability is necessary for BTO in the automotive sector, the co-location of suppliers may be less critical. Yet, it has also been shown that geographical proximity of suppliers affects the trade-off between product variety and operational performance when mitigated by modularity (Salvador et al. 2002). These differing views question the assumption for close proximity in JIT strategies as an enabler for BTO.

An alternative perspective argues that proximity provides additional benefits to JIT capability such as the development of knowledge-sharing (Dyer and Singh 1998). Work relating to industry clusters indicates that the sharing of knowledge within these industrial groupings provides for the development of specific capabilities (Saxenian 1994). It follows, therefore, that the capability for JIT could be enhanced by close proximity of supplier to supplier, and supplier to OEM, whereby tacit knowledge is transferred between firms within the industrial cluster, or in this case the supplier park.

21.2.4.5 Disturbances in the Supply Chain

There are many causes of supply chain disturbance that in turn affect the reliability of delivery. One of the causes of supply chain disturbance is where distant suppliers are more likely to experience disruptions in delivery, for instance, problems experienced as a result of the transport infrastructure or from extreme weather conditions (Svensson 2000). This has been described as one of the primary factors affecting the adoption of supplier parks (Cullen 2002). Yet, disturbances are not only limited to transport problems and can include events at supplier production sites such as strikes and machine breakdowns (Svensson 2000). It is unclear from existing research whether bringing suppliers close to their customer manufacturing sites does indeed reduce these types of disturbances overall.

To understand the role of supplier parks in BTO strategy we develop a conceptual framework to structure our inquiry, based on the drivers, factors, and outcomes. The "outcome" of developing supplier parks for BTO can be thought of as being dependent upon drivers and moderated by the factors described earlier. We argue that the drivers centre on the primary requirements of BTO, i.e. product mix and volume flexibility, with the addition of funding incentives that affect the decision to set up a supplier park. Factors moderate the relationship between drivers and outcomes and help explain why some supplier parks clearly support BTO and others appear not to support BTO.

21.3 Data Collection

This study adopts an exploratory case study (Marshall and Rossman 1989; Yin 1994) to investigate whether supplier parks are imperative for BTO in the automotive industry. While there are already several studies that describe the effects of BTO in the automotive sector (Holweg and Pil 2004), this research aims to explore how the phenomenon interacts with the supply chain using supplier parks as the unit of analysis. It adopts a multiple case approach that uses the rationale of theoretical replication, not statistical sampling logic, where each case is selected so that it "either predicts similar results, or produces contrasting results for predictable reasons" Yin (1994, p. 46). While ideally all 23 supplier parks in Europe would be investigated, limited resources and our exploratory approach meant that eight supplier parks were chosen as representing one or more factors from the conceptual framework, i.e. drivers and general characteristics of the customer (volume or premium manufacturers), that affect support for BTO (Table 21.1).

The study divides the cases into supplier parks that enable BTO, those with potential, and those that do not. It concludes with a matrix showing the relative position of all eight cases in terms of their capability to provide support for BTO. The number of cases adopted here is consistent with good practice in case research

Table 21.1 Case selection criteria

Organisation/location	Drivers	Other characteristics
1 Seat/Exel Logistics, Abrera, Spain	Government funding	Volume manufacturer
2 Ford Motor Co Ltd, Bridgend, UK	Product mix	Volume manufacturer
3 General Motors, Ellesmere Port, UK	Product mix	Volume manufacturer
4 Volvo Car Corp, Ghent, Belgium	Product/volume flexible Government funding	Premium manufacturer
5 Jaguar Cars Ltd, Halewood, UK	Government funding	Premium manufacturer
6 Audi AG, Ingolstadt, Germany	Government funding	Premium manufacturer
7 MG Rover Group, Longbridge, UK	Volume flexible	Volume manufacturer
8 Volvo Car Corp, Torslanda, Sweden	Product/volume flexible	Premium manufacturer

where "the ability to conduct six to ten case studies" is analogous to conducting a similar number of experiments on related topics.

The idea to investigate supplier parks emerged from an earlier research programme, "3DayCar", which studied the implications of introducing customer order fulfilment into the United Kingdom. 3DayCar shows that on average, 40 days are needed in the UK to build and deliver a new vehicle, from order entry at the dealership to final customer delivery. Yet, only 1 day is actually spent building the vehicle (Holweg and Pil 2001). During a visit to the DaimlerChrysler Smart factory in May 2000, the response by personnel on the site suggested that the proximity of suppliers in relation to vehicle manufacture is a significant factor that may improve BTO capability. This visit piloted the research and stimulated the development of the conceptual framework. Construct validity was addressed by discussing draft interview and case reports with research participants and adjusting these on the basis of their comments (Yin 1994). Thirty semi-structured interviews were conducted over an 18-month period involving 17 site visits across Europe. Finally, the research was disseminated by presenting the results at academic conferences as well as industry seminars organised by the Society of Motor Manufacturers and Traders (SMMT), the International Motor Vehicle Programme (IMVP) and other international conferences (such as Miemczyk et al. 2004).

21.4 The Reality of Automotive Supplier Parks

This section briefly describes the findings from the eight sites across Europe. The findings are summarised in Table 21.2 by the number of suppliers located on the supplier park, the number of vehicle models it serves, the distance in kilometres from the OEM assembly plant, annual OEM production volume, supplier park age, and country of location.

Table 21.2 Key figures for the supplier parks studied. *OEM* original equipment manufacturer

	Number of suppliers	Number of models	Distance to OEM (km)	Volume (per annum)	Age (years)	Country
Seat, Abrera	32	6	2.5	426,675	10	Spain
Ford, Bridgend	1	3	0.5	1,075,000	1	UK
GM, Ellesmere	4	2	1	350,000	3	UK
Volvo, Ghent	15	2	3	160,000	3	Belgium
Jaguar, Halewood	6	1	0.5	55,610	4	UK
Audi, Ingolstadt	11	2	0.5	308,594	6	Germany
MG Rover, Longbridge	3	4	0	163,144	2	UK
Volvo, Torslanda	15	4	3	170,000	4	Sweden

21.4.1 Supplier Park Descriptions

21.4.1.1 Seat, Abrera

The supplier park at Abrera, near Barcelona in Spain, is located 2.5 km from the Seat assembly plant. The park was established in 1992 when the main Seat assembly plant was moved from the suburbs of Barcelona to an industrial district 50 km away. This move coincided with the development of a supplier park. The site was financed by an investment company that rents the site to the users of the park (the suppliers and logistics providers). The area of the site was increased by 30% in 1998 to cope with an expansion in capacity at the vehicle assembly plant. The site now operates with 32 suppliers carrying out a number of operations including inventory management, consolidation, late configuration and assembly tasks, with all components being delivered in sequence to the plant by a third party logistics provider. The transportation is by truck with a 10-min journey time. Around 946 journeys are made per day delivering 63 component sets to three vehicle assembly lines.

21.4.1.2 Ford, Bridgend

The site at Bridgend, Wales was chosen specifically because it only assembles engines for Ford Motor Company and, more recent members of the Premier Automotive Group (PAG), such as Volvo and Land Rover. Many other Ford sites have associated supplier parks, e.g. Valencia in Spain, Cologne in Germany, and Bridgend plant managers view this as an important part of their own strategy to cope with increasing pressures from Ford and other PAG customers. The site faces many of the issues that traditional vehicle assembly plant supplier parks face, such as increasing volumes and variety, the opportunity for government funding, as well as critical supplier issues affecting competitiveness, such as the need for global sourcing. While construction of the park infrastructure has been completed and the plant is currently receiving deliveries from one supplier, Bridgend is still negotiating with other suppliers involved with JIT delivery to ascertain mutually beneficial conditions for their re-location.

21.4.1.3 GM, Ellesmere Port

The Ellesmere Port supplier park is another recent introduction following a re-organisation of the sequencing operation in 2001. The Ellesmere Port plant assembles two models, the Astra and Vectra for Vauxhall (UK) and Opel (Europe) brands. The introduction of a new model led to a reclassification of this facility to "flex-plant" in order to cope with demand variability in the European market; hence, the re-organisation of inbound logistics and supply. Originally, two suppliers were located close to the plant, followed by the introduction of a new consolidation and sequencing centre. The park includes a whole range of activities from light assembly and late configuration, to sequencing and warehousing. There are now four suppliers onsite, including a third party logistics provider (3PL). One supplier and 3PL handle the sequencing for the other suppliers, as well as sequencing inbound deliveries from suppliers located across the UK and Europe.

21.4.1.4 Volvo, Ghent

Established in 1999, the supplier park supporting the Ghent assembly plant supplies components and modules in sequence to the Volvo assembly plant. The suppliers are dispersed over an area between 1.5 and 3 km from the plant. The OEM plant assembles two different models with an annual target volume of 160,000 cars. There are 15 suppliers at the park supplying modules ranging from headliners and seats to tailgates and bumpers. The site was developed by a property services company with 10% of the investment costs met by Volvo and suppliers. Trucks are used to transport the goods to the assembly plant with around 175 deliveries per day. The supply of goods is organised by Volvo and line-side inventory is also financed by the OEM.

21.4.1.5 Jaguar, Halewood

The development of the supplier park at Halewood coincided with the ending of Ford Escort production and the beginning of Jaguar X-Type production in 1999. The site is not dedicated exclusively to automotive suppliers, as a pharmaceutical firm also occupies the facility. The transition to Jaguar production led to a major reduction in capacity needed at the plant, leading to a re-organisation of the production layout and a reduction in the labour force. The park itself employs

850 personnel through the automotive suppliers. The area is designated "objective one", which means that it qualified for European structural funding. This was used to pay for much of the development of the supplier park.

21.4.1.6 Audi, Ingolstadt

The Audi supplier park at Ingolstadt is an established site with 11 suppliers on site and a range of activities being carried out at the site. The site was developed to cope with an increase in both vehicle production volume and product variants, with Audi/VW adopting a module and platform strategy to decrease overall costs and increase flexibility. The site is 100% funded by the local government, who lease it to suppliers and Audi. The site also houses general consolidation activities from a range of automotive suppliers. Ingolstadt is limited in its capacity for the final assembly of vehicles; hence, some painted bodies are shipped to other locations in Germany and Hungary for this final stage of production. The supplier park is intended to assist this operation, in addition to sequencing parts for final assembly onsite.

21.4.1.7 MG Rover, Longbridge

The development of this supplier park was initiated by the re-structuring of manufacturing at Longbridge. When BMW relinquished control of Rover Group Birmingham in 2001, its new flagship model the "Rover 75" had to be relocated from its Oxford production site to Longbridge. The previous production location already had a number of co-located suppliers assisting with in-sequence delivery; thus, the intention was to replicate this at the new site. Longbridge was also undergoing change and spare capacity led to the availability of space for suppliers within the assembly site itself. However, only three suppliers followed the Rover 75 to its new location. These firms supply the "75" assembly line, with three other vehicle models being served by more distant suppliers. The original production ethos of the 75 was to build to order; hence, the use of a supplier park and sequencing centre was central to this strategy.

21.4.1.8 Volvo, Torslanda

Established in 1999, the supplier park at Arendal in Sweden supplies components and modules in sequence to the Torslanda Volvo assembly plant 3 km away. The OEM plant assembles four different models with an annual target volume of 170,000 cars. Similar to the Volvo supplier park at Ghent, there are some 15 suppliers at the park supplying modules including headliners, seats, tailgates and bumpers. The site was also developed by a property services company, with 10% of the investment costs met by Volvo and suppliers. Here, around 192 deliveries take place per day. Again, the supply of goods is organised by Volvo with line-side inventory also financed by the OEM. The sequencing system is run by the suppliers on the park and provides signals every minute for a 4-h delivery horizon.

21.5 Discussion

The data from the cases require codification or a common classification to enable cross-case comparison. The conceptual framework guides the classification of case data into drivers, factors and outcomes. Drivers are classified in terms of the need for product mix flexibility, the need for volume flexibility, and the availability of public funding. Factors are variables that moderate the relationship between drivers and outcomes. The combination of factors and their specific attributes vary case by case. The attributes of start-up costs are high or low (where high relates to significant new developments in infrastructure and low represents re-use of existing facilities and significant external funding)[1]. The attributes of asset specificity are primarily high or low, but include site, plant and personnel-related assets, and the risk of strategic inflexibility (again high and low referring to the two extremes of specific modules such as cockpits or unspecific products, e.g. nuts and bolts). Two elements of institutional norms are encountered here and focus on union resistance and mimetic behaviour that follows perceived industry best practice. The attributes of JIT capability centre on who holds the requisite skills and competencies to co-ordinate sequenced in-line supply. Supply chain disturbance is perceived as either present or absent in terms of the impact on component supply. This research finds cases of transportation disturbance and attempts to mitigate this through late configuration. Some additional factors are also presented in Table 21.3 that fall outside of the original classification, e.g. change in corporate strategy.

The outcomes from the research emerge in the form of different supplier park types in terms of scale, proximity, and capability to enable supply chain flexibility e.g. "large-scale distant", "small scale onsite" (Table 21.3). The cross-case analysis reveals that there are differences in the characteristics of supplier parks in this study. From a physical perspective, the supplier parks vary in size and location in relation to the vehicle assembly plant they serve. The parks also appear to differ with respect to how they enable BTO, and how the moderating factors affect the drivers for the development of supplier parks. The analysis now examines supplier parks that enable BTO, supplier parks with the potential to enable BTO, and supplier parks that do neither.

---

[1] Using a high/low measure was necessary due to the lack of available objective measures to distinguish between the factors.

Table 21.3 Summary of the key factors affecting the outcome of each supplier park. *SILS* sequenced in-line supply, *BTO* build-to-order, *JIT* just-in-time, *OEM* original equipment manufacturer, *SC* supply chain, *SP* supplier park

Case	Drivers	Factors	Outcome
Volvo, Torslanda	Product mix flexibility Volume flexibility	Low start-up costs Risk of strategic inflexibility 80% of supplied value delivered by SILS	A large-scale distant supplier park critical to sustaining BTO Inflexibility ameliorated by obligational contractual relationships
Volvo, Ghent	Product mix flexibility Volume flexibility	Low start-up costs Risk of strategic inflexibility 80% of supplied value delivered by SILS	A large-scale distant supplier park critical to sustaining BTO Inflexibility ameliorated by obligational contractual relationships
Seat, Abrera	Product mix flexibility Volume flexibility	Low start-up costs Low asset specificity Union resistance JIT capability held by third-party provider Distance allows capacity flexibility	A large-scale distant third party-controlled supplier park Separate location from the OEM assembly plant means capacity can be expanded
Audi, Ingolstadt	Public funding Product mix flexibility Volume flexibility	Low start-up costs Personnel and plant asset specificity Union resistance JIT capability held by OEM	A large scale adjacent supplier park that addresses the recent increase in component variants and volume Core role in both reducing logistics costs and overcoming capacity constraints
Jaguar, Halewood	Public funding Volume flexibility – (reduced during development)	Low start-up costs Union resistance JIT capability held by OEM, but long call-off lead time	A large scale, mixed use industry park, the result of a change in manufacturing strategy by Ford Europe No drivers for BTO, hence no benefits
MG Rover, Long-bridge	Product mix flexibility Volume flexibility – (reduced during development)	Low start-up costs Low site, personnel, and plant asset specificity Perceived need to maintain JIT capability Disturbance perceived as risk (transportation)	A small-scale, on-site dedicated supplier park capable of supporting BTO Insufficient number of suppliers to enable BTO Reduction in overall production volume means BTO is low priority
GM, Ellesmere Port	Volume flexibility	Supplier develops the JIT capability Risk of strategic inflexibility Disturbance perceived as risk (mitigated through late configuration)	A small-scale, adjacent supplier park capable of limited support by minimising the effects of SC disturbance through late configuration

Table 21.3 *Continued*

Case	Drivers	Factors	Outcome
Ford, Bridgend	Public funding Product mix flexibility	Low start-up costs High asset specificity for suppliers (primary manufacturing) Replicating industry best practice on SP JIT capability held by OEM Disturbance perceived as risk (transportation)	A small-scale shared industry park with one supplier and one logistics provider (low-level specific assets to Ford) despite the prospect of supporting BTO

21.5.1 Supplier Parks That Enable BTO

The Volvo, Audi and Seat supplier parks are all large in scale with a significant amount of supplied value routed through the parks (80% in the case of Volvo). They are also distant from the assembly plant, providing for some capacity flexibility if expansion is required; except in one case, Audi, which has limits to capacity variation due to union resistance. Each park had low start-up costs as a result of external funding. Asset specificity is viewed as high, but in the case of Volvo, obligational contractual relationships allow the risk of opportunism and strategic inflexibility to be reduced, supporting Dyer's (1996) statement that the gains of specialisation can outweigh the costs. The drivers for these parks come from the need to provide volume and mix flexibility in the supply chain, to reflect flexibility in the assembly plant, for which Volvo are particularly known in their BTO strategy. Capability in JIT operations in these cases is held by the OEM (with a strong control and coordination role). Yet, this is not the case at Seat, where much of the capability is held by the third-party logistics provider.

Overall, these parks enable BTO at the assembly plant because of the need for volume and mix flexibility. There is the additional benefit of low start-up costs, and the potential for strategic inflexibility is moderated by favourable supplier relations. This results in large-scale supplier park operations with high levels of outsourced in-sequence component supply.

21.5.2 Supplier Parks with the Potential to Enable BTO

Supplier parks with the potential to enable BTO include MG Rover Longbridge and GM Ellesmere Port. These are small-scale adjacent or onsite parks of insufficient size to provide significant support for the BTO strategy. Moreover, the reasoning for their introduction is driven by either the need to use spare capacity (in the case of MG Rover) or the need to provide spare capacity for a sister facility in continental Europe (in the case of GM). The start-up costs were not financed by an external body, which in turn affects the economics of locating suppliers close to the plant. However, other moderating factors also provide reasons why these sites might provide the potential for supporting BTO. The capability to provide in-sequence supply is being developed at both sites, thus supporting a process to provide product mix flexibility, especially as late configuration is introduced at both sites. Furthermore, the reduction of supply chain disturbance as a result of unreliable transportation provides conditions that are supportive of BTO.

The principles behind these small-scale cases in theory support BTO at the vehicle assembly plants. However, despite their potential they are inhibited by the lack of scale of the operations; in one case, this was the result of falling vehicle sales, and in the other due to sharing production with a sister site.

21.5.3 Supplier Parks That Do Not Enable BTO

The evidence from the last two cases, Ford and Jaguar, suggests that these supplier parks do not enable BTO and have limited potential to do so in the future. In both cases the strongest driver appears to be the availability of external funding for the required supplier park infrastructure. For Jaguar, the original intention to build a supplier park was to provide volume flexibility, but a change in manufacturing strategy at the European level has removed this requirement. Yet, the park was established, suppliers co-located and in-sequence supply initiated. The long call-off lead times of 12 h removes the urgency of in-sequence supply, as this period of notice does not require close supplier location. In the case of Ford, while park infrastructure has materialised, only one supplier has located to Bridgend and a viable business case still has to be made to the other partners. Evidence shows that this site demonstrated mimetic behaviour in that other Ford plants had already implemented supplier parks and reported performance improvements. While the plant was driven by a need for product mix flexibility (the number of engine variants produced had rapidly increased), the benefits of co-located suppliers to improve JIT capability and BTO were unclear. The types of suppliers would also produce high opportunism and strategic inflexibility issues, especially where primary manufacturing such as forging, casting, or machining were needed, as significant investment in supplier plant and personnel would be required.

Hence, for these last two cases the drivers for BTO were largely absent. In the case of Jaguar the supplier park was not implemented to provide BTO advantages, and space on the park was utilised by non-automotive companies, removing the potential volume flexibility advantage. The Ford engine supplier park has only partly materialised, despite external funding. The decision process for Ford may need re-aligning towards the benefits for BTO and an examination of whether engines could and should be produced JIT.

21.5.4 Supplier Parks: Imperative for Build-to-Order?

Three supplier park types have emerged from the cross-case analysis supported by Table 21.3. First are supplier parks that enable BTO because they are large scale, catering for volume and product mix flexibility. Start-up costs are often minimised through public and private funding. Second are supplier parks that in theory support BTO, but are small scale, where drivers for either volume or product mix are lacking. Third are also small or underdeveloped supplier parks that do not support BTO because of overall weak drivers for flexibility and recent changes in manufacturing strategy. This paper identifies a pattern among supplier park types, proximity, strategic BTO flexibility and scale.

The analysis demonstrates that large-scale parks that enable BTO are associated with being "distant" (more than 1 km) from the OEM assembly plant. Supplier parks that are geographically distant offer greater opportunity for expansion than onsite or adjacent parks, and hence are more flexible. These parks are driven by both volume and product mix flexibility, and combine several moderating factors that enable BTO. The parks that do not enable BTO or only possess the potential to do so are adjacent or onsite and are limited by the constraints of surrounding OEM infrastructure. According to this research adjacent and onsite parks are driven – at most – by either volume or product mix. Several moderating factors also need be considered in cases where the capability for BTO was less evident. For instance, changes in corporate strategy, union resistance to changes in working practice, and the difficulties of persuading suppliers to invest in an appropriate level of asset specificity.

Returning to the research question, this chapter indicates that only certain types of supplier parks are imperative for BTO, described here as distant from the OEM assembly plant, providing strategic BTO flexibility, and possessing sufficient scale. The combination of factors and drivers that lead to this type are volume flexibility, product mix flexibility, low start-up costs and managed asset specificity.

21.6 Conclusions

This study finds that there are a number of different types of supplier park; yet, only some of these have the characteristics to enable BTO. These are large-scale sites, 1 km or more distant from the OEM assembly plant, and provide both volume and product mix flexibility. The supplier parks that do not enable BTO are small scale, and provide volume and product mix flexibility only to a limited degree. These parks are characterised as onsite or adjacent to the OEM assembly plant.

In terms of research limitations, this is a European study where concepts such as the availability of public funding may be idiosyncratic to this region. While eight cases out of a European total of 23 is a good representation, a wider study across the total population including the US, South America and Japan might include additional variables. Case study methods are appropriate for explanatory research; yet, further theory testing could be enhanced through cluster analysis techniques covering supplier parks world-wide.

It is important to note that change is the "normal" state for the automotive industry and this could have ramifications for the factors that influence decisions on future supplier park development. The increasing trend in modularisation and supplier alliances (e.g. Hella, Behr and Plastic Omnium developing complete front-end modules) suggests an increased position of power for first-tier suppliers. Their desire for scale effects could further restrict co-location opportunities. This could also lessen the effect of specific technologies (assets), as more components of modules are shared over more end-products. A further interesting development is the desire for some vehicle manufacturers to remove nearly all short-term variability in their supplier schedules (through better schedule reliability). This development could potentially reverse the trend in co-locating suppliers altogether, where suppliers no longer need to respond to short-term variability in material volume and mix requirements.

This study has significant implications for theory. Contrary to received wisdom that supplier parks have developed because of the disruption caused by extended supply chains (i.e. Korean fuel pumps shipped to the UK), we find that they are adopted for a variety of reasons. This includes the availability of public funding, corporate re-structuring and the result of changes in strategy, in addition to the more apparent need for volume and product mix flexibility. Supplier parks become imperative for BTO in cases of increasing demand to deliver high product variety, the ability to cope with fluctuation in volume and the capability to respond to short order lead times. This research supports Dyer's (1996) view that competitive advantage in the form of BTO is contingent on the type of activity and degree of interdependence between OEMs and suppliers to achieve flexibility. A surprise finding is that supplier parks in close proximity to the OEM do not necessarily foster closer working relationships and knowledge sharing for BTO. Hence, more distant supplier parks are better placed to enable BTO than onsite or adjacent parks. This questions the "closer is better" hypothesis of Saxenian (1994).

Returning to the conceptual model raises the issue of how well the model reflected the divergence in supplier park approaches. The model's use of drivers appears to adequately predict supplier park development, although the mix of each set of drivers is variable across each case. Furthermore, those parks that appear to support BTO do so as a result of a combination of factors related to improving JIT capability, reducing supply chain disturbances and require integration due to the level of specific assets (e.g. supplier's products and processes that cannot easily be switched to other OEMs). The start-up costs and institutional norms appear to moderate whether supplier parks can actually be implemented, for example, where costs are supported by an external agency and unionisation does not limit managerial choices on activities that take place in supplier parks. Company strategic direction should also be included as a further moderating factor (either positive or negative) in the model, to account for issues that emerged from the cases. Hence, factors can be categorised into "barriers to implementation" and "enablers of

BTO". Further work should test these categories of factors across different cultural contexts to assess generalisability. BTO remains an under-defined construct, and while it was not the focus of this work to develop this, continued research should focus on defining BTO and its role in sustaining competitive advantage. More generally, research should attempt to empirically link the benefits of proximity (in terms of information and knowledge sharing, reduction in supply disturbance, etc.) with the overall performance of the supply chain as well as the performance of individual firms in that chain. While there are clear potential benefits for OEMs, suppliers inevitably suffer by losing economies of scale; hence, studying the effects at all levels of the supply chain is key. Furthermore, the effect of information systems may moderate the positive impact of supplier parks, supporting the work of Wafa et al. (1996), de-linking the need for proximity in JIT systems.

An important learning outcome for practitioners highlights the issue of where supplier parks have not fulfilled their promise to support BTO. Managers at these supplier parks need to focus on:

• Whether there is demand for flexibility in the first place
• Whether the site is not simply a convenient use of spare capacity
• Where there is genuine demand for BTO, is there adequate support from top management?

Supplier parks with the potential for BTO need to build scale by encouraging suppliers to locate with appropriate levels of asset specificity managed through trust-based, obligational supplier relationships (Sako 1990). Only by considering the supplier park in the context of a long-term vision and as part of a dedicated strategy towards building to order can VMs and suppliers expect to realise superior levels of performance.

References

Baldwin CY, Clark KB (1997) Managing in an age of modularity. Harv Bus Rev 75:84–93
Berry W, Cooper M (1999) Manufacturing flexibility: methods for measuring the impact of product variety on performance. J Oper Manag 17:163–179
Chew E (2003) Carmakers reap benefits of supplier park concept. Automot News Eur
Collins R, Bechler K, Pires S (1997) Outsourcing in the automotive industry: from JIT to modular consortia. Eur Manag J 15:498–508
Cullen T (2002) Carmakers split on supplier parks. Automot News Eur
Doran D (2001) Synchronous supply: an automotive case study. Eur Bus Rev 13:114–120
Doran D (2003) Supply chain implications modularization. Int J Oper Prod Manag 23:316–326
Dyer JH (1996) Specialized supplier networks as a source of competitive advantage: evidence from the auto industry. Strat Manag J 17:271–291
Dyer JH, Singh H (1998) The relational view: cooperative strategy and sources of interorganizational competitive advantage. Acad Manag Rev 24:660–679
Fisher ML (1997) What is the right supply chain for your product? Harv Bus Rev 75:105–116
Fredriksson P (2002) Modular assembly in the car industry – an analysis of organizational forms' influence on performance. Eur J Purch Supply Manag 8:221–233

Gunasekaran (2005) Build-to-order supply chain management: a literature review and framework for development. J Oper Manag 23:423–451

Holweg M, Greenwood A (2001) Product variety, life cycles and rate of innovation – trends in the UK automotive industry. World Automot Manuf

Holweg M, Miemczyk J (2002) Logistics in the "three-day car" age: assessing the responsiveness of vehicle distributions logistics in the UK. Int J Phys Distrib Logist Manag 32:829–850

Holweg M, Pil FK (2001) Successful build-to-order strategies start with the customer. Sloan Manag Rev 74–83

Holweg M, Pil FK (2004) The second century: reconnecting customers and value chain through build-to-order. MIT Press, Cambridge

Howard M, Vidgen R, Powell P (2003) Overcoming stakeholder barriers in the automotive industry. J Inform Tech 18:27–43

Hsuan J (1999) Impacts of supplier-buyer relationships on modularization in new product development. Eur J Purch Supply Manag 5:197–209

Jack E, Raturi A (2002) Sources of volume flexibility and their impact on performance. J Oper Manag 20:519–549

Kast FE, Rosenzweig JE (1981) Organization and management – a systems and contingency approach. McGraw Hill, London

Klein B, Crawford R, Alchian A (1986) Vertical integration, appropriable rents and competitive contracting process. In: Putterman L, Kroszner R (eds) The economic nature of the firm. Cambridge University Press, Cambridge

Kochan A (2002) Lean production helps logistics firms. Automot News Eur

Larsson A (2002) The development and regional significance of the automotive industry: supplier parks in Western Europe. Int J Urban Reg Res 26:767–784

Marinin KJ, Davis TRV (2002) Modular assembly in international automotive manufacturing. Int J Automot Techn Manag 2:353–362

Marshall C, Rossman BG (1989) Designing qualitative research. Sage, London

Miemczyk J, Howard M, Graves A (2004) Supplier parks in the European auto industry. In: Al VE (ed) 11th European Operations Management Association International Conference. Insead, France

Millington AI, Millington CES, Cowburn M (1998) Local assembly units in the motor components industry. Int J Oper Prod Manag 18:180–194

Sako M (1990) Prices, quality and trust: inter-firm relations in Britain and Japan. Cambridge University Press, Cambridge

Sako M, Murray F (1999) Modules in design, production and use: implications for the global automotive industry. Paper prepared for the IMVP Annual Forum. MIT, Boston

Sako M, Warburton M (1999) Modularization and outsourcing project: preliminary report of the European Research Team. Paper prepared for the IMVP Annual Forum. MIT, Boston

Salvador F, Forza C, Rungtusanatham M (2002) Modularity, product variety, production volume, and component sourcing: theorizing beyond generic prescriptions. J Oper Manag 20:549–576

Sanchez R, Mahoney JT (1996) Modularity, flexibility, and knowledge management in product and organization design. Strat Manag J 17 (Winter special issue):63–76

Saxenian A (1994) Regional advantage. Harvard University Press, Cambridge

Schonberger RJ, Gilbert JP (1983) Just-in-time purchasing: a challenge for U.S. industry. Calif Manag Rev 1:54–68

Shah R, Ward P (2003) Lean manufacturing: context, practice bundles and performance. J Oper Manag 21:129–150

Slack N (1991) The manufacturing advantage. Mercury Books, London

Stuart I, McCutcheon D, Handfield RB, McLachlin R, Samson D (2002) Effective case research in operations management: a process perspective. J Oper Manag 20:419–433

Svensson G (2000) A conceptual framework for the analysis of vulnerability in supply chains. Int J Phys Distrib Logist Manag 30:731–749

Upton D (1994) The management of manufacturing flexibility. Calif Manag Rev 36:72

Von Corswant F, Fredriksson P (2002) Sourcing trends in the car industry. Int J Oper Prod Manag 22:741–758

Voss C, Tsikriktsis N, Frohlich M (2002) Case research in operations management. Int J Oper Manag 22:195–219

Wafa MA, Yasin MM, Swinehart K (1996) The impact of supplier proximity in JIT success: an informational perspective. Int J Phys Distrib Logist Manag 26:23–34

Ward PT, Duray R (2000) Manufacturing strategy in context: environment, competitive strategy and manufacturing strategy. J Oper Manag 18:123–138

Williamson OE (1979) Transaction-cost economics: the governance of contractual relations. J Law Econ 22:233–261

Yin RK (1994) Case study research: design and methods. Sage, Thousand Oaks

Chapter 22
Managing the Transition to the "5-Day Car" in Europe

Gareth Stone, Valerie Crute and Andrew Graves

School of Management, University of Bath, Bath, UK

Abstract. As industry executives acknowledge, at present there is little evidence of a relationship between car production and the actual orders being generated in sales rooms. This reality has significant implications for customer satisfaction, the industry cost base and future competitiveness. Many executives think that build-to-order (BTO) concepts and practices offer the European automotive industry one of the best opportunities for survival. However, there are significant challenges to be overcome to achieve the BTO transition. In this chapter, key enablers and barriers that will govern the supply chain transition to a BTO business model for the European automotive industry are explored. Research was undertaken at a number of dissemination events regarding the implementation of BTO concepts, attracting top executives from around Europe, and has provided valuable insight into the possibility of an industry transition to BTO. The findings of this study are presented and the chapter goes on to discuss some of the key factors that industry leaders will need to consider in managing the transition to the "5-day car" in Europe including building on existing examples of good practice; training and re-educating industry leaders and the workforce in BTO principles and practices; clear planning and objective setting; enhancing supply chain learning; aligning performance and accounting practices; and communicating and benchmarking progress.

22.1 Introduction

This chapter describes some of the factors that need to be considered in managing the European car industry transition to the new paradigm of building cars to customer order using the "5-day car" business model. Research in this arena goes back as far as the 1980s, when Graves (1987) posited: "the current EC strategy of creating a pan-European market will require new forms of collaboration and alliances in production, component supply, technology development and distribution". The chapter will then consider the

383

challenges facing the automotive industry in terms of three linked and corresponding factors: the social, economic and environmental aspects of the industry. Initial
findings of a dissemination process are reported that took the form of a questionnaire administered at a number of organised events across Europe where top
automotive executives were asked to provide considered feedback on a range of
BTO-enabling technologies. Clearly, there is a strong need to research and understand both the potential obstacles and enablers in managing the BTO transition.
This chapter goes on to discuss some of the factors identified in research on the
adoption of promising practice and the relevance of such practices for managing
the BTO transition.

22.2 Why Build-to-Order?

In his book *The Empty Raincoat* Charles Handy claims that it is critical for organisations to start investing in new concepts and practices before the current ones reach
full maturity (Handy 1995). Handy referred to the concept as "S Curves". To abandon something that is successful and still growing for something uncertain is never
easy and this certainly seems to be apparent in the automotive industry today. However, if the industry were to wait until the current "S Curve" starts to decline much
further, then this may turn out to be the worst time to start developing a new business model, simply because the industry would have diminished resources to start
afresh. Unfortunately, the European industry is conservative, notoriously difficult
to change and has already in fact started on a new strategy to combat cost reduction
challenges. By increasing the pace of its march towards low wage economies, top
management believes it can remain competitive. This may provide short-term cost
reduction through lower wage bills, but raw materials are bought and priced on
a global market. This simply reflects the nature of the industry, which is typified by
ruthless competition on price, soaring raw material and energy costs and aggravated
further by a management hierarchy infatuated with cost management and obsessed
with restructuring manufacturing processes at the micro level. The approach truly is
short-sighted, as it will almost certainly decimate production in the domestic markets of Europe rapidly. In order to protect the European economy from this gradual
atrophy, new approaches that allow costs to be controlled whilst maintaining the
domestic markets are urgently required (Stone et al. 2007; Holweg and Pil 2004).
The 5-day car provides a viable competitive alternative that, through its structure,
requires a domestic capability to provide localised assembly.

A sustainable automotive production industry requires us to pay close regard to
certain economic, environmental and social challenges and constraints. Furthermore, in order to enhance decision-making and control of supply it requires an
element of what Beer describes as "controlled convergence" (Beer 1994) of the
total life cycle of vehicles, which includes design, production, distribution, use,

and increasingly, end-of-life management. Currently, Europe has the capacity to produce over six million more cars than can be sold. This over-production has led to over-capacity, with the ultimate effect of a downward pressure on prices. Over-capacity is not spread evenly across plants, OEMs or geographically across regions. However, plant closure on the scale witnessed today in order to reduce capacity, has a significant, adverse and enduring effect on brand diversity, the local populations, and ultimately, GDP.

Significant reductions in material and energy consumption have put further pressure on the industry to reduce waste and pollution, whilst at the same time increasing economic performance. There are some difficult challenges involved with transferring to more sustainable methods of vehicle production and use, which will require political, economic, social, environmental, technological and legislative change. The EU's ILIPT project has presented a range of tools, actions and protocols that may result in progress towards these goals, based on the evolution of existing technology and approaches described throughout this book. However, in the longer term there is a requirement for improved understanding of the scale and type of change required, at a system level, and for the associated implications for technology, industry and society to be clearly understood. Indeed, many European automotive industry experts are fully aware that wholesale change in the way that business is done is urgently required. The financial burdens induced by the requirement to meet increasingly sophisticated customer expectations are crippling the industry. Matthias Holweg, in a previous chapter, has observed that the mass production model, which has been with us for the best part of the last century, is clearly no longer viable in business environments typified by turbulence and flux. However, "the mass customisation paradigm" (Pine 1993) that has captured the imagination recently is also proving difficult to operationalise, even at local levels. Lean production has helped the situation to a degree by enabling the industry to sustain itself in relatively recent years, but the end result is that manufacturers are still losing money. It has been nearly two decades since lean production set out to employ the new promising practices based on the Toyota production system first captured in the now seminal book *The Machine That Changed the World* (Womack et al. 1991). Today, more than ever, the industry exists within a climate of over-capacity and escalating costs and it now appears that the industry has simply become more adept at sealing its own fate. Of course, many managers may already understand that this is untenable, but whilst their efforts have delivered considerable improvements in manufacturing efficiency, they have been largely ineffective in raising profitability; a direct outcome of a deliberate and possibly misplaced focus on factory processes.

A serious consideration for the industry, in terms of reducing the so-called "carbon footprint" is to reduce the number of movements each part or sub-assembly makes on its journey through what has become an overly complex and ramified supply chain. A common method, adopted by OEMs, to optimise supply by reducing the complexity and rationalising supply routes is to rely on logistics service providers (LSPs). Unfortunately, there is a limit to which the LSPs can stretch the efficiency potential without the full commitment of the OEM to support the use of equipment needed for delivering cars. There have been a number of calls recently to reduce the aptly named "empty kilometres" that road carriers travel in order to attempt an effective use of fuel and driving time approved under EU directives.

As Thelen (2007) indicates, the road networks are becoming less cost-effective as a means of reaching the market, and greater pressure is being put on ocean carriers to move products around the globe. However, this too is now reaching full capacity. The capacity shortage has been evident for several years and there have been several automotive solutions to allow containers to be utilised on both legs of a journey. Although innovative, these concepts are insignificant in the wider picture and are not in themselves sustainable, certainly not to support a global industry. Furthermore, as Thelen points out, the relocation of production sites to low=cost economies has only served to add greater pressure to container capacity. Building more carriers, adding more car decks and even lengthening existing vessels are all under consideration, but for the most part, the carriers' companies already squeeze their existing capacity in order to turn a profit. Whilst some might look to introducing new car carriers as a solution, the lead time for new vessels is far too long to meet the requirement of the current situation. Perhaps it is time to realise that shipping millions of parts around the world may be cheap today, but it is not the wise business opportunity it once appeared. Local production for local markets will eventually be seen as the responsible business model of the medium-term future.

The automotive industry is often thought of as the "engine of Europe" because of its economic and social importance and the historical role it has played in the development of Europe to date. Many social and economic observers look to the automotive industry as a barometer of social and economic activity. As one of Europe's major industries, it employs 2.3 million directly and it has been estimated that 12.6 million people are employed in the various supply networks across Europe (ACEA 2007).

It has long been understood that individual Governments should provide business with regulatory environments that enable companies and individual plants to be competitive and attract investment. Where plant closures do occur it is their responsibility to manage the social costs and economic regeneration. Some leading UK research by the SMMT and cited in a recent House of Commons report, indicates that competition on the basis of low-cost, semi-skilled workers and low value-added processes will be increasingly challenged and become less viable, as globalisation and increased economic integration brings lower cost countries in to challenge established markets in Europe. For example, despite a backdrop of increasing energy and raw material prices, the charges levied by automotive component suppliers have fallen in real terms by as much as 20% in order to remain competitive and attractive to manufacturers (HC399 2007). Flexible manufacturing systems that are capable of producing low volumes are required in order to allow manufacture with minimal inventory. Allied with competitive forces the industry is recognising the need for highly modular structures that can be applied to a spectrum of platforms. Another important requirement that has been recognised by the industry is the need for low investment vehicle programmes. The development of more cost-effective materials and processes along with new coatings options are also necessary to maintain and improve profitability. All that has been discussed in the chapters of this book so far can be considered as "available heuristics" for the automotive industry. Such solutions will support the industry in the transition to BTO, but there is clearly a need for a more coherent consensus building amongst the stakeholders that govern the industry. The industry, collectively, needs to be willing and able to search for feasible strategies to deal with an uncertain, complex and perhaps controversial planning environment. Yet, it must also be able to reassure itself that there is enough demand to meet the need for a "sea change" in manufacturing (Stone et al. 2007).

22.3 The Demand for Build-to-Order

Probably the most important factor driving change in the automotive industry today is the competitive pressure that exists to reduce development and manufacturing cycle times and costs and to identify and improve value for the customer. The new manufacturing systems must therefore improve flexibility, responsiveness and quality in order to deliver greater profitability and return on capital employed. In order to achieve this, the industry must improve its ability to work cohesively to change the industry structure and consolidate supply chains. The industry therefore requires flexibility and the ability to sustain inter-firm collaboration in supply networks.

Advances in communication technologies have led to a situation where we are witnessing new heights in the demand for mobility and change in the patterns of working and life style. This is allied with a significant demographic change. Car customers now demand more choice, which has directly led to a proliferation in the diversity of models and ranges available on the market today, complete with the latest developments in satellite navigation and blue-tooth technology. This new range of vehicles requires extremely flexible and on-demand manufacturing processes that are capable of evolution and can change to meet fashion and lifestyle imperatives, not to mention the requirements of omni-present changing EU regulation. Europe has a well-motivated and skilled workforce in place. Even so, the workers and unions have been confronted with a series of challenges in recent years. Economic restructuring brought about by regional integration, combined with market saturation and over-production, has led to job losses. At the same time and as a response to a changing business environment, managers have placed greater pressure on workers to increase productivity and product quality. Whilst this has been a useful intervention in the short-run, the possibility of improving slowly diminishes and is no longer sufficient alone.

22.4 Supply Chain Readiness

Today, most of the major vehicle manufacturers need to manage increased product complexity and reduced lifecycle time, whilst having to develop and manufacture a larger number of variants within shorter time-to-market cycles. The solution to this problem requires a dramatic shift in thinking, away from production-push and economy of scale, towards customer-pull and economies of scope. This involves reconsidering vehicles in terms of commonality; the process starts at the design stage. Features, such as body construction and vehicle complexity, can affect total lead-time, not just in production, but across the supply chain. This underpins and broadens the importance of vehicle design. Vehicle construction and the links with variety-driven vehicle complexity and process reliability have a wide-reaching impact stretching throughout the value stream. The scope of vehicle design appears to be increasing, and now includes not only the product characteristics, but also the means of its delivery.

Fierce competition and shorter product life cycles; hence, an increased number of new models and variants, have forced companies to focus on the flexibility of their systems to be able to respond to changing market demands swiftly. Flexibility in manufacturing is a major competitive advantage. The automotive supply chain will need to plan and give significant time to undertaking the changes necessary for the BTO transformation. Chakrabarti and Rubenstein (1976) suggest that management can institutionalise new practices and processes through the provision of "organisational slack". This fosters innovation since it allows individuals to spend a certain amount of time concentrating on un-programmed innovative activities. Re-engineering processes take time and patience and lack of time is often a negative factor. In addition, a degree of unlearning may also be required. A precondition for undertaking change is the need for flexibility and adaptability to change (Ascari 1995). Managers need to understand the value of experiential learning for the adoption of supply chain learning. On a practical level, it is suggested that over time, extra resource is used to "close out" improvement activities, perhaps through engaging additional "floating' staff (Bateman and Rich 2003). Such support will be necessary in the automotive context, where some employees are already suffering "initiative fatigue" and are feeling burned out by successive programme change initiatives.

22.5 Resistance to Change

Resistance to changes in practice by suppliers or customers is reported to be one of the barriers to successful implementation of major change initiatives (Drew 1994). Such barriers may also affect the transition to BTO. One way to mitigate this was highlighted by Hanson et al. (1994) who showed that companies that moved most rapidly to adopt new work practices from 1989 to 1994 formed learning relationships with other companies. Selto et al. (1995) also report that a common barrier from their interviews was negative relationships with suppliers. Panizzolo (1998), looking at the adoption of lean practices in Italy, found that firms find it more difficult to fully adopt innovative practices that concern the management of external relationships with suppliers and customers than internally oriented innovative practices. Indeed, Cagliano et al. (2001) similarly emphasise the need for small to medium-sized enterprises (SMEs) to move beyond traditional technical excellence or operational flexibility to meet the changing needs of their customer companies. In the context of the BTO transition, a common vision will be necessary for supply chain transformation and this will involve significant supply chain learning (SCL).

Bessant et al. (2003) utilised a case-based approach supported by a telephone survey in the UK. They discovered that UK supply chain management programmes do not yet incorporate SCL. Where SCL does occur, it is mostly limited to first-tier suppliers (or customers), and very seldom involves structured processes of learning from suppliers or customers. Boddy et al. (1998), again in the UK, implemented a postal survey of firms implementing supply chain partnering and found over half (54%) had been unsuccessful. In order to achieve greater understanding of SCL the term "absorptive capacity" was coined and defined as a firm's general ability to value, assimilate, and commercialise new, external knowledge (Cohen and Levinthal 1990). They suggested that an organisation's absorptive capacity tends to develop cumulatively, be path-dependent, and builds on prior investments in its members' individual absorptive capacity. Such issues will need to addressed, to eliminate barriers to learning in the BTO transition.

22.6 Company Ownership and Overcapacity as Potential Barriers

Several studies have suggested that there may be significant differences between European and foreign owned companies in both their performance and ability to adopt major changes in practices and processes. Such differences may be important to consider in managing the BTO transition, since such factors may be a barrier to new practice adoption. The literature trail begins with Hanson and Voss (1995), who in a survey of European manufacturers, suggested that across Europe, foreign-owned companies fared considerably better than those that were domestically owned, when it comes to new practice adoption. They posited that Japanese-owned sites are particularly outstanding at both practice and performance, and that American-owned companies had much better practice and somewhat better performance on average. On the other hand, companies with domestic parents had below average scores on both practice and performance. In their 1996 study, Collins et al. (1996) found that the origin of the parent company can influence results. The study showed that Swiss-owned plants achieved lower levels of overall practice and performance than non-Swiss and their results were consistent with the "Made in Europe" study, where foreign-owned companies fared considerably better than domestically owned companies in each country studied (Voss and Blackmon 1996).

The following year a European study found that many European industries already have vast over-capacity (Schmemner 1997). More than half the European companies in a recent IMD-sponsored survey reported that they were "suffering from over-capacity". Removing over-capacity is often a problem in Europe. Government mandates on severance pay and compensation to affected local communities make it more expensive to close facilities in Europe, and prevailing social mores make it a solution of last resort. The international dataset of Lowe et al. (1997) shows that part of the explanation for the higher performance of top plants lies with some combination of scale (volumes), capacity utilisation, and automation. Using data from the "Made in Europe" survey Voss and Blackmon (1996) support suggestions that manufacturing sites were more likely to have adopted world-class manufacturing practices if they had overseas parent companies. Japanese ownership was associated with the highest level of practices adoption, followed by North American and then other European ownership.

22.7 The Industry Executives' Perspective

It is clear that there may be many challenges faced by industry leaders in implementing BTO concepts and practices. A major review of the literature on the "adoption of promising practice" (Leseure et al. 2004) provides the framework for an analysis of industry executives' perceptions of the BTO transition. The research findings, presented in the following sections, are based on data collection and discussions held at 22 dissemination events throughout Europe where key industry executives were asked to comment on a range of BTO-inspired concepts. Responses were provided by top automotive producers and suppliers including OEMs such as BMW, Audi, Daimler and VW, along with suppliers including Siemens, Hella, Robert Bosch, Iveco Magirus Dräxlmaier HIB, Alcan Singen, ZF Lemförder and LSP Ferrostaal. A semi-structured questionnaire made available in both English and German language was used to guide interviews and a structured questionnaire was also used to collect data from participants.

22.8 Consultation Findings

Build-to-order has been heralded by a number of commentators as a robust solution to the industry's current malaise. Our survey of 55 invited executives indicated that the average level of production directly linked to a confirmed vehicle purchase was 58% in production sites today. However, this figure is somewhat misleading, since the confirmed vehicle purchase figures do not relate directly to end-customer vehicle purchases. Clearly, Europe is not capable of building a short lead time vehicle under current conditions. BTO, then, offers a vision for an industry in which the advantages of building cars to order has not gone unnoticed in recent years. Some 80.9% of the executives consulted in the dissemination of BTO concepts expressed that they believed that the majority (78% average) of cars should only be built to a confirmed customer order in a BTO system. The remaining 22% of finished vehicles being assembled for fleet markets etc., where variety is less problematic.

Prior studies have indicated that customers value and even desire BTO capacity and 63% of the executives approached in the study agreed that build-to-order may be "the most promising opportunity for the European automotive industry", with just 16% of those asked providing a negative or non-committal response. However, this general support for BTO cars as a concept comes with a powerful caveat; order to delivery must be achieved rapidly. It appears that late configuration of finished products would be insufficient. Therefore, a consequence of this time-based demand is that it requires major value adding activities to take place in facilities close to the market, requiring the industry to find new ways of driving cost out of the current structures. Executives were asked about the extent to which BTO is possible today for OEMs? The responses fell mainly into two distinct camps. Those who believed it was possible with some modifications to current processes amounted to 39.2% and those who thought it was possible today, but required major changes to the entire system accounted for 51.2% of the sample. Less than 5% thought implementing BTO would be practically impossible. Indeed, the findings of the consultation study suggest that most European manufactures have already undertaken some form of trial based on BTO concepts. Our survey also shows that half (50.1%) the executives we questioned were already aware of good examples of successful implementations of BTO concepts from within the industry, or else good practice necessary for implementation of BTO (36.5%). Interestingly, few were aware of the barriers in implementation of BTO that cause failure (13.5%), even though a thorough investigation into the likely barriers would provide key fundamental strategic decision criteria.

The validity of incentive schemes for BTO was also examined. Findings suggest that incentive schemes are still relevant for BTO today albeit at a lower level than may be justified for a mass production paradigm. Just 2% of executives gave incentive schemes a low priority, with an opposing 38.5% rating them as highly valued. Of significance was the finding that the majority (59.6%) thought that the incentive schemes were only of medium importance to help drive the adoption and implementation of initial change in a BTO scenario.

Academics and practitioners have been trying to get to grips with the exact drivers for change in business since the 1970s at least. Our most recent research suggests that transition to a BTO model may be driven by the suppliers who will emerge as the "pivotal" firms in the BTO transition process. Our survey shows that in general, most executives believe that BTO implementation would be impossible (58.2%) without supplier involvement. A further 27.3% felt that the process would be greatly improved if suppliers were involved in the entire process, whilst just 14.5% thought that supplier involvement would improve the implementation moderately; none of or respondents suggested that supplier involvement would not improve the implementation process at all.

Many executives thought access to sufficient resources (40.0%) was the most important factor in implementing BTO, with 22.0% suggesting time and a similar number (18.0%) suggesting confidence in the process and its leadership as being the most important success factor. In our survey, 31% of executives suggested that administration was the major impact factor for the cost of introducing new practices, whilst 22% stated that facilities would be a major cost impact factor. Additionally, 23.6% of executives placed re-training as a key factor impacting on the cost of implementing new practices.

The findings (72.2%) suggest that the size of a company influences its ability to adopt new practices and that large companies are less able to adopt new practices. Just 7.4% held the opposing view and 9.3% thought it made little difference today. From these observations, executives were additionally questioned on the ability of domestically owned companies to achieve the levels of change required to fulfil the ILIPT BTO concept and examine what barriers may be faced that are explicitly linked to the nationality of ownership. Stone et al. (2006) identified and proposed a range of possible drivers for the ultimate transition to a BTO business model. Some of these propositions were examined and the top three most likely outcomes amongst our industry executive perceptions were that 21.8% believed that the industry will eventually react to a crisis and be forced to adopt BTO to reduce cost structures, whilst only 10.9% believed that an enterprising new entrant might "steal a march" on the competition by rapidly launching a BTO car with a short lead time in Europe that would capture market share and therefore be difficult to emulate. Another popular speculation (23.6%) was that investment banks may ultimately require radical changes or that the final customer may exert pressure on the system to change (21%). Of course, there were some sceptical executives (10.9%) who believed that BTO may never happen.

22.9 Enablers to Effective Transition

As this research has shown, top industry executives in the automotive sector have strongly recognised the advantages of building cars to order in terms of both reducing the industry cost base and securing future competitive advantage. They likewise recognise the need for significant change in industry structure, processes and practices to achieve BTO goals and performance benefits. There is a clear understanding of the need to closely involve suppliers in such a transition; however, there seems to be a strong need to research and understand both the potential obstacles and the enablers in managing the BTO transition. In the following sections, some of the factors identified in previous research on the adoption of promising practice are summarised and their relevance to the BTO transition is considered.

22.10 Building on Existing Good Practice in BTO

Chakrabarti and Rubenstein (1976) demonstrated fairly early on the significance of the maturity of an innovation on the degree of success of adoption. This supports the idea that well-known, tested practices are easier to understand and implement. Indeed, Drew (1994) suggests that the lack of knowledge and skills was the third most important barrier to successful adoption of promising practice. Davies and Kochhar (1999) provided evidence from case studies and literature that indicated that failure to select practices based on a structured approach can lead to malpractices, fire-fighting, and sub-optimisation of performance. One reason for this may be that there is no real understanding of the practices being recommended as solutions. Taylor and Wright (2003) suggest that often misunderstandings of new practice lead to implementation problems and that misunderstanding of the processes is often more prevalent among less successful firms. Whilst Bateman and Rich (2003) point to a lack of internal support as being detrimental to the implementation of process improvement initiatives, they state that this lack of support is explained by the lack of understanding of the initiatives and the processes at stake by the parties expected to provide support.

All of the studies mentioned stress the need for developing understanding and experience in new practices before widespread adoption can be undertaken. Considering the scale of changes required for a transition to BTO, as outlined in the chapters of this book, it is likely that managers will similarly lack experience in the necessary BTO processes and practices. It is positive and appropriate, therefore, that companies are gaining this experience through trials based on BTO concepts, as reported in the preceding sections. Building on existing good practice will help companies to clarify their BTO vision and the strategies needed to make the transition. The development of company experience in BTO principles and practices will also help companies to customise their transition in such a way that fits the specific context and experience of both their own company and their supply chain.

22.11 Training and Education

There will likewise be a need to train and educate industry managers and leaders, and subsequently the remaining workforce in BTO rationale, concepts and processes. Re-training is often a key factor in decisions regarding a change of practice and can become expensive, as previous research involving significant transitions has indicated. In research carried out by Dixon et al. (1994) in the USA, the authors report that extensive employee training efforts took place prior to and during a total quality management (TQM) transformation. In one case, 5% of the total operating budget was spent on training. Ahire and Ravichandran (2001) suggest that technical training of employees was the second most important factor for transformation. Kassicieh and Yourstone (1998) tested the significance of "extent of training" and concluded that the extent of training correlated with cost reduction and increases in profits, but not with employee morale. McLachlin (1997), in case-based research in Canada, also found through pattern matching that the importance of the provision of training was supported. Provision of training was found to be a necessary condition for employee involvement, JIT flow and for JIT quality. However, an Italian survey somewhat controversially revealed that the provision of education and training to employees correlates with customer satisfaction, but found no correlations for planning, inventory management, improved efficiency, organisational climate, know-how and competence (Petroni 2002). In attempting to drive the transition to BTO, automotive industry leaders will need to be clear on both the necessary training, and "re-education" needs, of their own business and their supply chain.

22.12 Clear Planning and Objective Setting

The complex changes required in the BTO transition will need significant and careful planning to ensure that resources are available and interdependencies are considered. Previous research results have highlighted clear planning and objective setting as critical to the successful transition to new ways of working. For example, lack of clear planning results in poor implementation during major benchmarking implementations (Davies and Kochhar 1999). Research in the UK points to a lack of resources as a major barrier to successful transition and suggests that this is not necessarily associated with direct financial resources (i.e. available budget), but with issues such as access to production equipment and human resources. They also suggest that lack of resources is not necessarily affected by the size of the firm, but is correlated to the complexity and "interconnectedness" of the operations of the firm.

There have been a number of research papers on the subject of objective setting. The provision of clear objectives is reported to be an enabler of supply chain learning (Bessant et al. 2003). Guimaraes (1999) added that developing a rough-cut design to identify major issues early and determining all set-up details before implementation were crucial success factors. In a US review of 23 BPR projects, Dixon et al. (1994) suggested that objectives are often fuzzy initially because of the complexity of implementing major process changes in BPR projects. They highlight the need to clarify both objectives and the measures that will be used to determine successful achievement of those objectives. This is clearly a challenging issue for companies. In an international meta-analysis of 460 papers, Longbottom

(2000) found very little evidence to show that organisations are identifying and prioritising projects based on their corporate and strategic objectives. This leads to sub-optimal results and in some instances misuse of resources. The changes required to implement BTO, as outlined in this book, represent a significant paradigm shift for automotive companies and their suppliers. Careful planning of the BTO transition will be critical to success.

22.13 Aligning Key Performance Indicators

In recent years, both practitioners and researchers have emphasised the need to move beyond financial measures of operations and to incorporate a much wider variety of non-financial metrics into an organisation's performance reporting and reward systems. Now, in the intensely competitive global automotive industry, many European firms that are involved in automotive supply chains are in the process of reviewing their performance measures to enhance productivity and ensure that they remain competitive. There is palpable recognition of the need to benchmark their own performance against the best in the world and to adopt principles of best practice. As a result, a number of initiatives have been developed through partnerships and industry-led programmes, collaborative projects and further development in workforce practices. Companies are currently taking radical steps to redesign their performance measurement systems. However, OEMs are now using a variety of methods to deploy improvement throughout their supply chains. In addition, different measurement approaches are used by OEMs for the purpose of supplier assessment. This causes confusion among suppliers as to which approach to follow and creates waste within the overall system. As the BTO transition is planned, companies need to be aware of the potential for such confusion and waste for suppliers. An industry-led approach to identifying good practice in performance measurement and accounting for BTO may generate many benefits.

Leseure et al. (2004) noted that there is often disappointing coverage of the role of management accounting and performance data in the process of adopting promising practices. There is clearly a need in the BTO transition to develop appropriate performance and accounting measures that will drive the right behaviours. The ILIPT consortium has gone some way to addressing this issue by producing a framework report. Stone et al. (2006) illuminates some of the requirements for accounting and performance measurement systems by arguing for the development of key performance indicators (KPIs), cost models and simulation tools. Increasing variety traditionally leads to increased costs and inevitably lowers performance. Therefore, it is important to balance variety against improved revenue. Leseure et al. remark that a previously overlooked aspect is the development of robust complexity cost models. ILIPT has completed some initial work towards building a cost model, using a case study approach and drawing data directly from an automotive production plant.

The model has been developed based on an analysis of cost patterns against increasing complexity. Both information and material flows have been included. Preliminary analysis of predicted compared with actual cost data shows high correlation and the case study company is already employing the model as a guide during scenario planning in order to support the transition. Dynamic evaluation methodologies are being developed to enable analyses of structural and process changes performed for BTO in a number of scenarios across the supply chain. This methodology will form the basis of a prototype model to validate and demonstrate the validity of the BTO concepts. To validate and measure the transition to a BTO system a set of key performance measures are being developed to track progress. The key performance indicators fall into four groups: finance, process, structure and resource. The measures all integrate within the proposed process and product structures and form part of the process simulation models (Stone et al. 2006).

22.14 Enhancing Supply Chain Learning

In a conceptual paper published in the mid 1990s, Bessant et al. (1996) drew on a 5-year empirical study into the adoption of continuous improvement (CI) programmes within European enterprises to generate lessons in transition management. The research team concluded that "a disturbing number of CI programmes fail, mostly through decay rather than sudden decline. Those which survive only do so because of active and continuing efforts to energise, nurse, guide and shape – in other words, as the result of a difficult learning process around and acquiring this new capability." Davies and Kochhar (1999) undertook structured interviews in the UK with managers from aerospace and automotive companies focussed on the benchmarking of manufacturing planning and control systems. They found that even when best practices are accepted in principle, implementation does not follow – results of benchmarking are not put into practice. In other words, learning is not effectively taking place at either the company or the supply chain level. A fire-fighting mentality was the most important barrier to learning, which may hint at the fact that there is sometimes too much of a gap between current and best practice to even attempt implementation, at least in the short term. A further point is raised by Schmemner (1997) who posits that "the competitiveness of Western European manufacturing is slowly being eroded and European firms must act decisively to halt this decline". He goes on to suggest that costs in Europe are high and getting higher and that productivity is not keeping pace. European manufacturers have adopted some of the ideas that have proved useful to manufacturers elsewhere around the globe, but they need to do more to improve. To ensure an effective and sustainable transition to BTO, learning and development must take place at all levels in the supply chain. Factors that drive the introduction of supply chain learning include technological drivers and sustained pressure on pricing that force the supply chain to address waste and cost reduction (Bessant et al. 2003). Such pressures are clearly present in the BTO transition.

22.15 Communication and Benchmarking Progress

The role of clear communication and feedback on the success of improvement efforts has been found to be important at all levels – individual, organisational, and for the supply chain – in sustaining progress for transformation initiatives. In a US study by Kumar and Chandra (2001), the findings showed that in order for workers to identify problems and opportunities, and coordinate their efforts, management needs to provide them with feedback information in the form of manufacturing performance measures. It would seem that this holds true for both individuals and broader organisations. Bessant et al. (2003) detailed six supply chain cases where they found that holding extensive review meetings, to be enablers of supply chain learning. Drew (1994) posited that poor communication blocks effective BPR implementation and Guimaraes (1999) also recommends regular scheduled meetings between the project manager and each level of the project structure to drive significant changes. In leading the BTO transition, industry leaders will need to be aware of the imperative to provide feedback on progress and to develop benchmarks of BTO best practice.

Lessons can be learned from earlier efforts to drive industry change, where common measures were developed enabling progress to be benchmarked. The Society of Motor Manufacturers and Traders (SMMT) Industry Forum introduced a set of seven measures of competitiveness in 1998. These measures were endorsed by the UK automotive industry and used to monitor the impact of process improvement, which in turn was expected to result in improved competitiveness in the UK. Also in 1998, a set of ten key performance indicators were introduced for the construction sector for the purposes of:

- Benchmarking against others in the industry
- Assessing companies using a broader range of parameters

The Society for British Aerospace Companies (SBAC) also introduced a set of six key performance measures as an aerospace industry standard in 2000 (Ward and Graves 2001). Industry leaders in the BTO transition may also usefully consider introducing and promoting a range of appropriate measures to drive and measure the progress of BTO adoption in the industry. However, a word of warning was issued by Davies and Kochhar (1999), who suggest that benchmarking may not yield expected outcomes, if there is a lack of commitment to the findings and to the implementation process. The transition to BTO requires firm industry buy-in and a commitment to follow through in adapting to BTO concepts and practices.

22.16 Conclusion

The transition to BTO clearly needs flexibility in supply and strong leadership from the OEM. The industry executives in our study have clearly recognised the need to undertake significant change in the direction of BTO principles and practices. There is a commitment to making this change that will lead to reductions in the automotive cost base and will secure a competitive future for both OEMs and their supply chains. The leaders for change are likely to be the OEMs, but sustainability of the transition processes will be the responsibility of the full supply chain including innovative and flexible suppliers.

It is likely that the BTO transition will come about as part of a "controlled convergence" of the various elements that projects such as ILIPT have espoused. This chapter has highlighted some of the issues that may need to be addressed to successfully manage this important transition including building on existing examples of good practice; training and re-educating industry leaders and the workforce in BTO principles and practices; clear planning and objective setting; enhancing supply chain learning; aligning performance and accounting practices; and communicating and benchmarking progress.

References

ACEA (2007) Economic Report Q1. Available via
   www.acea.be/index.php/news/news_detail/acea_economic_report_q1_2007/
Ahire SL, Ravichandran T (2001) An innovation diffusion model of TQM implementation. IEEE
   Trans Eng Manag 48:445–464
Ascari A, Rock M, Dutta S (1995) Reengineering and organizational change: lessons from a comparative analysis of company experiences. Eur Manag J 13:1–30
Bateman N, Rich N (2003) Companies' perceptions of inhibitors and enablers for process improvement activities. Int J Oper Prod Manag 23(2):185–199
Beer S (1994) Decision and control: the meaning of operational research cybernetics. Wiley, Chichester
Bessant J, Caffyn S, Gilbert J (1996) Learning to manage innovation. Techn Anal Strat Manag 8:59–70
Bessant J, Kaplinsky R, Lamming R (2003) Putting supply chain learning into practice. Int J Oper Prod Manag 23(2):167–181
Boddy D, Cahill C, Charles M, Fraser-Kraus H, Macbeth D (1998) Success and failure in implementing supply chain partnering: an empirical study. Eur J Purch Supply Manag 4(2–3):143–151
Cagliano R, Blackmon K, Voss C (2001) Small firms under the MICROSCOPE: international differences in production/operations management practices and performance. Int Manuf Syst 12:469
Chakrabarti AK, Rubenstein AH (1976). Interorganizational transfer of technology – a study of adoption of NASA innovations. IEEE Trans Eng Manag 23:20–34
Cohen WM, Levinthal D (1990) Absorptive capacity: a new perspective on learning and innovation. Admin Sci Q 35(1):128–152
Collins R, Cordon C, Julien D (1996) Lessons from the 'made in Switzerland' study: what makes a world-class manufacturer? Eur Manag J 14:576–589
Davies AJ, Kochhar AK (1999) Why British companies don't do effective benchmarking. Integrated Manuf Systems 10:2–32
Dixon JR, Arnold P, Heineke J, Kim JS, Mulligan P (1994) Business process reengineering – improving in new strategic directions. Calif Manag Rev 36:93–108
Drew S (1994) BPR in financial services: factors for success. Long Range Planning 27(5):25–41

Graves AP (1987): comparative trends in automotive research and development. International Motor Vehicle Programme, Working Paper. MIT, Cambridge

Guimaraes T (1999) Field testing of the proposed predictors of BPR success in manufacturing firms. J Manuf Syst 18:53–65

Handy C (1995) The empty raincoat. Arrow Books, London

Hanson P, Voss C (1995) Benchmarking best practices in European manufacturing sites. Bus Process Re-Engineering Manag J 1(1):60

Hanson P, Voss C, Blackmon K, Oak B (1994) Made in Europe: a four nation study. IBM, London

HC399 (2007) Success and failure in the automotive car manufacturing industry. Fourth report of sessions 2006–2007. Stationery Office

Holweg M, Pil FP (2004) The second century: reconnecting customer and value chain through build-to-order; moving beyond mass and lean production in the auto industry. MIT Press

Kassicieh SK, Yourstone SA (1998) Training, performance evaluation, rewards, and TQM implementation success. J Qual Manag 3:25–37

Kumar S, Chandra C (2001) Enhancing the effectiveness of benchmarking in manufacturing organizations. Ind Manag Data Syst 101(2):80–89

Leseure M, Bauer J, Bird K, Neely A, Denyer D (2004) Adoption of promising practice: a systematic review of the evidence. Available via http://www.aimresearch.org

Longbottom D (2000) Benchmarking in the UK: an empirical study of practitioners and academics. Benchmarking 7(2):98–117

Lowe J, Delbridge R, Oliver N (1997) High performance manufacturing: evidence from the automotive components industry. Organ Stud 18:783

McLachlin R (1997) Management initiatives and just-in-time manufacturing. J Oper Manag 15:271–292

Panizzolo R (1998) Managing innovation in SMEs: a multiple case analysis of the adoption and implementation of product and process design technologies. Small Bus Econ 11:25–42

Petroni R (2002) Critical factors of MRP implementation is small and medium-sized firms. Int J Oper Prod Manag 22(3):329–348

Pine BJ (1993) Mass customisation: the new frontier in business. Harvard Business School Press, Boston, p 44

Schmemner R (1997) The erosion in European manufacturing. Prod Oper Manag 6:110–115

Schon D (1967) Technology and social change. Delacorte, New York

Selto FH, Renner CJ, Young SM (1995) Assessing the organizational fit of a just-in-time manufacturing system – testing selection, interaction and systems models of contingency theory. Account Org Soc 20:665–684

Stone G, Miemczyk J, Esser R (2005) Making build to order a reality: the 5 Day Car Initiative. In: Strengthening competitiveness through production networks: a perspective from European ICT research projects in the field of 'Enterprise Networking'. European Commission Publications, Brussels

Stone G, Parry G, Graves A (2006) Transformation of the European automotive industry: the future of lean. J Financ Transformation Finance Factory18:10–15

Stone G, Parry G, Graves, A, Esser R (2007) Requirement for a sea-change in European car production, In: Rabe M, Mihók (eds) New technologies for the intelligent design and operation of manufacturing networks. Fraunhofer IRB, Stuttgart

Taylor WA, Wright GH (2003) A longitudinal study of TQM implementation: factors influencing success and failure. Omega 31(2):97–111

Thelen W (2007) The space race. Automot Logist September/October:38–40

Voss C, Blackmon K (1996) The impact of national and parent company origin on world class manufacturing: findings from Britain and Germany. Int J Oper Prod Manag 16(11):98

Ward Y, Graves A (2001) Lean performance measurement for aerospace. Int J Aero Manag 1(1):85–96

Womack J, Jones D, Roo D (1991) The machine that changed the world: the story of lean production. Harper Perennial, New York

Part VI
The Road Ahead

Chapter 23
The Road to the 5-Day Car

Glenn Parry and Andrew Graves

School of Management, University of Bath, Bath, UK

Abstract. The automotive industry has continuously managed to rise to the challenges it has faced through increases in competitive pressures, increased productivity, quality, cost and the rapid development of technology. The industry has been led by those best able to develop and implement new process paradigms, from Ford's mass production through to Toyota and Lean production. The next industry leader may be the first to implement a fully integrated Build to Order network and free the capital employed in the current process. As well as competitive pressures, external social and governmental pressures may also drive this transformation. Environmental pollution and climate change are coupled with governments legislating to reduce road traffic congestion and improve safety whilst maintaining individual mobility. The BTO paradigm maximises the efficiency of supply chains and removes unwanted vehicles and unnecessary transportation costs from the system, reducing their associated congestion and carbon emissions. It may well offer European producers the ability to deliver customer value and socially responsible mobility for the 21st century.

23.1 Introduction

The European automotive industry has, over the past 20 years, faced several severe challenges to its prosperity and survival. First was the need to close the gap with the Japanese lean producers, particularly with regard to the transplant operations in the US and the UK. This has been largely achieved, not only through copying best-practice, but by developing lean and focussed strategies that have produced products that are now viewed as world leaders. Few analysts would have predicted in the early 1990s that Nissan would be rescued by Renault. The European industry has benefited further from structural changes to meet the global nature of the industry, thus overthrowing the "fortress Europe" mentality of the 1970s and 1980s. Realising they could no longer hide behind trade barriers and tariffs they built a pan-European market that has required new forms of collaboration and alliances in production, component supply and technology development. Many auto firms have now built a global base for future investment. Producers such as BMW and Mercedes-Benz are operating factory facilities in the US, whilst other VMs and suppliers have developed significant investments in China, India and Eastern Europe. The financial success of companies such as BMW, Porsche and their suppliers, has been spectacular over the past decade, as they have dominated their market segments.

As we move towards implementation, BTO may bring first mover advantage. Henry Ford rewrote the history book in the early part of the 20th century by radically re-thinking factory operations and economies of scale – through mass production. Most craft producers went out of business during the following decade. Since the 1970s, Toyota has led the world with their production system, outperforming the traditional mass producers with a 2:1 advantage in productivity and quality. While the move to BTO is likely to be achieved in stages, the next industry leader may be the first to implement a fully integrated BTO network. This BTO network will reap the benefits of satisfied customers, reductions in the cost base, and improved future competitiveness. The free capital the system would generate would provide unprecedented returns for investors – a move that would draw in further investment, giving the OEM the finances to develop vehicles that delight their customers; a virtuous circle that could accelerate the fortunes of the early adopters. Already, some of the OEMs have instigated BTO-type operations within their networks and feasibility trials are ongoing, but full-scale adoption of a BTO approach may require the intervention of agents of change. These may originate from investment banks, OEMs, the supply chain, the dealer network or even the customer. The customer gets their "bespoke" vehicle within a short timeframe. Companies hold less risk in terms of stocks of finished goods and are able to free up their capital more quickly to invest in new products and processes. Financiers should see a greater return on capital employed. The concept offers advantages to all. The drivers towards BTO are, however, wider than financial, commercial and consumer interests, and include the need to address social and governmental challenges.

23.2 Social Pressure and Government Challenge

The European car industry now faces the most critical challenge since its inception. This challenge is driven by two powerful externalities:

• Environmental pollution and climate change
• Mobility and road traffic congestion

Ignoring these challenging drivers for change may result in firms being regulated out of business, often by legislators who are hostile to the very concept of individual mobility. Vehicle manufacturers, therefore, have to satisfy not only the "voice of the customer", building vehicles with lower prices, improved quality and the latest technology and design, but also the "voice of society", which increasingly demands high levels of social responsibility from VMs on behalf of its citizens.

The current solution for addressing environmental pollution involves building so-called "environmentally friendly" hybrid vehicles. While such technologies may go some way to providing innovative solutions, such vehicles are often constructed of components supplied from a dispersed global supply base and shipped half way around the globe. Such practices may become increasingly viewed as both wasteful and counter-productive. BTO concepts and practices offer a viable alternative, with a focus upon producing fewer vehicles that will still meet customer needs, together with shorter supply chains and building closer to the customer base. This results in fewer emissions from both the manufacturing base and less pollution from the logistics providers. In addition, new and technologically advanced solutions will be rapidly adopted to ameliorate both road traffic congestion and pollution. Flexibility of both the production processes and the end-product will enable consumers to customise their vehicles for different usage patterns, instead of purchasing several bespoke vehicles. New technology will be able to be retro-fitted to existing vehicles to upgrade capability at the customers' demand.

Meeting the social pressures for future mobility will require building a fundamentally new relationship among government, industry and the customer. As the authority that owns (or controls) the traffic infrastructure, the government plays a critical role in deciding national policy between public and private transportation. Government acts as the co-ordinator of regulations and protocols between countries and trading partners. Government may also play a catalytic role through the encouragement of collaborative R & D programmes such as ILIPT, in order to foster industrial synergies and encourage innovation where firms lack the resources to invest in long-term projects. It is evident that legislators by their actions or inactions will play a major role in deciding the future trajectory for this area of automotive research and development in the coming decade. The requirement for auto firms to be pro-active by developing closer collaboration with legislators is therefore essential in order to deliver a focussed strategy for global competitiveness for the automotive sector in the future.

23.3 Industry's Challenge

It is clear that through the implementation of lean production techniques, European VMs have produced vehicles with significantly reduced waste, both material and management, compared with the mass producers of 20 years ago. However, due to the "push" nature of production and the industry's obsession with market share and economies of scale, the sector has created massive over-production of vehicles in Europe. The full benefits of developing the "lean enterprise" have, therefore, still not been realised across the industry with regard to meeting or exceeding the demands of a less wasteful manufacturing base. Through BTO, the opportunity is available to fully utilise lean philosophies in order to dramatically reduce excess waste and thereby satisfy both the voice of society and that of the customer.

In addition, the European automobile industry is currently undergoing a technological transformation through a combination of "technology-push" and "demand-pull". As this book has shown, a range of new technologies and processes are being developed by traditional suppliers and new entrant firms, particularly in the areas of advanced materials and electronics. On the demand side, several factors are combining to stimulate innovation, such as changing consumer preferences regarding performance, style, crash protection, fuel economy and emissions. European VMs are therefore challenging the status quo on every technological front. However, their ability to develop revolutionary new concepts is restricted in two key areas. First, the investment banks – arguably the VMs' most important suppliers – require greater confidence in the returns generated if they are to invest at the required levels. This is due to the uncertainty of future returns available from a sector that, although faster at delivering products than the aerospace or construction sectors, lags well behind the fast clockspeed industries such as information systems or microchip producers. Therefore, firms such as Ford or VW have to compete, not just against other auto firms for capital and the best managers and engineers, but also against firms such as Intel, Microsoft and telecommunications companies. This is a serious disadvantage and is only partly overcome by government-funded research and development programmes.

The second issue relates to technology-push and the fundamentally conservative nature of the industry, driven by extensive "sunk costs" that have historically dominated production. Likewise, at the roots of the automotive industry, sectors such as petrochemicals and the steel industry have until recently been relatively slow at developing new and innovative technologies with regard to alternative materials and fuels. Radical technical change has historically been hindered by market uncertainties and technological problems that have mitigated against the rapid deployment of revolutionary concepts and solutions. Nevertheless, as the solutions offered in this book illustrate, change is underway. ThyssenKrupp Steel, for example, have taken a lead and are now committed to a radical re-think of car body variants that are built from common parts and require low tool investment. The ModCar body consists of only four modules, which will radically reduce manufacturing costs and complexity and allow the development of new types of vehicles to meet increased customer demands.

The chapters in this book have sought to support this industry transition by providing an overview of a new paradigm for automotive manufacture, build-to order. The contributors each provide different perspectives that build up the broader picture of a BTO supply concept that may provide competitive advantage to those able to implement this approach. Market developments have been highlighted and the need to understand what value individual companies bring to the value stream through their core competences has been demonstrated. We have examined ways of re-imagining the product through modularity, bringing late configuration and greater adaptability. With the integration of the supply chain and interoperability of IT systems customers can be linked directly to the supply chain and build process. Process modelling has demonstrated the feasibility of the BTO concepts, which have been examined by industry experts for critique and validation. Industry leaders have recognised the need to undertake significant change, and we are now moving towards adoption of BTO principles and practices in an industry that is often described as conservative, but which has a history of change.

23.4 Conclusion

This book and the ILIPT 5-day car project, offers a road map of possible engineering and managerial paths that could build upon the success of the European industry's capacity to survive, by reinventing itself over the past 60 years in the face of increased competition from overseas. The industry's competitive advantage has been largely its world-class technological and design capacity and its willingness to become a global leader in the development of ground-breaking processes and practices. It is, therefore, arguably better placed than many of its competitors to make the next significant transformation in order to meet the challenges set out in this chapter.

Build-to-order offers the chance for automotive companies to eliminate overcapacity and realise the true potential of lean production via the pursuit of perfection through the removal of waste and maximisation of customer value. BTO may well offer the global automotive industry, and particularly the European producers, the ability to pursue true technological collaboration between all stakeholders and provide socially responsible mobility for the 21st century.

About the Editors

Dr Glenn Parry is a Senior Fellow at the University of Bath's School of Management. He is part of the Core Team for the ILIPT project and Theme III leader. His research is industrially focused and he has managed and contributed towards consortia from the automotive, aerospace and construction sectors. He has worked for British Steel in radiation cured coatings, TWI developing organo-ceramic coatings, LEK Consulting as a strategy analyst and WMG as part of the UK Lean Aerospace Initiative focussed on ERP and core competence. His current research interests include the move to service provision, costing and ERP management. Project involvement includes BAE-S^{4}T, Lean Flight Initiative, Supply Chain 21 and Agile Construction Initiative. He holds a PhD in Materials Science from Cambridge University, a BSc in Chemistry & Business and an MPhil in Materials Science from Swansea University and a Certificate in Teaching in Higher Education and a Diploma in Psychotherapy from Warwick University. He can be contacted on parryglenn@hotmail.com.

Professor Andrew Graves holds a Doctorate in Science & Technology Policy from the University of Sussex and began his career in the UK automotive and aerospace industries in the 1960s, specialising in Manufacturing Techniques and Supply Chain Management. After serving with the military, he spent 12 years in Grand Prix racing as a technician and manager with several leading Formula 1 teams and over the past 20 years has worked in academia participating in global research programmes in Lean Manufacturing. He worked as a Senior Research Fellow at the Science Policy Research Unit at the University of Sussex from 1985, where he undertook research with MIT's International Motor Vehicle Program (IMVP) on R & D and technology issues. In 1994 he moved to the University of Bath's School of Management as Co-Director of the IMVP. He now holds the Chair of Technology Management. Recent research initiatives focus on transferring lean techniques from the automotive industry to other sectors and include the following projects: the UK Lean Aerospace Initiative, the Agile Construction Initiative, the Lean Flight Initiative, 5DayCar/ILIPT, the Supply Chain 21 Initiative (SC21).

About the Authors

Kati Brauer, doctoral candidate at 4flow research, 4flow AG, Berlin, Germany. Kati studied Business Administration and Mechanical Engineering at the Technical University of Freiberg (Germany) and the Polytechnical University of Valencia (Spain) and acquired her Master's Degree (Diploma) in 2007. In her diploma thesis, she developed a model for evaluating automotive network configurations. Her research interests include supply chain management, production and logistics planning. Currently, she is pursuing her Ph.D. thesis in the field of integrated scheduling and transportation planning and is actively involved in consulting projects in the field of logistics.

Michael Berger holds a diploma in Electrical Engineering and a Ph.D. in Computer Science from Dresden University of Technology. He has been working for more than 14 years in Computer Science research, specialising in Distributed Systems, Mobile and Ubiquitous Systems, and Intelligent and Multi-Agent Systems research. Dr. Berger has been a member of Siemens Corporate Technology in Munich since 1997 and is currently competence field leader in "Agent Technologies". Dr. Berger has many scientific publications, is actively involved in workshops and conferences, and is technical reviewer for the EU and lecturer at Dresden University of Technology and Ludwig-Maximilians-University in Munich.

Dr. Valerie Crute is a research fellow in the School of Management, University of Bath. She lectured for 7 years in the Department of Communication, University of Ulster, where she obtained a Ph.D. in Social Psychology, before specialising as a consultant in the management of organisational change. She managed the UK Lean Aerospace Initiative (UKLAI) at Bath, a unique research collaboration between four leading UK universities and over 30 manufacturing companies. Her research over 10 years has examined factors influencing successful organisational transformation including managerial, cultural and communication issues both within and across collaborating organisations.

411

Jan-Gregor Fischer, Siemens AG, holds a Master's degree in Computer Science from the Technical University of Munich, focussed on IT in mechanical engineering and Ubiquitous Computing & Augmented Reality (AR). His Master's Thesis comprised the development of a generic ontology-based software platform. In 2001, he joined c.baX GmbH, Munich, as a software engineer specialising in the development of custom data acquisition software. Since 2006, he has been working on ambient intelligence and agent technologies, semantic data interoperability and reasoning with the Intelligent Autonomous Systems research group at Siemens Corporate Technology (CT) in Munich.

Philipp Gneiting, Daimler AG, Group Research & Advanced Engineering, Materials, Manufacturing and Concepts. Philipp holds a Master's degree in Business Administration from the University of Mannheim, focussing on industrial and operations management. In his thesis, he developed mathematical methods for optimising distributed decision processes in automotive production networks. At the time of writing, Mr. Gneiting is currently pursuing his Ph.D. thesis in the field of modular automobile production in cooperation with the Daimler Research centre in Ulm. Additionally, he is actively working on projects at Daimler in the area of inter-enterprise collaboration, production network optimisation and evaluation of modular concepts.

Dr.-Ing. Maik Gude, Technische Universität Dresden, Institute of Lightweight Structures and Polymer Technology, Head Scientist. Dr. Gude holds a diploma degree in Applied Mathematics from the Technische Universität Clausthal and a Ph.D. in Mechanical Engineering from the Technische Universität Dresden. His research focus is the development of material and structural models and the design of lightweight structures made of advanced materials. Dr. Gude is involved in a number of fundamental and applied R & D projects, especially those aimed at novel composite materials and structures.

Alexandra Güttner, Daimler AG, Group Research & Advanced Engineering; Materials, Manufacturing and Concepts. Alexandra is currently pursuing her Master's degree in Industrial Engineering and Management at the University of Karlsruhe with Daimler Research & Development, Ulm, focussing on complexity and variety management. In her thesis, she is developing a mathematical methodology for calculating the floor space required by highly varied products subject to the number of variants or complexity drivers. Additionally, she is contributing to the work of the EU ILIPT project.

Prof. Dr.-Ing. Bernd Hellingrath holds a Master's Degree in Computer Science from the University of Dortmund. He received his Ph.D. in 2001 from the faculty of Mechanical Engineering at the same university. In 1988, he started working at the Fraunhofer Institut Materialfluss und Logistik (IML) as a researcher focussing on the application of Artificial Intelligence technologies within logistics. Since 1995, he has been head of the main Enterprise Modelling department at Fraunhofer IML. In 2006, he became a full professor at the faculty of Business Computing at the University of Paderborn. His major working areas lie within the development of methodologies for the design, planning and execution of processes within production and logistics networks, the application of ICT systems supporting supply chain management processes and the usage of simulation technologies for the design and operation of logistics systems.

Matthias Holweg, M.Sc., Ph.D., is a Senior Lecturer in Operations Management and the Director of the Centre for Competitiveness and Innovation at the Judge Business School, University of Cambridge. He is a principal investigator on several research projects, including at MIT's International Motor Vehicle Program (IMVP), where his research focusses on the dynamics of competition and patterns of evolution of the global automotive industry. Prior to joining the faculty at Cambridge, Matthias held positions at MIT's Center for Technology, Policy and Industrial Development and at the Lean Enterprise Research Centre at Cardiff Business School. He can be contacted at m.holweg@jbs.cam.ac.uk.

Dr. Mickey Howard is Senior Lecturer in Operations Management at the School of Management, University of Bath. His research examines product-service innovation and through-life capability management in the automotive, aerospace, and retail sectors. Mickey complements this with private advisory work for organisations such as BAE Systems, Volvo Cars and the Ministry of Defence. He regularly publishes in the *International Journal of Operations & Production Management* and the *Journal of Purchasing & Supply Management*. He guest lectures at business schools such as Harvard, Copenhagen, and Audencia Nantes, and has recently been awarded a Chartered Institute of Purchasing & Supply research scholarship.

Prof. Dr.-Ing. habil. Werner Hufenbach, Technische Universität Dresden, Director of the Institute of Lightweight Structures and Polymer Technology. Prof. Hufenbach holds a diploma degree in Forming Technology, a Ph.D. in Mechanical Engineering and a habilitation degree in Applied Mechanics and Material Mechanics from the Technische Universität Clausthal. He became a Professor at the Technische Universität Clausthal and in 1993 he was appointed Director of the Institute of Lightweight Structures and Polymer Technology at the Technische Universität Dresden. His in-depth research and development activity is in the area of lightweight structures engineering. He is well known for his contributions in the fields of multi-materials design and function-integrated lightweight structures.

Thomas Huth, FEV Motorentechnik GmbH, Engine Management and Electronics. Thomas holds a diploma in mechanical engineering from the University of Applied Sciences in Aachen, focussing on the electro-mechanical valve train

(EMVT). At FEV he works in the field of EMVT development. Mr. Huth is studying for a Ph.D. at the institute of combustion engines, University of Aachen.

Katja Klingebiel, ebp consulting GmbH. Katja Klingebiel holds a Diploma degree in Business Mathematics from the University of Dortmund. She is working as project manager in the area of design, evaluation and implementation of logistics, supply chain management and RF-ID processes for the automotive industry and is carrying out management training for several automotive companies. Additionally, she is currently pursuing her Ph.D. thesis in the field of build-to-order concepts, thereby focussing on identification and evaluation of costs and the benefits of these concepts for automotive networks.

Jörg Mandel, group manager for network logistics at the Fraunhofer Institute for Manufacturing Engineering and Automation IPA in Stuttgart, Department of Enterprise Logistics. Jörg completed his Master's study at the University of Stuttgart and since 2000 has worked as a scientist at Fraunhofer Institute for Manufacturing Engineering and Automation in the fields of modelling, visualisation and optimisation of production and logistics networks. In different projects, he analyses and optimises the process chain of automotive suppliers using JIT and JIS concepts for OEMs in Germany. Jörg Mandel is currently pursuing his Ph.D. thesis in the field of build-to-order processes in collaborative supply chains.

Joe Miemczyk is Assistant Professor in Operations and Supply Chain Management at Audencia, Nantes, France. He teaches on the MBA, International Management and Grande Ecole programmes and is part of the MASC Institute, dedicated to the study of purchasing and supply chains. Previously, he was a business process engineer for quality and environmental management at the Unipart Group of Companies. He started his academic career with the 3DayCar programme and was most recently theme leader on the European ILIPT project. Joe has published widely on automotive supply chains in journals such as the *Journal of Business Logistics* and the *International Journal of Operations and Production Management*, and has presented at numerous conferences, including the European Operations Management conference and the Logistics Research Network.

Stefanie Ost, Fraunhofer Institute for Manufacturing Engineering and Automation. Mrs. Ost holds a Diploma in Industrial Engineering from the Technical University of Chemnitz, focussing on production management and logistics. In her thesis, she developed a performance measurement system for a company in the mechanical engineering industry. Currently, Mrs. Ost is working as a research fellow in the Department of Corporate Logistics at the Fraunhofer Institute for Manufacturing Engineering and Automation in Stuttgart. She actively participates in projects dealing with topics including efficient business processes for supply chain management, concepts for network planning and control and inter-enterprise collaboration.

Andreas Reichhart, Judge Business School, University of Cambridge. Andreas holds a Ph.D. in Management Studies from Judge Business School, where he studied the flexibility of automotive supply chains. He regularly publishes papers in academic journals as well as trade press articles on topics as diverse as "co-located supplier clusters", "lean distribution" and "supply chain flexibility". Currently, Andreas works for McKinsey & Company, where he helps international clients increase the efficiency of their supply chains.

Jens K. Roehrich is a doctoral researcher at the Centre for Research in Strategic Purchasing and Supply (CRiSPS), School of Management, University of Bath, UK. He holds a dual degree in industrial engineering and business from Germany and a Masters Degree in operations management from the University of Bath, UK. Jens has been appointed a full member of the Chartered Institute of Purchasing & Supply (MCIPS), UK. His current research themes evolve around the dynamic governance of inter-organisational relationships inherent in long-term product-service supply arrangements reflecting a multi-disciplinary research perspective.

Heinrich Schleich is Professor for Production Management at the Leuphana University of Lueneburg, Germany. He does scientific research with focus on production and supply chain optimisation, which includes the EU ILIPT programme. He is director of Institute for Production Technologies (IPRO) and director of the MBA program "Manufacturing Management". Heinrich Schleich has an industrial background, working for more than 20 years in the mechanical engineering industry and as a management consultant, with a focus on operations. He holds a Master's degree (Diploma) and a doctorate (Dr.-Ing.), both from the Technical University of Aachen, Germany.

Thomas Seidel, Head of 4flow research, 4flow AG, Berlin (Germany). Thomas Seidel is responsible for the research department of 4flow AG, which proactively develops innovations in supply chain management. He holds a Master's Degree (Diploma) in Business Administration and Mechanical Engineering, which he acquired at the Technical University of Berlin. His studies also led him to the Technical University of Kaiserslautern (Germany), the University of Calgary (Canada) and to the University of Western Sydney (Australia). As a consultant, Thomas has advised customers on issues of order fulfilment and logistics processes. 4flow research projects focus on supply chain design and corresponding management information systems.

Jens Schaffer is a researcher in the field of Production Management in the Department of Automation Technology at the Leuphana University of Lueneburg, Germany. He also works as a consultant to the automotive industry for YesCon Engineering. He has an industrial background, with 6 years' experience as a project manager in international automotive development projects. Jens Schaffer holds

Master's degrees in Mechanical Engineering (plastics technology) and in Business Administration. His activities in research focus on variant management in industrial processes and the related cost structures. As a consultant he optimises and manages product development and production processes for a variety of customers.

Thomas Sommer-Dittrich, Daimler AG, Group Research & Advanced Engineering, Materials, Manufacturing and Concepts. Thomas studied applied engineering and logistics at Berlin Technical University and Jiao Tong University Shanghai, China. As a Research Fellow he focussed on logistics management, especially flexible logistics for disassembly factories and networks, as well as on logistics technologies, e.g. automated guided vehicles. He worked as a consultant for advanced technologies and management in the European automotive industry, with a focus on sustainability, flexibility and complexity management. Within Daimler's Strategic Manufacturing Field, Thomas Sommer-Dittrich is responsible for new concepts for Excellence in Production Planning and Operations (ExPO). The research also includes work packages of the Modular Car (ModCar) and Flexible Supply Networks (FlexNet) within the EC-funded ILIPT project. Thomas is author of several publications, mainly in the field of flexible logistics in production and disassembly and AGV.

Gareth Stone is a researcher at the Lean and Agile Research Centre based at the University of Bath, School of Management. Gareth has an industrial background, working for 15 years in the aerospace industry. His main research areas are associated with the exploration of logistical and policy implications of automotive build-to-order concepts across Europe. Currently, he is developing transition strategies for the European ILIPT Project. He has recently been involved with key performance metrics for the UK aerospace industry and the development of promising agile practice within UK construction. He has lectured in Operations Management at Southampton Business School and has undertaken a 3-year project with the UK Lean Aerospace Initiative.

Michael Toth, Fraunhofer Institute for Material Flow and Logistics (IML), Enterprise Modelling and Supply Chain Management Group. Michael holds a Diploma degree in Information Science and Business Administration from the University of Paderborn, focussing on logistics and supply chain management. At the time of writing, Mr. Toth is pursuing his Ph.D. thesis in the field of demand and capacity management for automotive networks. Additionally, he is actively working on projects at Fraunhofer IML in the area of inter-enterprise collaboration, supply network simulation, virtual engineering and evaluation of supply chain strategies. He is also a consultant in different industrial projects for the automotive and engineering industry.

Andreas Untiedt, ThyssenKrupp Steel AG, holds a Bachelor degree for Mechanical Engineering from the University of Glamorgan as well as a diploma from the University of Applied Sciences and Arts, Hannover, focussing on thermo-dynamics and combustion engines. With the experience from the areas of designing automotive engine components and coordinating development of modular product structures in the field of automotive chassis between OEM and supplier, he is now working on innovative product development for the automotive industry in the Lightweight Construction & Innovation Center Auto department of ThyssenKrupp Steel.

Markus Witthaut leads the Supply Chain Management group at the Fraunhofer Institute for Material Flow and Logistics IML, Dortmund, Germany. He holds a Diploma – the German equivalent to a Master's degree – in Computer Science from the University of Dortmund. Since joining Fraunhofer IML in 1992, he has been involved in many industrial and research projects in the logistics domain. His main research areas are the development and application of new business processes and supporting information technology for the optimisation of supply chains.

ILIPT Members

4flow AG

Thomas Seidel
Stefan Wolff
Kati Brauer
Leiv Klarmann
Ines Nadj

BMW Motoren GmbH

Wilhelm Niedermayr
J. Prenninger
Martina Handstanger
Alois Rohrauer
Christian Tschurtschenthaler
Anita Jürets

Ceramicx

Frank Wilson
Margaret O'Brien
John White

CLEPA

Bjoern Hedlund
Lars Holmqvist
Gloria Pellischek

Daimler AG

Thomas Sommer-Dittrich
Alexandra Güttner
Klaus Fuerderer
Philipp Gneiting
Robert Sommer
Stephan Bürkner
Danuta Wowreczko
Ralph Rinkl
Hartwig Baumgärtel

	Markus Kern
	Shahram Hami-Nobari
	Alfred Katzenbach
Dana	Sanjay Walia
Debonding Limited	Peter S. Bain
	Giovanni Manfrè
	Matthew Machon
De-Bonding Ltd c/o MG Consult srl (Italy)	Sara Sogari
EBp Consult GmbH	Katja Klingebiel
	Kasra Nayabi
	Hannes Winkler
	Frank Gehr
	Jochen Braun
EFTEC AG	Rolf Holderegger
	Jesus Garcia
	Heinrich Sommer
EU Commission & Reviewers	Florent Frederix
	John Harris
	Keith Ridgeway
	Christoph Mandl
	Enzo Carrubba
	Angel Ortiz
Ferrostaal	Diana Volz
	Stefanie Jancar
	Nino Sebastian Müssig
	Hinnerk Pflüger
	Ralf Becker
	Markus Schlueter
FEV Motorentechnik GmbH	Thomas Huth
	Gerhard Lepperhoff
	Thomas Crott
	Olaf Elsen
	Thomas Esch
Fraunhofer Gesellschaft	Bernd Hellingrath
	Markus Witthaut
	Michael Toth
	Jörg Mandel

	Stefanie Ost
	Silvia Körber
	Peter Dürr
	Marco Motta
	Axel Wagenitz
	Andre Nickl
	Stephan Schüle
	Kathrin Hesse
	Matthias Keller
	Axel Kuhn
	Thomas Meise
	Tobias Hegmanns
	Mehmet Kürümlüoglu
	Joachim Lentes
	Meike Dittmar
	Christian Mazzocco
	Gabriela Janusz
	Jörg Kimmich
	Walter Schneider
	Jürgen Bischoff
Freeglass	Benyahia Rym
	Luc-Henry Blanche
	Sascha Ruckwied
	Doris Gampp
Hella	Andrej Marek
	Mare Kysel
	Martina Karafiatova
	Petr Roubicek
JMA	Gerd Fricke
Lear Automotive (EEDS)	Alberto Garcia
	Yann Darroman
	Juan Carlos Alonso
Plastic Omnium	David Barlow
	Neil Houghton
Platos GmbH	Jan-Christoph Maass
	Karl-Werner Witte
	Jürgen Kauth
PUC-Rio	Luiz Felipe Scavarda

Saint-Gobain Glass	Alexandre Richard
Saint-Gobain Sekurit	Jean-Louis Bravet
Siemens AG	Michael Berger
	Jan-Gregor Fischer
	Caroline Wagner-Winter
	Michael Pirker
	Jochen Geiger
Siemens VDO	Karl Josef Kerperin
	Holmer-Geert Grundmann
	Stephan Hoessl
	Gregor Duda
	Tamara Hüttinger
	Andreas Vollmer
	Karlheinz Dietz
	Christine Eisenmann
	Janet Loschinski
	Isabel Siegel
	Christiane Markert
	Jörg Kiefer
St. Petersburg Institute for informatics and automation of the Russian Academy of Sciences	Alexander V. Smirnov Nikolay Shilov
ThyssenKrupp AG	René Esser
	Andreas Breidenbach
	Andreas Untiedt
	Ralf Sünkel
	Klaus Wolf
	Jan Kurzok
	Sabine Hofmann
	Rob Theunissen
TRW	Christoph Hartwig
	Olaf Josef
TU Dresden	Maik Gude
	Werner Hufenbach
	Anne Höner
	Barbara Röllig
	Andreas Ulbricht
	Rene Fuessel

University of Bath

Andrew Graves
Glenn Parry
Gareth Stone
Valerie Crute
Joe Miemczyk
Jens Roehrich
Mickey Howard
Jan Legge

University of Cambridge

Matthias Holweg
Andreas Reichhart
Eui Hong
Colin Roberts

University of Lüneburg

Jens Schaffer
Heinrich Schleich
Kerstin Vollmer

University of Patras

Sotiris Makris
Nikos Giannelos
Zois Papachatzakis
George Papanikolopoulos
Dimitris Mourtzis

VDI/VDE-IT

Helmut Kergel
Sabine Globisch
Volker Härtwig
Peggy Neick
Marcus Smolarek
Anette Hilbert
Lysann Müller
Jürgen Valldorf
Wolfgang Gessner

Glossary

ATO	Amend-to-order: Products are amended during production or during pre-production planning to match customer orders. If no customer order is allocated to a product, it is still manufactured.
BTO	Build-to-order: Refers to a demand-driven production approach where a product is scheduled and built in response to a confirmed order received for it from a final customer. The final customer refers to a known individual owner and excludes all orders by the OEM, national sales companies (NSC), car dealers, fleet orders or other intermediaries in the supply chain. BTO excludes the order amendment function, whereby forecast orders in the pipeline are amended to customer requirements, as this is another level of sophistication for a BTS system.
BTS	Build-to-stock: Refers to products that are built before a final purchaser has been identified, with production volume driven by historical demand information.
CAE	Computer-aided engineering
EAI	Enterprise application integration
EMVT	Electro-mechanical valve train enables the independent control of single valves by separate electronic operations.
First-tier supplier	Company that has a contract or purchase agreement to provide products and/or services directly to an OEM.
FlexNet	Flexible supply network, the name given to Theme II of the ILIPT project.
IBP	Independent body panels
ICDP	International Car Distribution Programme

ICT	Information and Communication Technology
ILIPT	Intelligent Logistics for Innovative Product Technologies. The name of the project that addresses the conceptual and practical aspects of the delivery to the customer of a bespoke product only several days after placing the order. This 4-year, €16 million joint European Commission and industry-funded project will reach completion in the summer of 2008.
IMVP	International Motor Vehicle Program
IntePro	Integration of Complex Product Processes, the name given to Theme III of the ILIPT project.
IS	Information system
IT	Information technology
JIS	Just-in-sequence: A progression of just-in-time where the products delivered arrive in the correct order, to match the manufacturing order sequence.
JIT	Just-in-time: An inventory system designed so that materials or supplies arrive at a facility just when they are needed.
KPI	Key performance indicator
LERC	Lean Enterprise Research Centre
LSP	Logistics service provider
LTO	Locate-to-order: Refers to the use of systems that give detail of stock to allocate orders to goods held.
MIT	Massachusetts Institute of Technology
ModCar	The name given to the ILIPT modular concept car and Theme I of the ILIPT project.
MTF	Make-to-forecast
MVVT	Engines using a mechanical variable valve train (MVVT) mechanically adjust valve timings on both the intake and outlet camshafts as a function of gas pedal position and engine speed.
NE	New entrants
OEM	Original equipment manufacturer: Refers to the car production companies, e.g. DaimlerChrysler, BMS etc.
RCAR	Research Council for Automobile Repairs
SBAC	Society of British Aerospace Companies

SCL	Supply chain learning
NSC	National sales companies
Second-tier Supplier	Company who provides products and/or services directly to a first-tier supplier, even though they may hold contracts or purchase agreements with an OEM.
SMMT	Society of Motor Manufacturers and Traders
Supply Chain	The supply chain includes all the activities involved in delivering a product from raw material to the customer, including sourcing, manufacture and assembly, warehousing, distribution and inventory tracking, order management and their associated information systems.
TPS	Toyota production system
VM	Vehicle manufacturer
VOB	Virtual order bank: an integrated order management and scheduling system that is used to link customer demand directly with capacity available in the supplier network, enabling the synchronisation of the network in real-time.

Index